Critical Research on Sexism and Racism in STEM Fields

Ursula Thomas
Georgia Perimeter College, USA

Jill Drake
University of West Georgia, USA

A volume in the Advances in Religious and Cultural Studies (ARCS) Book Series

Information Science REFERENCE
An Imprint of IGI Global

Published in the United States of America by
Information Science Reference (an imprint of IGI Global)
701 E. Chocolate Avenue
Hershey PA, USA 17033
Tel: 717-533-8845
Fax: 717-533-8661
E-mail: cust@igi-global.com
Web site: http://www.igi-global.com

Library of Congress Cataloging-in-Publication Data

Names: Thomas, Ursula, 1972- editor. | Drake, Jill, 1968- editor.
Title: Critical research on sexism and racism in STEM fields / Ursula Thomas and Jill Drake, editors.
Description: Hershey, PA : Information Science Reference, 2016. | Includes bibliographical references and index.
Identifiers: LCCN 2016003163| ISBN 9781522501749 (hardcover) | ISBN 9781522501756 (ebook)
Subjects: LCSH: Sexism in education. | Racism in education. | Women in science. | Science--Study and teaching--Social aspects.
Classification: LCC LC212.8 .C75 2016 | DDC 370.81--dc23 LC record available at https://lccn.loc.gov/2016003163

This book is published in the IGI Global book series Advances in Religious and Cultural Studies (ARCS) (ISSN: Pending; eISSN: Pending)

British Cataloguing in Publication Data
A Cataloguing in Publication record for this book is available from the British Library.

Advances in Religious and Cultural Studies (ARCS) Book Series

ISSN: Pending
EISSN: Pending

Mission

In the era of globalization, the diversity of the world and various cultures becomes apparent as cross-cultural interactions turn into a daily occurrence for individuals in all professions. Understanding these differences is necessary in order to promote effective partnerships and interactions between those from different religious and cultural backgrounds.

The **Advances in Religious and Cultural Studies (ARCS)** book series brings together a collection of scholarly publications on topics pertaining to religious beliefs, culture, population studies, and sociology. Books published within this series are ideal for professionals, theorists, researchers, and students seeking the latest research on collective human behavior in terms of religion, social structure, and cultural identity and practice.

Coverage

- Impact of Religion on Society
- Gender
- Globalization and Culture
- Sociology
- Social Stratification and Classes
- Cultural Identity
- Human Rights and Ethics
- Cults and Religious Movements
- Cross-Cultural Interaction
- Group Behavior

IGI Global is currently accepting manuscripts for publication within this series. To submit a proposal for a volume in this series, please contact our Acquisition Editors at Acquisitions@igi-global.com or visit: http://www.igi-global.com/publish/.

The Advances in Religious and Cultural Studies (ARCS) Book Series (ISSN Pending) is published by IGI Global, 701 E. Chocolate Avenue, Hershey, PA 17033-1240, USA, www.igi-global.com. This series is composed of titles available for purchase individually; each title is edited to be contextually exclusive from any other title within the series. For pricing and ordering information please visit http://www.igi-global.com/book-series/advances-religious-cultural-studies/84269. Postmaster: Send all address changes to above address.

Titles in this Series

For a list of additional titles in this series, please visit: www.igi-global.com

Handbook of Research on Chaos and Complexity Theory in the Social Sciences
Şefika Şule Erçetin (Hacettepe University, Turkey) and Hüseyin Bağcı (Middle East Technical University, Turkey)
Information Science Reference • copyright 2016 • 458pp • H/C (ISBN: 9781522501480) • US $305.00 (our price)

Combating Violent Extremism and Radicalization in the Digital Era
Majeed Khader (Home Team Behavioural Sciences Centre, Ministry of Home Affairs, Singapore) Loo Seng Neo (Home Team Behavioural Sciences Centre, Ministry of Home Affairs, Singapore) Gabriel Ong (Home Team Behavioural Sciences Centre, Ministry of Home Affairs, Singapore) Eunice Tan Mingyi (Home Team Behavioural Sciences Centre, Ministry of Home Affairs, Singapore) and Jeffery Chin (Home Team Behavioural Sciences Centre, Ministry of Home Affairs, Singapore)
Information Science Reference • copyright 2016 • 582pp • H/C (ISBN: 9781522501565) • US $190.00 (our price)

Handbook of Research on Race, Gender, and the Fight for Equality
Julie Prescott (University of Bolton, UK)
Information Science Reference • copyright 2016 • 744pp • H/C (ISBN: 9781522500476) • US $320.00 (our price)

Islamic Economy and Social Mobility Cultural and Religious Considerations
Hasan Shahpari (Community College of Philadelphia, USA) and Tahereh Alavi Hojjat (DeSales University, USA)
Information Science Reference • copyright 2016 • 372pp • H/C (ISBN: 9781466697317) • US $190.00 (our price)

Nationalism, Cultural Indoctrination, and Economic Prosperity in the Digital Age
Bryan Christiansen (PryMarke, LLC, USA) and Joyce Koeman (University of Leuven, Belgium)
Information Science Reference • copyright 2015 • 364pp • H/C (ISBN: 9781466674929) • US $205.00 (our price)

Contemporary Social Issues in East Asian Societies Examining the Spectrum of Public and Private Spheres
Mika Markus Merviö (Kibi International University, Japan)
Information Science Reference • copyright 2014 • 349pp • H/C (ISBN: 9781466650312) • US $175.00 (our price)

New Media and Communication Across Religions and Cultures
Isaac Nahon-Serfaty (University of Ottawa, Canada) and Rukhsana Ahmed (University of Ottawa, Canada)
Information Science Reference • copyright 2014 • 328pp • H/C (ISBN: 9781466650350) • US $195.00 (our price)

IGI GLOBAL
DISSEMINATOR OF KNOWLEDGE
www.igi-global.com

701 E. Chocolate Ave., Hershey, PA 17033
Order online at www.igi-global.com or call 717-533-8845 x100
To place a standing order for titles released in this series, contact: cust@igi-global.com
Mon-Fri 8:00 am - 5:00 pm (est) or fax 24 hours a day 717-533-8661

Table of Contents

Detailed Table of Contents

Understanding why women are underrepresented in various Science, Technology, Engineering, and Mathematics fields remains an important area of research. In the United States and in many industrialized nations around the world, STEM professions remain male dominated. Explanations for why women are not participating STEM professions are many and diverse. The Ecology Systems Theory (EST) presents a lens through which the causes for the continued underrepresentation of women in STEM fields may be examined. EST is widely accepted theoretical framework for exploring the influences that contribute to the development of an individual. The study presented in this chapter explored the familial, educational, economic, and social experiences of 125 female participants working in a STEM field. Findings suggest there are influences at specific levels in EST that can and do affect the educational and career aspirations of women in relationship to STEM fields.

The engineering field, in particular, struggles to recruit and retain students, especially women of color. Thus, consideration of how academic environments, such as treatment by faculty and peers, interaction with faculty, and available resources for learning and tutoring, uniquely affect women of color is examined. Several theories, such as critical racial theory, intersectionality, and campus climate framework, highlight the importance of examining individual characteristics and details of the environmental context. This study used data from a sample of 850 women students in 120 U.S. engineering undergraduate programs from 31 four-year institutions. Black women engineering students experienced and perceived more differential treatment because of their race/ethnicity but interacted more with faculty than White women students. This study provides critical implications for policy and practice regarding how administrators and faculty members can design engineering programs to create better climate and offer resources for women of color students.

Cultural ideals of gender normativity creates stereotypes that lead to the identification of specific occupations as being male-oriented or female-oriented, further perpetuating institutional sexism. Research shows that stereotypical beliefs are pervasive and exist in all professions, including higher education. Women academics, especially in the STEM fields, are still present in low numbers and often have to sacrifice marriage and family for their careers due to the overarching masculine organizational structure that forms the basis of most higher education institutions. This chapter will discuss the history of women in academia, gender equity in higher education, and the consequences of gender normative language on women in academia.

This chapter documents the experiences of the ongoing journey of an African American female physicist. They correspond to those in documented studies of other African Americans and females in both the specific field of physics as well as the broader area encompassing Science, Technology, Engineering, and Mathematics (STEM). While there are some anomalies, when scaled with the norm of these groups, there is a thread of consistencies in the obstructions and difficulties that seem to be unique to mostly African Americans and on a smaller scale to White females. The intent of this writing is to shine a light on the status of affairs particularly in the scientific Ph.D. community, an area that many have felt was immune to the difficulties faced by African Americans on the lower end of society. It is evident that our society is neither "post-racial" nor "post-sexist", even on the higher intellectual turf.

This chapter explores issues that have influenced Black women in Canada and the Caribbean in their pursuit of advanced degrees and regular academic roles in Computer Science (CS) and related fields. The concern is with the low representation of Black women in such roles and the reasons why some women have chosen the field and others have fled from it. The issues are related to gender and race and have been gathered from the scientific literature on Computer Science Education. Black women in Canada and the Caribbean who are on paths towards and in regular academic roles in CS were surveyed about how these issues have influenced them and their peers for or against CS in general and more specifically advanced education in Computer Science, since these such choices for CS are a requirement for regular academic roles in CS. This work delivers insights into an understudied locale for an underrepresented group in a field critical to economic development in Canada and the Caribbean. Based on our findings, we make recommendations to increase the number of this underrepresented group.

There is a critical need to develop initiatives aimed at expanding and diversifying the healthcare workforce, beginning with medical education. This chapter addresses racial and gender disparities in academic medicine and biomedical research. Racial and ethnic minorities and women remain underrepresented in medicine, biomedical research, and healthcare leadership. These disparities are national issues and have far-reaching effects which translate into significant educational and healthcare-delivery disparities. Research has shown that health care professionals from underserved backgrounds are more likely than others to work in underserved areas which would address current shortages in health professionals for the medically underserved. The authors describe their theory for the existence and persistence of these disparities. They offer evidence of these disparities while concluding with current initiatives to address these disparities, calling for innovative approaches to training underrepresented minorities and women as physicians and biomedical research scientists.

This project examines the critical issues of stereotypes and gender bias that face students, particularly women, of color in higher education and into their careers, as well as the lack of representation of women of color in higher education STEM related disciplines. Interviews with three, African American females from their start as undergraduate students in Biology, into graduate programs, and eventually with careers as science educators in public science education classrooms, as well as a review of current literature, highlights the need for women of color within STEM disciplines, as well as strides that have been made toward increasing the number of both females and males of color in STEM related areas.

The existing literature does not focus on risk-taking differences within females and how these differences may influence academic choices in Science, Technology, Engineering, and Mathematics (STEM) fields. The current project examined differences in academic risk-taking between STEM and non-STEM female

The discussion promoted in this text is part of the inquiry of PhD in Education in progress since 2012 at Fluminense Federal University (Niterói, Brazil). The above-mentioned doctoral research is dedicated to the study of educational trajectories of prestigious black professionals in Brazil. The main objective of this text is to bring up a reflection on the studies about the relation between prestigious college courses and race/skin color in Brazil carried out in the contemporaneousness. For this purpose, it is necessary to analyze the process of selection of the professional careers that demand the ownership of the bachelor's degree (higher education). This text is supported by studies about social stratification. Bibliographical inquiry was the methodology elected for the development of the present text.

Surgery is a medical specialty that has a rich history of rigorous training, and the development of a young surgeon requires both individual study and external reinforcement to ensure competency. Although, women are entering medical school at increasing numbers, they only make up to 19% of American surgeons. Minority women represent an even smaller number. Recent studies document that minorities report challenges during surgical residency that may inhibit successful surgical training. These challenges are explored as they relate to limited mentoring and collegial isolation while training in trauma surgery.

Preface

THE PERVASIVE AND PERSISTENT CHALLENGE OF RACIAL AND GENDER DISCRIMINATION IN STEM SETTINGS

In a recently released report for the National Science Foundation entitled, *Double Jeopardy? Gender Bias Against Women of Color in Science,* 60 female scientists were interviewed about their treatment in their work settings, 100% reported encountering gender bias (Williams, Phillips, & Hall, 2014). Among African American scientists interviewed for the report, 48% reported being mistaken for administrative or custodial staff at their own place of employment. This study and many others soundly indicate that gender-based and racial-based discrimination are persistent problems haunting both educational and professional STEM settings (Ceci, Williams, & Barnett, 2009; Espinosa, 2011; Freeman, 2004; Hollenshead & Thomas, 2001). With mounting studies revealing the pervasiveness of discrimination in STEM settings, the role of gender bias and racial discrimination play in the underrepresentation of woman in STEM fields has gained a great deal of attention as the shortage of STEM professional amasses. It is broadly accepted across STEM fields that addressing issues that negatively impact the recruitment and retention of women in STEM field is important not only to address the shortage of STEM professionals but to also uphold commitments to equality of opportunity and the expansion of human capital and productivity. The consequences of racial-based and gender-based discrimination in STEM fields has brought the importance of addressing these issues to the fore. In response, sectors of both government and industry have poured enormous amounts of money into programs and research aimed at addressing the underrepresentation of women and minorities to STEM professions. Comprehensive reform initiatives range from programs designed to combat STEM stereotypes held by teenage girls through mentoring such as National Aeronautics and Space Administration launches NASA G.I.R.L.S. online mentoring program to multi-million dollar grants such as the Robert Noyce Scholarship Program awarded to universities to increase participation in STEM careers.

THE STATE OF INTERVENTIONS

Despite ongoing efforts to end gender and racial gaps STEM professions, discrimination in work and educational settings still exists. Long-standing structural barriers and inappropriate organizational norms continue to subject women to disparaging practices. Many of the work place initiatives fall short of developing adequate measures to address not only subtle, but even blatant discriminatory behaviors

(Ceci, Williams, & Barnett, 2009; Xu, 2008). Undeniably, drastic changes are needed in both work and educational settings to end the underrepresentation of women and minorities in STEM professions. Changes must take place to not only alleviate to the current shortage STEM workers, but to ensure there are adequate STEM professionals in the workforce pipeline. According to U.S. Census Bureau, in 2011 racial and ethnic minorities comprised 50.4 percent of the children in the United States under the age of 1. In 2020, the majority of the population under the age of 18 will be an ethnic or racial minority (Colby & Ortman, 2014). If current practices continue, more than half of the potential STEM work force could be the target of gender and/or racial discrimination leading to a reduce recruitment and retention of STEM professional. Ending or at least reducing the minority and female gap in STEM fields necessitates an end to hostile and prejudiced environments that discourage women of color from entering or succeeding in STEM settings. Not only does discrimination hurt women, but many aspects of organizational effectiveness in work and school settings suffer as well. Loss is produced in the form of decreased productivity, increased attrition, and civil litigation.

SEARCHING FOR A SOLUTION

What can be done to prevent or reduce gender and racial discrimination in STEM educational and profession settings? Gender-based and racial-based discrimination stems from a complex mixture of and structural and cultural factors that are rooted and social norms as well as racial and gender stereotypes. At the core of the problem are social beliefs that shape societal perceptions regarding who is a scientist or a mathematician or an engineer. Compounding the problem, many of the individuals engaging in gender-based and racial-based discriminatory practices are not aware of their biases. Discriminatory work place behaviors are carried out in a status quo manner and as a result STEM work environments maintain their male-dominated status quo. Because many unknowingly behave in prejudiced ways, combating racial and gender bias extends beyond anti-discrimination policies. After all, there is no reason to change one's behaviors, if one believes he is already in compliance with such policies. Therefore, one of the first steps in eradicating racial and gender-based discrimination is to educate those acting in discriminatory ways about the words and actions that convey prejudice. Women, especially women of color working in STEM professions have a range of experiences that they bring to the conversation when examining words and actions that convey prejudice. The sharing of their personal stories can be an effective instrument for increasing wrongdoers' awareness of the pervasiveness discriminatory practices as well as the plethora of negative consequences associated with those practices.

THE PURPOSE OF THE BOOK

This book conveys the pathways and journeys of women on color in regards to their experiences as STEM professionals. Research based on those experiences as well as deliberations from the literature on gender relations and power for women of color is shared. Both the nature and significance of racial and gender discrimination in STEM settings are explored. The book also draws on the stories of women of color not only to highlight the prejudice encountered, but to also share pathways for overcoming prejudicial treatment. The chapter authors for this text are employed in a variety of settings including business, education, and medical settings. The chapter authors offer unique perspectives on gender and racial discrimination with regard to their specific settings. By including chapters about women of color in variety of STEM settings, including medical settings, this book contributes to the existing literature on women of color in STEM in a distinct manner.

Organization of the Book

The book is organized into 13 chapters. A brief description of each of the chapters follows:

Chapter 1: Grace under Fire – Examining the Underrepresentation and Persistence of Women in STEM Professions from an Ecological Systems Theory Perspective

This chapter explored a myriad of experiences that were communicated as contributing to the selection and maintenance of a STEM career for the female participants.

Chapter 2: Constructing Conducive Environment for Women of Color in Engineering Undergraduate Education

This chapter provides crucial implications for policy and practices concerning how administrators and faculty members can design engineering programs to create a healthier climate and to propose resources for women of color.

Chapter 3: Barriers to Success – Disadvantages of Gender Normative Language to Women in STEM

This chapter will talk about the account of women in academia, gender equity in higher education, and the cost of a gender normative language on women in academia.

Chapter 4: Overcoming the Myriad of Obstacles – The Ongoing Journey as a Female African American Physicist

This chapter documents the experiences of the ongoing journey of an African American female physicist. The target of this writing is to reflect a light on the status of affairs particularly in the terminal degree population.

Chapter 5: Standing Alone in Computer Science Education – A Story of Black Women in Academia

This chapter delves into problems that have plagued Black women in Canada and the Caribbean in their quest of advanced degrees and standard academic roles in Computer Science (CS) and related fields.

Chapter 6: Race and Gender Inequalities in Medicine and Biomedical Research

This chapter attends to racial and gender inequality in academic medicine and biomedical research. Racial and minorities and women remain sorely lacking in medicine, biomedical research, and healthcare administration.

Chapter 7: African American Women in STEM Education – The Cycle of Microaggressions from P-12 Classrooms to Higher Education and Back

This chapter examines the critical issues of stereotypes and gender bias that confront students, for the most part women, of color in higher education and into their careers, along with the deficiency of representation of women of color in higher education STEM related branches of learning.

Chapter 8: Are Female STEM Majors Academic Risk Takers?

The current project examined differences in academic risk-taking between STEM and non-STEM female students.

Chapter 9: Fortitude – A Study of African Americans in Surgery in New York City

Covert racism is ever-present in the field of medicine and medical education. The most heavily impacted are African American and Caribbean American females and males. This chapter details the actions of whites allies, and a dedication to maintain a continued watchful stance for this protected population.

Chapter 10: Black Women and Science in Higher Education – Not Much Has Changed: Still the Struggle

This chapter gives an account on the experiences of a female STEM faculty member's academic journey, and then entrance into the workforce as she negotiates the high and low points of being a black woman in a science field, as a science educator that will mentor other black women scientists, while also trying to maintain a feasible family life.

Chapter 11: STEM Fields and Ethnic Women in New Zealand – Issues and Opportunities

This chapter examines how public policy engages women in STEM and how it can enact gender issues.

Chapter 12: Prestigious College Courses and Race/Skin Color – Some Thoughts about Higher Education in Brazil

This text examines the context that exists between the opportunity for students of color in Brazil and the advanced college courses offered to them.

Chapter 13: Becoming a Surgeon

This chapter documents the life and experiences of an African American female surgeon as she navigates institutional racism and sexism through a reflective narrative methodology. This chapter also documents her resolve as she continues in her career.

Ursula Thomas
Georgia Perimeter College, USA

Jill Drake
University of West Georgia, USA

REFERENCES

Ceci, S. J., Williams, W. M., & Barnett, S. M. (2009). Women's underrepresentation in science: Sociocultural and biological considerations. *Psychological Bulletin*, *135*(2), 218–261. doi:10.1037/a0014412 PMID:19254079

Colby, S. L., & Ortman, J. M. (2014). *Projections of the size and composition of the U.S. population: 2014 to 2060. Current Population Reports, 25-1143*. Washington, DC: U.S. Census Bureau.

Espinosa, L. (2011). Pipelines and pathways: Women of color in undergraduate STEM majors and the college experiences that contribute to persistence. *Harvard Educational Review*, *81*(2), 209–241. doi:10.17763/haer.81.2.92315ww157656k3u

Freeman, C. E. (2004). *Trends in educational equity of girls and women: 2004* (NCES Report No. 2005-016). Retrieved from U.S. Department of Education, Institute of Education Sciences, National Center for Education Statistics website: http://nces.ed.gov/pubs2005/2005016.pdf

Williams, J. C., Phillips, K. W., & Hall, E. V. (2014). *Double jeopardy? Gender bias against women of color in science*. Report for the National Science Foundation. Retrieved from www.worklifelaw.org

Xu, Y. J. (2008). Gender disparity in STEM disciplines: A study of faculty attrition and turnover intentions. *Research in Higher Education*, *49*(7), 607–624. doi:10.1007/s11162-008-9097-4

Chapter 1
Grace under Fire:
Examining the Underrepresentation and Persistence of Women in STEM Professions from an Ecological Systems Theory Perspective

Ursula Thomas
Georgia Perimeter College, USA

Jill Drake
University of West Georgia, USA

ABSTRACT

Understanding why women are underrepresented in various Science, Technology, Engineering, and Mathematics fields remains an important area of research. In the United States and in many industrialized nations around the world, STEM professions remain male dominated. Explanations for why women are not participating STEM professions are many and diverse. The Ecology Systems Theory (EST) presents a lens through which the causes for the continued underrepresentation of women in STEM fields may be examined. EST is widely accepted theoretical framework for exploring the influences that contribute to the development of an individual. The study presented in this chapter explored the familial, educational, economic, and social experiences of 125 female participants working in a STEM field. Findings suggest there are influences at specific levels in EST that can and do affect the educational and career aspirations of women in relationship to STEM fields.

INTRODUCTION

The importance of recruiting and retaining women in science, technology, engineering, and mathematics (STEM) fields is a long-standing, widely accepted notion in the United States as well in many other nations around the world (American Association of University Women, 2010; Ceci, S., Williams, W., & Barnett, S., 2009; Kleinhubbert, 2013; Gayles, 2012). As far back as fifty years ago, researchers began

DOI: 10.4018/978-1-5225-0174-9.ch001

writing about the lack of women in science (Rossi, 1965). Recently, the call to recruit and retain women in STEM careers has reached all-time highs. The looming shortfall of professionals with STEM expertise has spurred the release of journal articles, books, and reports stressing the importance of increasing the number of females and minorities who major and then choose a STEM career (Committee on Equal Opportunities in Science and Engineering 2004; National Science Foundation (NSF) 2006a, b; Southern Education Foundation, 2005; Wagner, 2008; Bellanca, 2010). According to the Bureau of Labor & Statistics (2010), by 2018 computer and mathematical science occupations are projected to add 785,700 new jobs to the U.S. economy. Jobs in these fields is projected to grow at double the average pace for almost all other occupations. The projected demand for a workforce with degrees or credentials in a STEM field, coupled with the expected retirement of many current STEM professionals, makes recruiting and retaining women of particular importance (NSF, 2006b; Southern Education Foundation, 2005).

The Untapped Female Workforce

More women are attending universities and colleges than ever before in the history of the United States. The number of females enrolled in post baccalaureate programs has exceeded that of males since 1988 (National Center for Educational Statistics, 2013). The gap between females and males enrolled in college is even greater among Hispanics and African Americans than their white counterparts, (Lopez & Gonzalez-Barrera, 2014). The ratio of males to females earning advanced degrees furthers reveals the untapped potential of females as STEM professionals. As of 2011, among adults 25 and older, 10.6 million U.S. women have master's degrees or higher, compared to 10.5 million men (National Center for Educational Statistics, 2015). Yet, amid the record-setting number of educated females who could potentially pursue a STEM career, there continues to be a formidable gap between women who earn STEM degrees and pursue a career in a STEM field as compared to their male counterparts (Freeman, 2004). Even more disheartening is the fact that for the past three decades policymakers, educators, researchers, and others have spent a great deal of time and resources implementing initiatives aimed at increasing the number of females in STEM careers. Numerous grants and programs exist to recruit more women into STEM careers. Some of the more well-known programs include: *Educate to Innovate*, *The Girl Scouts STEM Program*, *The SciGirls Seven*, and *Techbridge*. Yet, except for few fields such as biology and chemistry, little progress has been made in increasing the representation of women in STEM fields. Moreover, some fields, such as computer science and mathematics, have seen a decline in women earning bachelor degrees (AAUW, 2013; National Science Board, 2014)

The benefit of increasing the number of women who pursue and then attain a career in a STEM field is not limited to a larger more diverse workforce. The unique perspectives women possess can enrich and shape innovations made in STEM disciplines in distinctive ways. Case in point, recent improvements to Facebook's features such as the photo viewer and news feed were the result of suggestions from female engineers (Huhman, 2012). With STEM professionals solving problems and working on improvements that serve males and females, it is logical to include perspectives from both genders. When the perspectives of only one gender dominate the work of STEM professionals, the products or solutions presented may include incorrect assumptions or worthwhile innovations may be overlooked. It is not difficult recognize that society as a whole is "enriched when women fully contribute the advancement of science and technology" (Milgram, 2011, p. 5). The real question is not why do we need more women in participation in STEM professions, but how can we convince more women to choose a STEM career. This is a lingering question, which has yet to be fully answered.

Reasons for the Under-Representation of Women in STEM Fields

Many studies have examined factors believed to have contribute to the under-representation of women in STEM fields (Ceci, Williams, & Barnett, 2009; Isaac, 2001; National Science Foundation, 2011; Riegle-Crumb & Moore, 2013; Griffin, 2009; National Science Board, 2014; Xu, 2008; AAUW, 2013). The factors investigated cover a wide range of range environmental and gender-related characteristics. Neurological differences, educational experiences, socialization patterns, and self-efficacy are among the many explanations cited as gender specific characteristics that may be associated with the gender gap. One area that has not received a great deal of attention is the role family members have in the recruitment and retention of women in STEM professions. Many of the aforementioned factors are impacted by one's home environment as a child and an adult (Bronfenbrenner, 1979). Also, the manner in which one responds to the negative conditions found in a challenging, and perhaps discriminatory, work environment are connected to one's upbringing and home environment. Considering the well-established impact family can have on the development of an individual, it makes sense to explore the extent to which the interactions between family and an individual may influence one's interest and ability to obtain and then maintain a career in a STEM profession (Rozek, Hyde, Svoboda, Hulleman, & Harackiewicz, 2015; Sheldon & Epstein, 2005; Tziner, Loberman, Dekel, & Sharoni, 2012).

In this chapter, the professional and personal experiences of 125 women who have successfully acquired and maintained a STEM career are considered and examined through the lens of Ecological Systems Theory. Familial, educational, economic, and social factors that were reported as contributing to the selection and maintenance of a STEM career are examined. Explanations for the unique ways in which family member's contributed to these women's success as STEM professionals are discussed.

LITERATURE REVIEW AND THEORETICAL UNDERPINNINGS

Ecological Systems Theory

The Ecological Systems Theory (EST) developed by Urie Bronfenbrenner (1917-2005) presents a model for understanding the interaction between the sociocultural influences in one's environment and an individual's development. The EST asserts that human beings exist and develop within complex systems of environmental structures and relationships and that each of these systems in their own distinctive way exert influence on the development of an individual. (Tissington, 2008; Whiston & Cinamon, 2015). According to the Ecological Systems Theory, individuals exist in five nested layers of systems (see Figure 1). The five systems are the Microsystem, Mesosystem, Exosystem, Macrosystem, and the Chronosystem. The Microsystem, which is the inner most level of EST, refers to the bidirectional and reciprocal interactions occurring in an individual's immediate surroundings including physical, social, and psychological interactions. One's family members, peers or co-workers and close friends are a part of the Microsystem.

The Mesosystem consist of the broader social system and includes the connections between two or more microsystems. The extent to which a parent influence his or her child's teacher and vice versa is an example of a mutually influential relationship encompassed in this system. Encircling the Mesosystem is the Exosystem. The Exosystem is the circle of influence exerted on an individual from the corporate culture of one's work or a community's values or customs. In general, individuals have little

Figure 1. Illustration depicting the five levels of the Ecological Systems Theory by Urie Bronfenbrenner (Bronfenbrenner, U. (1994). Ecological models of human development. In T. Husen & T. N. Postlethwaite (Eds.), International Encyclopedia ofEducation (2nd Ed., Vol. 3, pp. 1643-1647). Oxford, England: Pergamon Press).

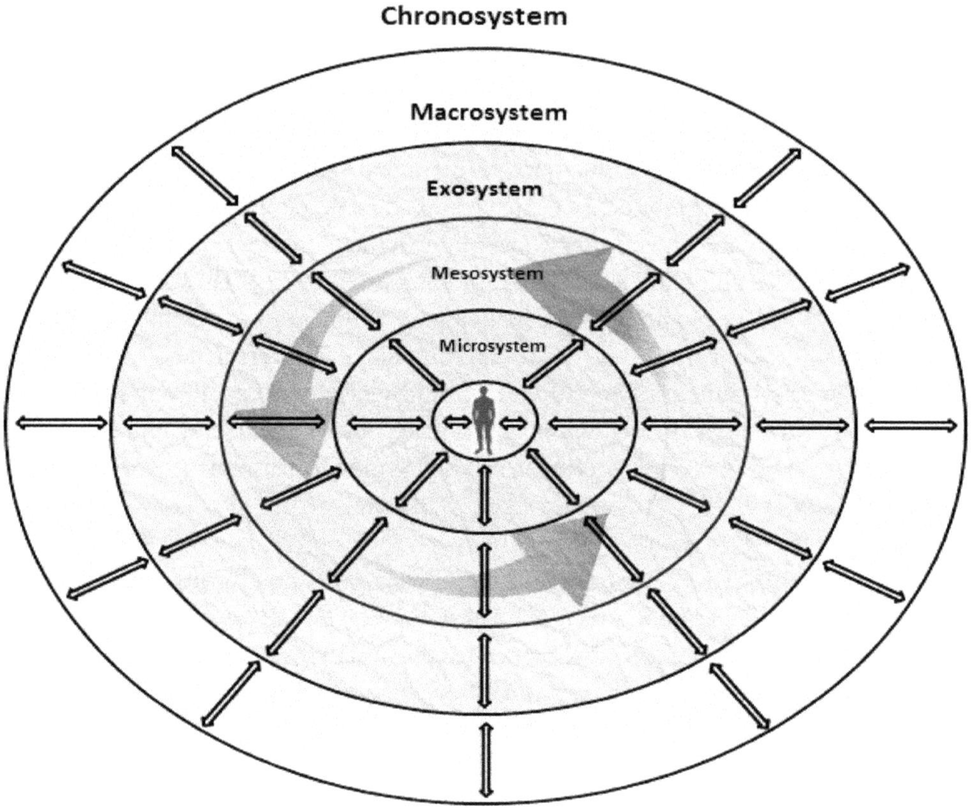

or no influence on the Exosystem, but this system characteristically "has a stronger and more direct influence on individuals" (Whiston & Cinamon, 2015, p. 46). Encompassing the Exosystem and all the systems nested within it is the Macrosystem. This system includes societal values, customs, and laws. When societal beliefs regarding the role of women in society influences a young girl's academic course taking patterns, this system is exerting influence the individual. The final system is the Chronosystem. This system explains the influence of the dimension of time on the individual and well as all the nested systems (Lakic, 2012).

When considering the Ecological Systems Theory, factors reported as contributing to the underrepresentation of women in STEM professions can convincingly be categorized into one of the five systems. Leaving a STEM position due to frustration over receiving less pay for the same level and quality of work, would be example of an interaction in the Exosystem in which the corporate environment influences administrators' decisions regarding salary. Leaving a STEM position because of constant pressure from one's spouse to be at home more, would be the result of interaction between the individual and his or her spouse at the Microsystem level. Recognizing where the interaction exists and how the interactions influence an individual's behaviors can be important in determining which of the current and proposed initiatives aimed at addressing the shortage of women in STEM could produce the best results. According to EST, the power of the individual to influence the system diminishes the further away from the system is form individual's immediate surroundings. Additionally, the timing in which proposed solutions are implemented matter as well. For example, ensuring women are not sexually harassed in the work place, does not greatly influence the number of female, high school seniors who decide to take AP STEM courses. However, increasing the number of tuition-free scholarships offer to college freshman majoring in a STEM discipline is more likely to exert influence on a high school student's course taking decisions. Hence, the right treatment has to be given at the right time in one's development for optimal influence. So, if the Ecological system(s) and the developmental period(s) can be identified which have the greatest impact on one's career aspirations can be discovered, then corresponding interventions delivered during the appropriate developmental period and in the precise ecological system can be carried out.

Career and Educational Influences at the Microsystem Level

The Microsystem level is considered to be the most influential level in Ecological Systems Theory in relationship to the development of an individual. Interactions between an individual and his or her family members, close friends, teachers, and peers or co-workers occur in the microsystem level. Numerous studies have shown that interactions between children and their parents influence educational and career decisions in regards STEM fields. Zeldin and Pajares (2000) found that participants' self-efficacy for math-related careers was shaped by positive perceptions of family members', teachers', and peers'. Ferry, Fouad, and Smith (2000) found that parental encouragement had "significant" direct effects on grades in math and science courses. Jeeyon and Mijin (2015) studied 207 Korean college students of which 127 were females and found that the better the quality of a student's attachment to his or her mother and father, the higher the student's self-efficacy in career decision making and the student's commitment to career choices. The findings of a longitudinal study of 281 adolescences, conducted by Germeijs and Verschueren (2009), suggest that "higher levels of perceived security with parents are beneficial to adolescents' process of choosing a major in higher education" (p.475). Tziner, Loberman, Dekel, and Sharoni (2012) observed "a positive correlation was found between job characteristics deemed important by the son/daughter and the characteristics they perceived to be descriptive of their parents' jobs" (p. 103). Rozek, Hyde, Svoboda, Hulleman, and Harackiewicz (2015) found that increasing a mother's STEM utility value (associated usefulness and merit of STEM) is associated with a significant indirect effect on the number of 12th grade STEM courses taken.

The extent to which teachers and their instructional practices impact learning is well-documented (O'Brien, Blodorn, Adams, Garcia, & Hammer, 2015; Vgotsky, 1978; Wentzel, 1998). Much like parents, teachers can and do influence educational and career aspirations in relationship to STEM fields. According to Muller, Katz, and Dance (1999), high schools that foster positive relationships between students and teachers have higher graduation rates and have students who score higher on math achievement tests (Muller et al. 1999). Nye, Konstantopoulos, and Hedges (2004) found that there are substantial differences among teachers' ability to produce achievement gains in their students in regards to mathematics achievement.

Intelligence is one of the many personal characteristics of an individual that is developed in the Microsystem level of EST. Family and the materials in one's immediate surroundings have also been shown to have on intelligence. The interaction between parents and their children is considered one of the largest influencers. The role of quality experiences occurring between an individual and his or her parents/guardians is so important that even changes in family structure can negatively impact on one's Intelligence Quotient (IQ). Black, Devereux, and Salvanes (2010) found that even sudden changes in family size, such as the birth of a twin could negatively impact an individual's IQ.

An individual's characteristics such as intelligence, aptitude, and gender are attributed to the underrepresentation of women in STEM professions (Freeman, 2004; Halpern,et al., 2007; Ceci, Williams, & Barnett, 2009). The excuse that men are "mathematically superior and innately better suited to STEM fields" remains a common belief despite a growing body of research indicating otherwise (Corbitt & Hill, 2010). Even when females show promise in relationship to STEM fields, intelligence is not considered to be a significant factor in the success. Murphy and Sullivan (1997) studied project SPLASH, a program for minority adolescent girls with significant promise in science and mathematics. In the SPLASH project, girls and parents were asked to identify the most substantial reason for why they or their daughters were selected as a participant in the program. These girls and their parents chose options related to effort rather than intelligence. But, 80% of the girls as with 79% of their parents selected "smart" among the other characteristics provided in a checklist of 10 descriptive traits about their daughter. Girls viewed their performance problems in content areas as a personal failure. The research found that boys preserve their self-esteem despite lack of achievement and participation in a content area. Granted, gender differences do exist in relationship to brain size, neurological functioning, and IQ, but the extent to which those differences contribute to the intellectual achievement and the underrepresentation of women in STEM professions continues to be a widely debated topic (Halpern, 2012). As a whole, males tend to have a brain that is 10% larger than their female counterparts (Giedd, Raznahan, Lenroot, & Mills, 2012). However, Lynn and Irwing (2004) found that there was little to no difference in intelligence between males and females. Males have been found to outperform females on spatial awareness and visualization assessments. Yet, females outperform males on some verbal assessments (Colom, Contreras, Arend, Leal, & Santacreu, 2004) Although school-aged males tend to outperform girls at the highest levels on science and mathematics standardized assessments, females earn better grades in mathematics and science courses (Mosatche, Matloff-Nieves, Kekelis, & Lawner, 2013).

Self-efficacy refers to an individual's belief in his or her ability to execute specific behaviors to attain a designated goal "under designated situational demands" (Bandura, 2007, p. 646). Self-efficacy and related personal qualities such as self-esteem are primarily developed in one's Microsystem and have been examined in relationship to the shortage of women in STEM fields. Unlike gender differences in regards to intelligence, research findings pertaining to the relationship between self-efficacy and educational attainment in STEM fields are more consistent and suggest that a positive correlation exists (Colbeck et al. 2001; Lent et al. 2005; Leslie et al. 1998; Rayman and Brett 1995; Zeldin and Pajares 2000*)*.Roué (2007) studied a group of 2800 girls taking part in a summer technology and engineering model program. The results show that the greatest forecaster of math and science interest was self-esteem comprising 36.4% of the variability in the interest scale. Self-efficacy was the second highest forecaster accounting for 26.5% of the variability in STEM participation. MacCorquodale (1984) studied the self-image ratings of more than two-thousand high school and middle school girls as compared to their level of interest in taking science classes. The findings of study suggest that the universally held image of the scientist are consistent with the self-image of students who have an interest in science to the degree that intelligence, self-confidence, independence and creativity are involved. Lent et al. determined that self-efficacy was positively correlated with females' interest in engineering-related activities and aspirations regarding earning a bachelor's degree in engineering. In fact, a higher level of self-efficacy has been associated with women remaining in a STEM career despite encountering an assortment of barriers.

Career and Educational Influences at the Mesosystem Level

According to Bronfenbrenner (1994), the Mesosystem involves the linkages between entities in the Microsystem such as family and school, and relationships between students and peers. Research exists supporting the idea that relationships between microsystems can and do influence career and educational attainment and aspirations in relationship to STEM fields (Carolan, 2012; Murphy & Sullivan, 1997; Parsons & Chandler, 1995). Carolan (2012) conducted a longitudinal study examining the relationship among high school size, social capital, and adolescents' mathematics achievement and found that when parents of adolescent friends know each other achievement in mathematics is significantly predicted. Parsons and Chandler (1995) studied the effects of mother-daughter science club, which was established to lessen adolescent girls' progressively more harmful attitudes toward mathematics and science. Towards the beginning of the study, among the girls expressing interest in science, their preference was for the life sciences. By the end of the project, a quarter of the girls who initially preferred life sciences, articulated an interest earth and physical sciences. The mothers' participation in the school-based program influenced their daughters' dispositions toward science fields. Many of the mothers, shared recollections of damaging experiences from their own school science courses and attempted to provide positive experiences that were more constructive for their daughters. Sheldon and Epstein (2005) found that "schools that effectively implemented activities that encouraged parents to participate with their children in home learning activities reported improved percentages of students who were proficient in mathematics from 1 year to the next"(p. 204). The extent to which school-based initiatives aimed at parents impact students' disposition and performances presents demonstrates how interactions at Mesosystem level can and do impact educational aspirations in relationship to STEM fields.

Wood's (2002) study looked at whether learning can take place if an individual does not participate fully in the socio-cultural environment. The goal of this research was to understand how socio-cultural factors affect a woman's choice to participate in science, technology, engineering, and mathematics. The study found that belonging and participation in line with socio-cultural groups became significant features when investigating learning in the field of engineering. This study is reflective of the mesosystem as it highlights peer relationships within a specific learning environment.

McPherson (2014) investigated how eight undergraduate African American women in science mathematics and engineering gained access to cultural capital and informal science learning opportunities throughout their educational journey. The study found that African-American women have access to cultural capital and informal science learning inside and outside of the home and school environments across preschool through college settings.

The linkages between church and family are accounted for at the Mesosystem levels. Studies exist which suggests achievement in STEM courses can be positively correlated with participation in religious activities. Jeynes (2009) conducted a study which examined the Bible literacy of students and their grade point averages (GPA). Jeynes found that Bible literacy scores positively correlated with higher GPAs. In a study that compared the achievement of students enrolled in public, charter and religious private schools, Jeynes found that "youth from religious private schools performed better academically than their counterparts in both public charter schools and traditional public schools, even when the results were adjusted to account for socioeconomic status, selectivity, race, and various other factors (Jeynes, 2013, para. 3).

Career and Educational Influences at the Exosystem Level

The Exosystem is the level of ecological systems theory that accounts for those environments and entities that influence the individual indirectly by affecting the people, places, and things in the meso- and microsystems. The manner in which the economic status of the community in which one's family lives impacts the educational and recreational programs available to an individual is an example of an Exosystem interaction. Educational policy and legislative reforms such as the adoption of Title IX, Individuals with Disabilities Education Act (IDEA), No Child Left Behind Act, and the Affordable Healthcare Act, are among the myriad of Exosystem reforms that exert influence on one's meso- and microsystem.

Legislative initiatives directed at increasing the number of women working in STEM fields are a part of the career and educational influences of the Exosystem. The number of initiatives aimed at increasing the representation of females in STEM career is numerous. Whether these initiatives are resulting in more women choosing a STEM career is difficult to establish. There are indicators that many of these programs are making a difference. The National Aeronautics and Space Administration (NASA) has created and supported more than 22 events and programs focused on Robotics and STEM. Of the high school females participating in the various NASA programs, 100% go to college, and 91% pursue STEM degrees. In 2013, the National Science Foundation award 55% of its Graduate Research Fellowships to women (White House Council on Women and Girls, 2014).

Career and Educational Influences at the Macrosystem Level

The Macrosystem refers to the level of EST that describes the outermost layer of an individual's environment. It explains those things that influence an individual life experiences and perspective as a result of belonging to a particular social, racial, political or cultural group and/or being born during a specific historical period. The political ideologies and customs of one's country are forces of influence at this level. The manner in which one's social class, ethnic group, and religious and political affiliations impact an individual's development are explained by the influence of the Macrosystem level on other EST levels. Many of the forces accounted for at this level are assumed to play major roles in the underrepresentation of women participating in STEM careers (American Association of University Women, 2010). Gender-role socialization, cultural values, environmental and social barriers, even economic conditions are among the causes thought to contribute to the underrepresentation of women in STEM fields. Inequities are apparent in social and occupational strata and are especially true for occupations that have been historically earmarked for males in the dominant culture. With only a few exceptions (e.g., Cullen, et al. 2004), research consistently shows that stereotype threat contributes to gaps in academic performance between women and men (Brown, et al. 2000; Brown & Josephs 1999; Gonzales, et al. 2002; O'Brien & Crandall 2003). Melymuka (2008) "found that 63% of women in science, engineering and technology have experienced sexual harassment" in one form or another (para.7). Sexual harassment, a lack mentors, a sense of isolation, and even unfair evaluations have been cited as work environment factors contributing to the lack of women in STEM fields (Porter, 2014). The Society of Asian Scientists and Engineers (2015) questioned several women in the STEM population and found a common set of negative work place themes to include; gender stereotyping, lack of female mentors, and absence of flexibility. Women, who have remained in STEM despite a number of reported issues, have needed to possess "some unusual patterns of socialization in order to reach their occupational destinations" (Rossi, 1965, p.1197). The forms of influence at the Macrosystem level are perhaps the most difficult to alter to address the underrepresentation of women in STEM fields. Yet, it is the influences at the Macrosystem level that likely generate the greatest challenges for women.

Career Aspiration Influences at the Chronosystem Level

The Chronosystem is comprised of the dimension of time. It explains the influence of both change and constancy in one's life. On the surface, it may not seem as though time would be an influential factor in women's educational and career aspirations in regards to STEM fields. Yet, the timing of major events and influences in a female's life can and do impact educational and career decisions. Women are very concerned about their internal "clock". Not wanted to delay getting married and bare children has been associated with when a woman pursues graduate degrees and many professions require degrees beyond the bachelor degree to advanced and in some cases retain a STEM career. When women delay baring children, they are faced with rearing young children in the most demanding portion of their career. For women who do not pursue a graduate degree until later, if they typically have children at a younger age resulting in them having teenagers or young adults when returning to school allowing them to have more time to dedicate to career and/or educational demands. However, woman earning their bachelors, masters, or doctorate degrees at the age of 45 or older are underemployed. So, females often must choose between familial stressors such as parenting young children or waning employment opportunities when pursuing educational and/or career goals albeit in STEM fields or other professional degrees.

THE CURRENT STUDY

Purpose

The purpose of the study was to explore the familial, educational, economic, and social experiences that may contribute to the selection and maintenance of a STEM career for a woman. Grounded theory was used as the unit of analysis. For this study, a research method that develops from patterns in the data collected is most appropriate. An inductive perspective in approach moves from a general to a more definitive meaning. This is based on three components to include concepts, categories, and propositions also known as hypotheses. The pinnacles of the analyses are the concepts. The theory is rooted in the crafting of a concept rather than in the raw data collected.

Strauss & Corbin (1998), authors of "Basics of Qualitative research: Grounded Theory Procedures and Techniques" are two of the model's utmost supporters, and demarcate it as follows: "The grounded theory approach is a qualitative research method that uses a systematic set of procedures to develop an inductively derived grounded theory about a phenomenon (p.12)." Grounded theory's purpose is to unpack or construct an explanation of a phenomenon to include labeling the key elements of certain phenomenon and further classify the connection or relationship of the labeled elements to the contextual factors and processes associated with the study (p.19). The objective is to move from the generic to definitive without deleting the characteristics that make a subject or issue exceptional.

Combining data and theory is the explicit goal of grounded theory. The collection of data, subsequent analysis, and tenets of theory conception are reciprocal and this is achieved rather explicitly in a procedural manner. It is paramount that the asking of questions and proposing comparisons are respectively identified to direct any form of analysis in the process of constructing a theory (p.23). An instance of this

Figure 2. Graph depicting the STEM careers represented in the population studied

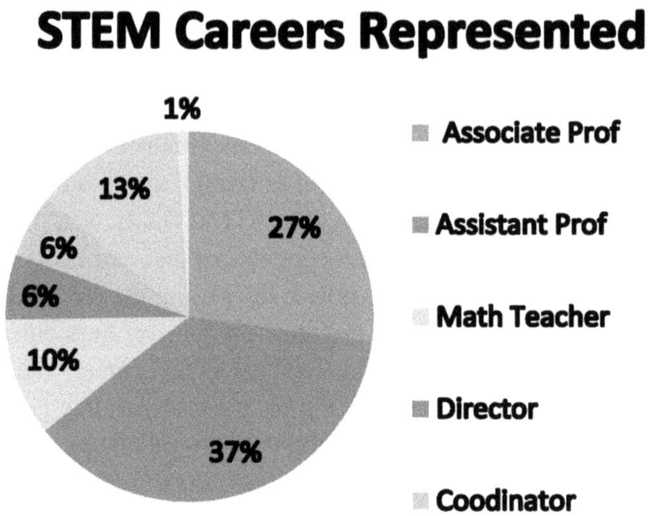

would be purposefully expressing that research questions be open ended and generic instead of hard-pat hypotheses in the theory that emerge and report as a phenomenon. This should as well be congruent to the participants.

Population

The participants for this study consisted of 125 women working STEM fields throughout the United States. While 167 participants began out survey used for this study, only 125 participants complete the entire survey. Eighty-four percent of the participants worked as an educator in some capacity either at the secondary or higher education level. The STEM positions of the participants included K- 12 teachers (10%), university faculty (64%), program directors/coordinators (12%), graduate students (13%), and one participant working in an educational leadership position (1%). The years of experiences were as follows: zero to four years of experience (10.6%); five to ten years of experience (23%); eleven to fifteen years of experience (15.5%); sixteen to twenty years of experience (9.9); and more than twenty years (41%) (see Figure 2).

Instrumentation

The survey instrument use in this study had ten items in total: four questions about demographic and six open-ended six questions (see Figure 3).

The survey was distributed to participants via e-mail. Participants complete the survey on line.

Survey data from six open-ended items was analyzed through open coding and axial coding. Phenomena, or the central ideas represented in the data, were identified in order to create concepts for categories. A basic code list was generated in order to analyze the open-ended survey data line by line. The data from the interviews were coded line by line in order to identify what basic codes group themselves into categories (Strauss & Corbin, 1998). The researcher used NU*DIST to assist in the coding process.

Figure 3. Open-ended questioned presented to the participants of this study

1. Do you believe that you possess particular character traits that enabled you to attain a position in your particular STEM field? If so, what are those traits?
2. Were there significant experiences in your K-12 or college education (positive or negative) that encouraged you to pursue a career in this STEM field?
3. What, if any, familial experiences positive or negative encouraged you to pursue a career in your particular STEM field?
4. What, if any economic considerations impacted your pursuit of a career in this field?
5. What factors personal, professional, and/or financial have enabled you to maintain a career your particular STEM field?
6. Thinking of the success you have experienced as a women in one of the STEM fields, what (if any) formal or informal supports (political, social, cultural, corporate, etc.) have contributed to your success?

Data

Questions 1. In response to question, "Do you believe that you possess particular character traits that enabled you to attain a position in your particular STEM field? If so, what are those traits?" The following themes emerged from the data analyses (Figure 4).

Question 2. Were there significant experiences in your K-12 or college education (positive or negative) that encouraged you to pursue a career in this STEM field?

Figure 4. Graph depicting the character traits identified as enabling participants to attain a STEM position

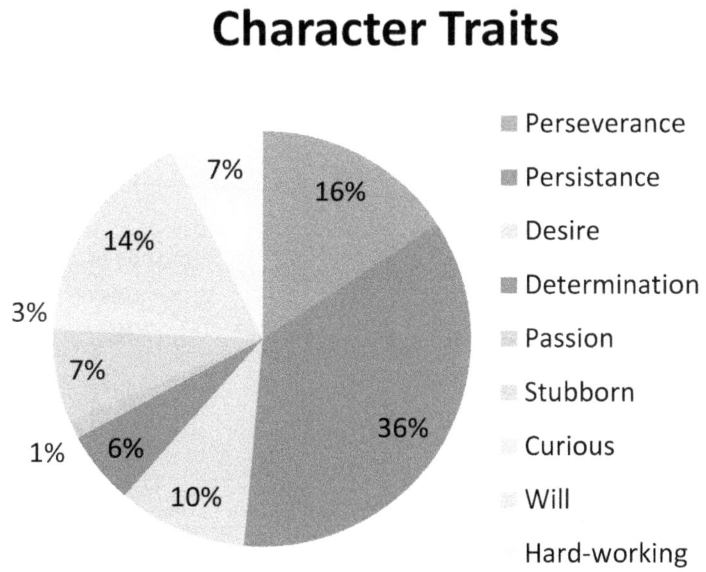

Figure 5. Graph depicting the significant experiences of STEM professionals

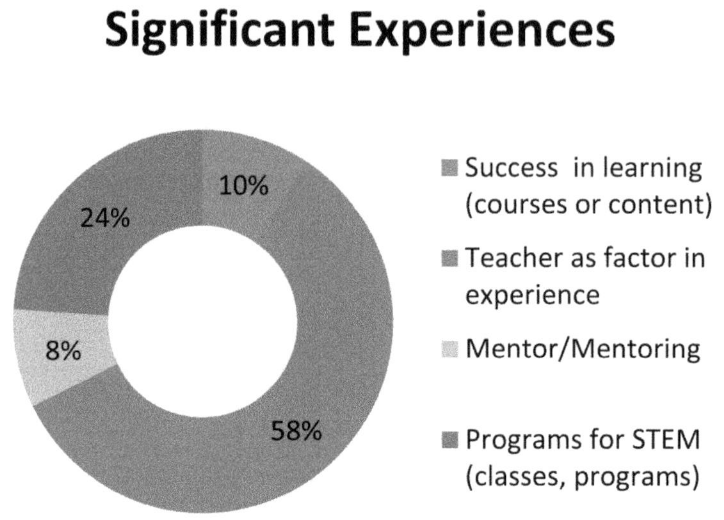

In regards to K-12 experiences, participants mentioned the following as contributing to their selecting a career in a STEM field: former teacher/professor (58%); participation in a STEM program or course (24%); experiencing success as a student (10%); and receiving mentoring (8%) (Figure 5).

Question 3. What, if any, familial experiences positive or negative encouraged you to pursue a career in your particular STEM field?).

Figure 6. Graph depicting the familial experiences of STEM professionals

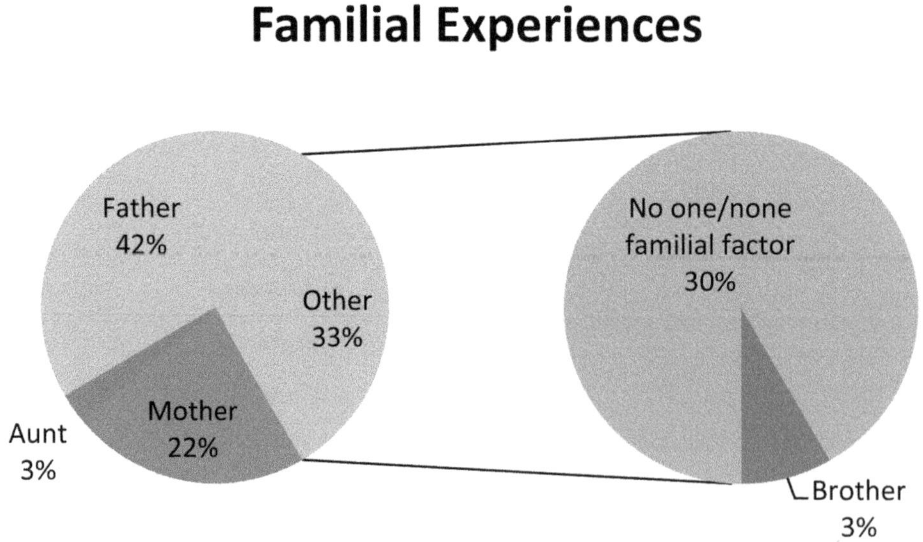

Figure 7. Graph depicting the economic considerations of STEM professionals

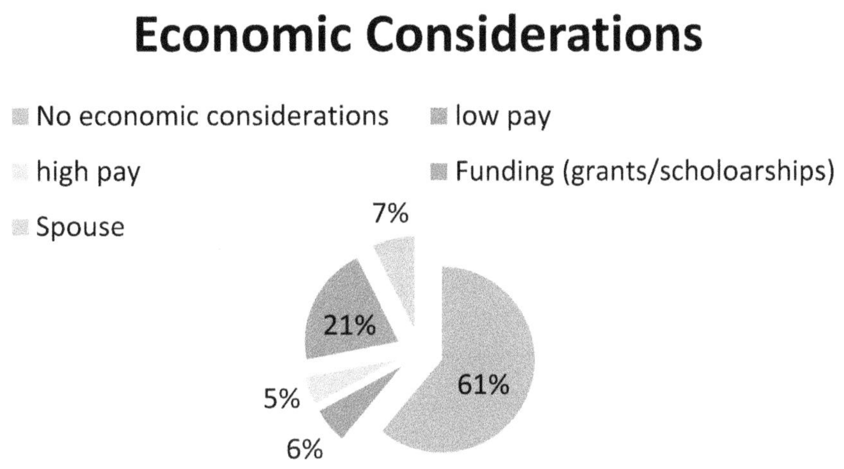

In regards to familial contributions, participants described their fathers (42%) over other family member (mothers, aunts, brother) as being a positive factor in their selecting a career in a STEM field (Figure 6).

Question 4. What, if any economic considerations impacted your pursuit of a career in this field? (see Figure 7)

In regards to the impact financial factors, participants overwhelmingly (61%) responded that no economic considerations impacted their decision to pursue and then acquire a position in a STEM field. Of the financial considerations that were mentioned included: scholarships/grants (21%); spousal support (7%); opportunity for a high paying job (5%); and low pay (6%).

Question 5. What factors personal, professional, and/or financial have enabled you to maintain a career your particular STEM field?

In regards to factors which have contributed to the participants maintaining a STEM career, the following were mentioned: spousal support (32%); external funding (26%); opportunities for development (14%); experiencing success (8%); personal passion/drive (7%); interest in the field (7%); and intellectual ability (6%).

Question 6. Thinking of the success you have experienced as a women in one of the STEM fields, what (if any) formal or informal supports (political, social, cultural, corporate, etc.) have contributed your success?

In regards to other informal supports the top three supports mentioned were: none (25%); mentors (20%); and parent or immediate family members (15%).

In regards to K-12 experiences, participants mentioned the following as contributing to their selecting a career in a STEM field: former teacher/professor (58%); participation in a STEM program or course (24%); experiencing success as a student (10%); and receiving mentoring (8%). In regards to familial contributions, participants described their fathers (42%) over other family member (mothers, aunts, brother) as being a positive factor in their selecting a career in a STEM field. Some participants describe their mother as attempting to negatively impact their pursuit of a STEM career.

In regards to the impact financial factors, participants overwhelmingly (61%) responded that no economic considerations impacted their decision to pursue and then acquire a position in a STEM field. Of the financial considerations that were mentioned included: scholarships/grants (21%); spousal support (7%); opportunity for a high paying job (5%); and low pay (6%). In regards to factors which have contributed to the participants maintaining a STEM career, the following were mentioned: spousal support (32%); external funding (26%); opportunities for development (14%); experiencing success (8%); personal passion/drive (7%); interest in the field (7%); and intellectual ability(6%). In regards to other informal supports the top three supports mentioned were none (25%); mentors (20%; and parent or immediate family members (15%).

Discussion

The following themes emerged from the data analyses. First, there were significant experiences in their K-12 or college education (positive or negative) that encouraged them to pursue a career in this STEM field. Secondly, formal or informal supports (political, social, cultural, corporate, etc.) have contributed to respondents' success. These two themes are highlighted and discussed in turn.

Significant Experiences, K-16 Experiences

In examining the respondents' comments, four overarching constructs emerge; success in learning through courses or content, teacher as a factor in their experiences, mentor/mentoring, and programs for STEM (programs and classes). The respondents experienced immediate success through coursework and content area courses primarily in the high school and undergraduate years. They cited success came is the form of challenges during problem solving and the opportunity to provide proof of being correct when challenged by others in class and in student teaching or tutoring others who were not as proficient in mathematics.

Overwhelmingly reported by respondents were teachers as a significant factor in their experiences. This construct generated the most responses from the respondents. When reconstructing the open codes, the data revealed that in terms of those "who" made the greatest impact on their experiences were female teachers during middle and high school, consistently. They were described as role models in addition to their strong teaching. When the respondents reported on "what" teachers did, they cited encouraging them to pursue teaching in STEM fields. "How" teachers were significant came through tutoring and mentoring as reported by the respondents. When exploring "why" the respondents chose teaching as an option, two very distinct responses arose. They report that though they had great female teachers in middle and high school. They report as many responses that poor or ineffective teachers were as big a factor as effective teachers. The clustered actions of teachers as significant factors include that some respondents wanted to be a teacher or desire to be a teacher. The consequences of teachers as a significant factor were reported as becoming a teacher to offset negative experiences.

Specific programs were reported as a significant factor in the K-16 experience. Respondents report that programs such as summer STEM programs, upward bound, and state scholar program were significant and consequently, pushed respondents toward graduate programs.

Formal or Informal Supports and Contributions to Success

During the analysis of this particular construct, the researchers chose to delineate formal and informal supports. After this action, the researchers proceed to apply the coding method used in the previous construct. In analyzing the data, the researchers found the following clusters: mentors, family/parents, teachers/professor, and programs, the university, professional organizations, culture, finances, funding and grants. The two clusters that reside in the formal and informal supports are mentors and teachers.

Formal supports include mentors, teachers, programs, the university, funding, grants and professional organizations. In a formal capacity, mentors were supportive through community-based programs and assigned university mentors. Because of an effective formal mentor, one respondent credits a formal mentor as responsible for her stayed in graduate school and completing a PhD program. When asked how formal mentors supported them, several respondents stated that they were provided membership into a community that valued the presence of women in mathematics. This community was identified as the university and professional organizations (master's program, math lead teacher district program and EDGE). Grants were also identified as instrumental in supporting the respondents. Teachers were identified as formal supports as they were attached to fellowships and programs.

Informal supports were identified as mentors, family, teachers, culture and professional organizations. According to the respondents, mentors informally surfaced as high school teachers, and colleagues. As respondents spoke about families, they cited that families provided spiritual and moral support. The data also reveals that several families placed value on math and career choice of the respondents.

Culture was a complicated theme in the data. Culture, according the respondents, manifested itself on a number of different levels: academic, familial, and political. Respondents share that the culture of the United States very individualistic and formal supports are critical because academia has its own culture. Traditionally that culture does not support women. The familial support was essential to continue into their career paths.

RESULTS AND IMPLICATIONS

The findings from this study suggest experiences as a female student at either the high school and/or the collegiate level substantially influence women's choosing a career in a STEM field. Fifty-eight percent of the participants indicated that a high school or college professor positively influenced their decision to pursue a career in a STEM field. This finding suggests that high school and university faculty play a major role in ensuring that women who enter their institutions develop an interest in a STEM career. Furthermore, the participants' comments suggest that efforts to educate STEM faculty on the classroom behaviors and activities that promote women's interest in pursuing a career in a STEM field is necessary.

A review of this literature suggests that the most prominent barriers are inadequate academic preparation by elementary and secondary schools, insufficient attention to the psychological barriers that limit persistence in STEM fields, and inadequate support by colleges and universities.

The participants of this study also communicated this importance of familial support in selecting and maintaining a career in a STEM field. Approximately, 64% of the participant indicated that a parent influenced their decision to pursue a career in a STEM field. This finding suggests that more research and resources may need to be directed towards educating parents about the ways in which they can support their daughters' interests in a STEM career.

Solutions and Recommendations

The recommendations are threefold and multifaceted coming from the latest research. Beninger (2014) recommends in the *Catalyst Report* that the field must retain the top talent they have recruited as well as helping women feel like they belong. She also states the field must support women's advancement through sponsorship and ensuring transparency to support women's advancement. The National Women's Law Center (2012) focuses on recommendations for action the utilizing Title IX to include strengthening enforcement of the Title IX. It also recommends that schools and universities need to conduct regular trainings for teachers, professors and administrators about addressing stereotypes, implicit bias and negative climate that may discourage or intimidate girls and women. Lifelong and Equal Academic Futures (LEAF) sponsored by the National Science Foundation (n.d.) proposed improvements to retain and advance women in STEM fields. The recommendations include the following: "(1) preparing graduate students to overcome career obstacles through mentorship and dialogues about career decisions and work-life balance; (2) targeting the subtle biases and stereotypes that hinder the advancement of women; and (3) empowering graduate students to improve universities and STEM workplaces to be more flexible and equitable." O'Brien, Blodorn, Adams, Garcia, and Hammer (2014) recommend an intersectional approach to understanding gender-STEM stereotypes and the participation of women in STEM will help develop a more complete scientific literature that accurately reflects the experiences of people from a variety of ethnic backgrounds.

FUTURE RESEARCH DIRECTIONS

As a solution, the authors of this study propose a framework reflective of an "integrative reconciliation" lifted for Ausubel's theory to engage the whole woman into the STEM field through another lens. A theoretical framework for such an undertaking would be womanist theory. It is imperative that we further construct a view or lens that drills deeper that the somewhat topical approaches offered in the introduction of this chapter. It is the hope of the researchers that future research would include a deeper examination of womanist theory as a perpetual lens for this work.

Recognized the by the American Heritage Dictionary in 1993 defined womanist theory as "having or expressing a belief in or respect for women and their talents and abilities beyond the boundaries of race and class; exhibiting feminism that is inclusive especially of Black American Culture". The terms womanist and womanism were quickly embraced by, and frequently used in depiction of, African American women's fight for self-determination and community, previously and currently.

This terminology was greeted by some audiences for having a more fervent sound than feminist. Alice Walker explained it in 1983 in saying, "Womanist is to feminist as purple to lavender." Womanist theory and womanism aided the discourse in giving body to the psychosocial experience of African American and other women of color who have always been at the head of change to depose the gender and race caste systems. These same women have often been marginalized or rendered invisible in history texts, the media, and mainstream movements led by white feminists or male civil rights leaders (Byrd, 2012). The current STEM field is the face such a caste system.

CONCLUSION

The finding of this study confirm the tenants of Ecological Systems Theory that there are influences across the various levels of EST that can positively or negatively influence educational and career decisions of women in regards to STEM fields. Considering the levels of EST seem to have varying degrees of influence on the decisions females make in regards to STEM education and STEM careers, initiatives aimed at addressing the underrepresentation of women in STEM careers can be strategically aligned with the EST levels with the most influence. Strategically, implementing initiatives may result in more effect use of public and private funding directed toward increasing the number of women in STEM fields. In the end, those STEM fields in which women are not well represented may have a diverse and more effective work force benefitting all who are served by those fields.

REFERENCES

American Association of University Women. (2008). *Where the girls are: The facts about gender equity in education*. Washington, DC: Author.

American Association of University Women. (2013). *Improve Girls' and Women's Opportunities in Science, Technology, Engineering, and Math*. Retrieved August 13, 2015, from http://www.aauw.org/files/2013/02/position-on-STEM-education-111.pdf

Ausubel, D. P. (1970). *The use of ideational organizers in science teaching*. Occasional paper 3. The Science Series of the Science and Mathematics Education Information Analysis Center.

Barile, J. J., Donohue, D. D., Anthony, E. E., Baker, A. A., Weaver, S. S., & Henrich, C. C. (2012). Teacher-Student Relationship Climate and School Outcomes: Implications for Educational Policy Initiatives. *Journal of Youth and Adolescence*, *41*(3), 256–267. doi:10.1007/s10964-011-9652-8 PMID:21404108

Bellanca, J., & Brant, J. (2010). *21st century skills: Rethinking how students learn*. Bloomington, IN: Solution Tree Press.

Beninger, A. (2014). *High potentials in tech-intensive industries: The gender divide in business roles*. Retrieved from Catalyst website: http://www.catalyst.org/system/files/high_potentials_in_techintensive_industries_the_gender_divide_in_business_roles_1.pdf

Bennett-Conroy, W. (2012). Engaging parents of eighth grade students in parent-teacher bidirectional communication. *School Community Journal*, *22*(2), 87–110.

Black, S. E., Devereux, P. J., & Salvanes, K. G. (2010). Small family, smart family? Family size and the IQ scores of young men. *The Journal of Human Resources*, *45*(1), 33–58. doi:10.3368/jhr.45.1.33

Bozeman, S. T., & Hughes, R. J. (2004). Improving the graduate school experience for women in mathematics: The EDGE program. *Journal of Women and Minorities in Science and Engineering*, *10*(3), 243–253. doi:10.1615/JWomenMinorScienEng.v10.i3.40

Bronfenbrenner, U. (1994). Ecological models of human development. In T. Husen & T. N. Postlethwaite (Eds.), *International Encyclopedia of Education* (2nd ed., Vol. 3, pp. 1643–1647). Oxford, UK: Pergamon Press.

Bureau of Labor and Statistics. (2010). *Occupational outlook handbook: Overview of the 2008-18 projections*. Washington, DC: Author.

Byrd, K. (2015). The relationship between media influence and ethnic identity development among low-income African American and white adolescent girls. *Dissertation Abstracts International. A, The Humanities and Social Sciences*, *75*(7-A(E)).

Ceci, S. J., Williams, W. M., & Barnett, S. M. (2009). Women's underrepresentation in science: Sociocultural and biological considerations. *Psychological Bulletin*, *135*(2), 218–261. doi:10.1037/a0014412 PMID:19254079

Chandler, F. T., & Parsons, S. (1995, April). *A case study of a mother/daughter science club*. Paper presented at the Annual Meeting National Association for Research in Science, San Francisco, CA.

Ching-Hua, M., Ying-Chu, H., & Tzu-Wei, F. (2012). The role of the mother-daughter relationship in Taiwanese college students' career self-efficacy. *Social Behavior & Personality: An International Journal*, *40*(9), 1511–1521. doi:10.2224/sbp.2012.40.9.1511

Colom, R., Contreras, M. J., Arend, I., Leal, O. G., & Santacreu, J. (2004). Sex differences in verbal reasoning are mediated by sex differences in spatial ability. *The Psychological Record*, *54*(3), 365–372.

Corbett, C., Hill, C., & St. Rose, A. (2008). *Where the girls are: The facts about gender equity in education*. Retrieved from American Association of University Women website: http://www.aauw.org/files/2013/02/Where-the-Girls-Are-The-Facts-About-Gender-Equity-in-Education.pdf

Carolan, B. V. (2012). An examination of the relationship among high school size, social capital, and adolescents' mathematics achievement. *Journal of Research on Adolescence*, *22*(3), 583-595.

Davis-Kean, P. E. (2005). The influence of parent education and family income on child achievement: The indirect role of parental expectations and the home environment. *Journal of Family Psychology*, *19*(2), 294–304. doi:10.1037/0893-3200.19.2.294 PMID:15982107

Diekman, A. B., Brown, E. R., Johnston, A. M., & Clark, E. K. (2010). Seeking congruity between goals and roles: A new look at why women opt out of science, technology, engineering, and mathematics careers. *Psychological Science*, *21*(8), 1051–1057. doi:10.1177/0956797610377342 PMID:20631322

Eliot, L. (2010). The truth about boys and girls. *Scientific American Mind*, *21*(2), 22–29. doi:10.1038/scientificamericanmind0510-22

Ferry, T. R., Fouad, N. A., & Smith, P. L. (2000). The role of family context in a social cognitive model for career-related choice behavior: A math and science perspective. *Journal of Vocational Behavior*, *57*(3), 348–364. doi:10.1006/jvbe.1999.1743

Freeman, C. E. (2004). *Trends in educational equity of girls and women: 2004* (NCES Report No. 2005-016). Retrieved from U.S. Department of Education, Institute of Education Sciences, National Center for Education Statistics website: http://nces.ed.gov/pubs2005/2005016.pdf

Garmon, L. B. (1983). Women in physics: Today and looking toward the future. In B. Lotze (Ed.), *Making contributions: An historical overview of women's role in physics* (pp. 91–119). College Park, MD: American Association of Physics Teachers.

Germeijs, V., & Verschueren, K. (2009). Adolescents' career decision-making process: Related to quality of attachment to parents? *Journal of Research on Adolescence, 19*(3), 459-483. doi:10.1111/j.1532-7795.2009.00603.x

Giedd, J. N., Raznahan, A., Mills, K. L., & Lenroot, R. K. (2012). Review: Magnetic resonance imaging of male/female differences in human adolescent brain anatomy. *Biology of Sex Differences*, *3*(1), 19–27. doi:10.1186/2042-6410-3-19 PMID:22908911

Glass, J. L., Sassler, S., Levitte, Y., & Michelmore, K. M. (2013). What's so special about STEM? A comparison of women's retention in STEM and professional occupations. *Social Forces*, *92*(2), 723–756. doi:10.1093/sf/sot092 PMID:25554713

Graff, E. J., Babcock, L., & Laschever, S. (2004). Women don't ask: Negotiation and the gender divide. *The Women's Review of Books*, *22*(2), 4. doi:10.2307/4024424

Halpern, D. F., Benbow, C. P., Geary, D. C., Gur, R. C., Hyde, J. S., & Gernsbacher, M. A. (2007). The science of sex differences in science and mathematics. *Psychological Science in the Public Interest*, *8*(1), 1–51. PMID:25530726

Hill, C., Corbett, C., & St. Rose, A. (2010). *Why so few? Women in science, technology, engineering, and mathematics.* Retrieved from American Association of University Women website: http://www.aauw.org/files/2013/02/Why-So-Few-Women-in-Science-Technology-Engineering-and-Mathematics.pdf

Huhman, H. (2012, June). STEM fields and the gender gap: Where are the women? *Forbes.* Retrieved from http://www.forbes.com/sites/work-in-progress/2012/06/20/stem-fields-and-the-gender-gap-where-are-the-women/

Isaacs, B. (2001). Mystery of the missing women engineers: A solution. *Journal of Professional Issues in Engineering Education and Practice*, *127*(2), 85–91. doi:10.1061/(ASCE)1052-3928(2001)127:2(85)

Jeeyon, L., & Mijin, K. (2015). Parental attachment, career decision self-efficacy, and commitment to the career choice among Korean college students. *Journal of Asia Pacific Counseling*, *5*(1), 39–52. doi:10.18401/2015.5.1.3.

Jeynes, W. H. (2009). The relationship between Biblical literacy, academic achievement, and school behavior among Christian- and public-school students. *Journal of Research on Christian Education*, *18*(1), 36–55. doi:10.1080/10656210902751826

Jeynes, W. (2013). The data are in: Religious private schools deserve a second look. *Public Discourse.* Retrieved from: http://www.thepublicdiscourse.com/2013/05/10218/

Kleinhubbert, G. (2013, September). Lab equality: Sciences struggle to attract young women. *Spiegel Online International.* Retrieved from: http://www.spiegel.de/international/germany/germany-boosts-efforts-to-get-women-in-stem-fields-a-924023.html

Laura, P., Lundy-Wagner, V., Drezner, N., Gasman, M., Yoon, S., Bose, E., & Gary, S. (2009). The contribution of HBCUS to the preparation of African American women for STEM careers: A case study. *Research in Higher Education*, *50*(1), 1–23. doi:10.1007/s11162-008-9110-y

Lopez, M. H., & Gonzalez-Barrera, A. (2014, March 6). Women's college enrollment gains leave men behind. *Pew Research Center: Fact Tank*. Retrieved from http://www.pewresearch.org/fact-tank/2014/03/06/womens-college-enrollment-gains-leave-men-behind/

Ma, X., & Klinger, D. A. (2000). Hierarchical linear modelling of student and school effects on academic achievement. *Canadian Journal of Education*, *25*(1), 41–55. doi:10.2307/1585867

MacCorquodale, P., & National Institute of Education (U.S.). (1984). *Self-image, science, and math: Does the image of the "scientist" keep girls and minorities from pursuing science and math?*. Washington, DC: National Institute of Education.

Mao, C.-H., Hsu, Y.-C., & Fang, T.-W. (2012, November26). The role of the mother-daughter relationship in Taiwanese college students' career self-efficacy. *Social Behavior and Personality*, *40*(9), 1511–1522. doi:10.2224/sbp.2012.40.9.1511

Maryland State Department of Education. (n.d.). *STEM education glossary*. Retrieved from http://mdk12.org/instruction/curriculum/stem/pdf/STEMEducationGlossary.pdf

McPherson, E. (2014). Informal learning in science, math, and engineering majors for African American female undergraduates. *Global Education Review*, *1*(4), 96–113.

Milgram, D. (2011). How to recruit women and girls to the science, technology, engineering, and math (STEM) classroom. *Technology & Engineering Teacher*, *71*(3), 4–11.

Mosatche, H. S., Matloff-Nieves, S., Kekelis, L., & Lawner, E. K. (2013). Effective STEM programs for adolescent girls: Three approaches and many lessons learned. *Afterschool Matters*, *17*, 17–25.

Muller, C., Katz, S., & Dance, L. J. (1999). Investing in teaching and learning: Dynamics of teacher–student relationship from each actor's perspective. *Urban Education*, *34*(3), 292–337. doi:10.1177/0042085999343003

National Science Board. (2006). *America's pressing challenge - building a stronger foundation: A companion to science and engineering indicators - 2006* (NSB-06-02). Retrieved from http://www.nsf.gov/statistics/nsb0602/nsb0602.pdf

National Science Board. (2014). Science and engineering indicators 2014. Arlington, VA: National Science Foundation (NSB 14-01).

National Science Foundation. (2006). *Investing in America's future: Strategic plan FY 2006–2011* (NSF 06-48). Retrieved from http://www.nsf.gov/pubs/2006/nsf0648/NSF-06-48.pdf

National Science Foundation, Division of Science Resource Statistics. (2004). *Women, minorities, and persons with disabilities in science and engineering: 2004 (NSF 04-317)*. Arlington, VA: Author.

National Science Foundation, Division of Science Resources Statistics. (2011). *Women, minorities, and persons with disabilities in science and engineering: 2011 (NSF 11-309)*. Arlington, VA: Author.

National Women's Law Center. (2012). *The next generation of Title IX: STEM–science, technology, engineering, and math.* Retrieved from http://www.nwlc.org/sites/default/files/pdfs/nwlcstem_titleix-factsheet.pdf

Nye, B., Konstantopoulos, S., & Hedges, L. (2004). How large are teacher effects? *Educational Evaluation and Policy Analysis, 26*(3), 237–257. doi:10.3102/01623737026003237

O'Brien, L. T., Blodorn, A., Adams, G., Garcia, D. M., & Hammer, E. (2015). Ethnic variation in gender-STEM stereotypes and STEM participation: An intersectional approach. *Cultural Diversity & Ethnic Minority Psychology, 21*(2), 169–180. doi:10.1037/a0037944 PMID:25244590

Perna, L., Lundy-Wagner, V., Drezner, N., Gasman, M., Yoon, S., Bose, E., & Gary, S. (2009, January1). The Contribution of HBCUS to the Preparation of African American Women for Stem Careers: A Case Study. *Research in Higher Education, 50*(1), 1–23. doi:10.1007/s11162-008-9110-y

Porter, J. (2014). Why are women leaving science, engineering, and tech jobs? *Fast Company.* Retrieved from: http://www.fastcompany.com/3037075/strong-female-lead/why-are-women-are-leaving-science-engineering-tech-jobs

Roue, L. C. (2007). Young women's perceptions of technology and engineering: Factors influencing their participation in math, science and technology. Menomonie, WI: University of Wisconsin--Stout.

Rossi, A. S. (1965). Women in science: Why so few? Social and psychological influences restrict women's choice and pursuit of careers in science. *Science, 148*(3674), 1196–1202. doi:10.1126/science.148.3674.1196 PMID:17748114

Rozek, C. S., Hyde, J. S., Svoboda, R. C., Hulleman, C. S., & Harackiewicz, J. M. (2015). Gender differences in the effects of a utility-value intervention to help parents motivate adolescents in mathematics and science. *Journal of Educational Psychology, 107*(1), 195–206. doi:10.1037/a0036981

Satterfield, D., Heisel, S., & Budischak, S. (2013). *Retaining women in STEM careers: Graduate students as the building blocks of change.* Unpublished manuscript. Retrieved from http://www.nsf.gov/news/special_reports/gradchallenge/images/winners/entries/second-place-parasite-ladies.pdf

Sheldon, S. B., & Epstein, J. L. (2005). Involvement Counts: Family and Community Partnerships and Mathematics Achievement. *The Journal of Educational Research, 98*(4), 196–206. doi:10.3200/JOER.98.4.196-207

Snyder, T. D., & Dillow, S. A. (2015). *Digest of Education Statistics 2013* (NCES 2015-011). Retrieved from U.S. Department of Education, Institute of Education Sciences, National Center for Education Statistics website: http://nces.ed.gov/pubs2015/2015011.pdf

Southern Education Foundation. (2005). *Igniting potential: Historically black colleges and universities and science, technology, engineering and mathematics.* Atlanta, GA: Author.

Stout, J. G., Dasgupta, N., Hunsinger, M., & McManus, M. A. (2011). STEMing the tide: Using ingroup experts to inoculate women's self-concept in science, technology, engineering, and mathematics (STEM). *Journal of Personality and Social Psychology, 100*(2), 255–270. doi:10.1037/a0021385 PMID:21142376

Strauss, A. L., & Corbin, J. M. (1998). *Basics of qualitative research: Techniques and procedures for developing grounded theory*. Thousand Oaks, CA: Sage Publications.

Tziner, A., Loberman, G., Dekel, Z., & Sharoni, G. (2012). The influence of the parent offspring relationship on young people's career preferences. *Revista de Psicología del Trabajo y de las Organizaciones*, *28*(2), 99–105. doi:10.5093/tr2012a8

Vygotsky, L. S. (1978). *Mind and society: The development of higher mental processes*. Cambridge, MA: Harvard University Press.

Wagner, T. (2008). *The global achievement gap*. New York, NY: Basic Books.

Wentzel, K. (1998). Social relationships and motivation in middle school: The role of parents, teachers, and peers. *Journal of Educational Psychology*, *90*(2), 202–209. doi:10.1037/0022-0663.90.2.202

White House Council on Women and Girls. (2014). *White House Council on Women and Girls: recent agency accomplishments*. Washington, DC: Government Printing Office Catalog.

Wood, S. L. (2002). *Becoming a Woman Engineer in the Community of Practice: Validity and Value in Engineering- Education Research*. Academic Press.

Xu, Y. J. (2008). Gender disparity in STEM disciplines: A study of faculty attrition and turnover intentions. *Research in Higher Education*, *49*(7), 607–624. doi:10.1007/s11162-008-9097-4

Chapter 2
Constructing Conducive Environment for Women of Color in Engineering Undergraduate Education

Hyun Kyoung Ro
Bowling Green State University, USA

Kadian McIntosh
University of Arizona, USA

ABSTRACT

The engineering field, in particular, struggles to recruit and retain students, especially women of color. Thus, consideration of how academic environments, such as treatment by faculty and peers, interaction with faculty, and available resources for learning and tutoring, uniquely affect women of color is examined. Several theories, such as critical racial theory, intersectionality, and campus climate framework, highlight the importance of examining individual characteristics and details of the environmental context. This study used data from a sample of 850 women students in 120 U.S. engineering undergraduate programs from 31 four-year institutions. Black women engineering students experienced and perceived more differential treatment because of their race/ethnicity but interacted more with faculty than White women students. This study provides critical implications for policy and practice regarding how administrators and faculty members can design engineering programs to create better climate and offer resources for women of color students.

INTRODUCTION

Women of color have been severely underrepresented in Science, Technology, Engineering, and Mathematics (STEM) fields. Despite the gender gap in U.S. colleges and universities closing as women become the majority in enrollment and attainment numbers in most fields (Burge, 2011), achievement levels are unequal among women college students in engineering (Huang, Taddese, & Walter, 2000). In a two

DOI: 10.4018/978-1-5225-0174-9.ch002

decade span (1983-2002), the percent of women holding engineering degrees in the United States only increased from 13.3 percent to 20.9 percent (National Science Board, 2009). Furthermore, the lack of racial/ethnic diversity within undergraduate engineering is of particular concern. From 2002 to 2012, the percentage of historically underrepresented minority students earning bachelor's degrees in engineering rose only slightly from 12.3% to 12.9%; the percentage of African Americans actually declined from 5.3% to 4.2%, and the percentage of Hispanic Americans rose from 7.3% to 9.3% (National Science Foundation, 2015). Given the low participation in engineering by women and racial minorities, a look at underrepresented women undergraduates' enrollment and degree completion in engineering is even more discouraging. The National Center for Education Statistics (2011) reported that 18.3 percent of all bachelor's degrees in engineering awarded in 2010 went to women. Of those, White women received 11.4 percent and Asian women 2.8 percent. As a comparison, together, African American, Latina, and Native American women completed 2.8 percent of degrees. It is obvious that efforts to attract and retain underrepresented women as well as White women to engineering programs are still necessary.

Research indicates that women of color have struggled partly because of being in a double bind – that is being both female and minority (Malcolm, Hall, and Brown, 1976) in terms of disparities in access, disadvantages in experiences, and differences in outcomes (Ong, Wright, Espinosa, and Orfield, 2011). Individual as well as institutional factors are associated with the struggles of women of color students in engineering fields. Many studies and policy reports focus on individual factors, for example, parental socioeconomic status or academic preparedness and their influence on students' access, persistence, and degree completion in STEM disciplines. Larger educational systems, such as STEM programs' climates and resources for academic success have been emphasized in research and policy reports because the chilly climate, lack of diversity, and homogenous environment are negatively associated with women and racial minority students' college experience and outcomes (Seymour and Hewitt, 1997).

Much of the literature has overlooked potentially informative aspects. First, many studies aggregate STEM fields although each discipline may each have their own unique academic policy and culture (Ferrini-Mundy & Güçler, 2009). Women of color in engineering might have different experiences and perceptions from those in physics because each department has its unique faculty cultures, academic and co-curricular programs, and organizational structures, policies, and practices (Terenzini and Reason, 2005, 2010; Lattuca, Terenzini, Harper, & Yin, 2010). Second, the constant need to call attention to access, persistence, and degree completion may distract from the equally important need to address climate with specificity to certain students' perception of treatment by others and available resources for academic success. While recent studies have focused on the access, persistence, and degree completion among specific students such as racial minority men and women students in undergraduate engineering (Lord et al., 2009; Lord et al., 2011; Ohland et al., 2011), they have examined these groups separately and not with a focus on the climate. Third, many STEM studies, including engineering education studies and climate studies focus on student experiences among the entire student body or only delineate between two groups (i.e. male/female, racial minority/non-racial minority) or racial categories. Researchers, however, have not focused much on minority women students' perception of climate in engineering or STEM disciplines, except for a few qualitative studies (e.g., Johnson, 2007, Johnson, 2012). This intersection between gender and race/ethnicity is critical in engineering fields because specific challenges such as mistreatment, discrimination, or access to resources might be missed when groups are examined in isolation (Crenshaw, 2014; Malcolm et al., 1976). Thus, based on the sociohistorical context and realities of inequitable treatment in engineering undergraduate programs (Malicky, 2003; Sosnowski, 2002; Varma & Hahn, 2007), this study chose to address the gap in research by examining women students' reports on climate experiences, such as treatment from their instructors and peers, witnessing offensive word usage, interaction with faculty members, and perception of career projections and the importance of learning/tutoring centers by sub-racial groups: Blacks, Latinas, Asians, and Whites.

BACKGROUND

Critical race theory has been employed in research for decades to examine the salience of identities, specifically race (Bell, 1992; Delgado & Stefancic, 2001; Ladson-Billings, 1998). These studies suggest that racial minorities do not navigate through an environment in which everyone has an equal chance of succeeding (Solorzano, Ceja, & Yosso, 2000). Racial minorities tend to have more negative experiences, feel more isolated, and persist at lower rates (Hurtado, Milem, Clayton-Pederson, & Allen, 1999). Hence, critical race theory insists on not only questioning the assumed context of meritocracy, but also speaking to "the specificity of individuals and groups in social and cultural contexts" (Ladson-Billings, 1998, p.10). Thus, it is important to not simply dismiss the low number of underrepresented students or their varying experiences to lack of qualifications or engagement, but to seek to understand the experiences women of color have within the college environment of their academic disciplines.

Similar to critical race theory, Ogbu (1981)'s cultural-ecological model highlights contextual differences and differentiated experiences. Ogbu utilized cultural ecology or the study of the relationship between the population and environment via cultural values and strategies (i.e. using resources to achieve goals) used to cope with environmental demands (Ogbu, 1981). Ogbu promoted using cross-cultural models in studying cross-cultural groups in order to understand the context of groups' specific influences and demands (Ogbu, 1981). Emphasizing community and institutional factors such as discriminatory structures and instructor-student dynamics (Foley, 2005), Ogbu (1974, 1981, 2003) as well as other ethnographers (Foley, 1990; McQuillan, 1998; Willis,1981), demonstrated that students may adapt or react by seeking alternative competencies or by disengaging. Ogbu's works also focused on the destructive legacy an oppressive system may have in influencing students' behaviors (Foley, 2005). Disadvantaged students have an agency which functions within an inequitable society that would necessitate a systematic social change as well as individual change (Hamann, 2004). While this body of literature was important in refuting claims that some students were simply unsuited for better schools (Ogbu, 1981), it did not efficiently address positive adaptations or classroom environments (Foley, 2005).

Both critical race theory and cultural ecology emphasize the importance of examining impacts on specific identities within certain environments. Thus, in focusing on women and underrepresented minority students, research should explore the role of multiple identities, especially gender and race/ethnicity based on the intersectionality approach suggested by Crenshaw (1989, 2014). Intersectionality promotes analyzing the significance of multiple group memberships and the outcomes these memberships produce. This knowledge would help to better serve the targeted intersection rather than relying on a single-axis framework (Crenshaw, 1989, 2014). Studies often, however, focused on one group (gender or race) and avoided discussion of the other identities (whether privileged or not) at play, and consequently may have reached different conclusions. Intersectionality research addresses this neglect and its resulting misconceptions (Cole, 2009).

As Cole (2009) suggested, researchers' should consider who is included in the categories examined, the role inequality played, and the patterns of similarities. By including the intersections of a population, more understanding can be gained about that specific group rather than relying on broader generalizations. Therefore, when examining engineering students' experiences, it is important to analyze not only female students or minority students, but also the intersection of students' race and gender. In addition, since intersectionality is situated within a specific context which lends power or advantage to certain group memberships (Cole, 2009), it is important to understand the context of engineering fields as a traditionally White male dominated environment. The environment in which students live and the experiences that they have are essential to comprehending the relationship between social categories, practices, and opportunities (Bonilla-Silva, 2006). Women of color as a whole or even specific races and ethnicities of women may have different experiences than White women or other racial minority women because of the environment in which those experiences take place. Finally, assessing similarities across groups is important in maintaining focus on sociostructural systems rather than individuals (Cole, 2009; Crenshaw, 1989, 1994). This framework could provide clearer understanding and better implications to engineering programs by focusing on actionable policies for women of color students.

Despite calls from critical race theorists and advocates for intersectionality, the intersection of race and gender, although essential in understanding how identity relates to experiences, is often overlooked in research and practice (Crenshaw, 2014). One of the limitations of most studies is that researchers did not disaggregate the experiences of female students by comparing White female and minority female students' experiences and perceptions. Espinosa (2011) notes that even when investigating all STEM education literature, there are few studies, especially empirical, about women of color. However, when examined, women of color are shown to have a distinct psychosocial position than other minorities or other women (Cole, 2009; Espinosa, 2011). Thus, women of color engineering students may also have different experiences and perceptions.

MAIN FOCUS OF THE CHAPTER

Student Experience and Climate

In addition to critical race theory and intersectionality, research examining or utilizing campus climate frameworks also helped to guide understanding of students' experiences. This understanding is often missing in engineering studies as they focus on preparation or persistence, but not what happens during college (Lord et al., 2009). Climate has been conceptualized as having multiple dimensions such as psychological (i.e. perceptions), behavioral (i.e. interactions), organizational/structural (i.e., resource allocations), historical (i.e. past policies), and compositional (i.e. representativeness) (Hurtado et al., 1998, 1999; Milem Chang, & antonio, 2005). This framework indicates the various ways in which students experience the climate. Research shows that climate can have profound impacts on students' cognitive, psychological, and developmental outcomes (Hurtado & Carter, 1997; Persaud & Salter, 2003; Whitt, Edison, Pascarella, Terenzini, & Nora, 1998). For example, Baker, Tancred, & Whitesides (2002) found that female engineering students who felt that their academic environment was uncomfortable were less likely to plan to attend graduate school. Moreover, many studies on climate have shown that students perceive their environment differently based on their background and experiences. Privileged groups tend to view the climate more favorably than disadvantaged groups (Milem, 2003; Rankin & Reason, 2005). Therefore, aspects within the climate need to be examined for several subgroups.

Perceptions of Environment

One important aspect of climate is the psychological dimension which takes into account how students perceive their environment and the way they are treated. Studies have found continued existence of a negative climate for women (Malicky, 2003). For example, women engineering students reported that their male peers behaved rudely towards them (Seymour & Hewitt, 1997). In addition, women tend to be more aware of other students' victimization (Haines, Wallace, & Cannon, 2001). Perhaps, then, women may be conscious of the mistreatment of other students subsequently influencing their view of the climate. Furthermore, race appeared to influence the character of relationships with both peers and faculty members (Justin-Johnson, 2004). Women of color have different experiences not only from other White women, but also from men of color (Varma & Hahn, 2007). Women of color might feel ignored, unsupported, and feel as though they are not wanted or even expected in the field (Ong, 2005; Varma, Prasad, & Kapur, 2006). They also experience microaggressions which are both gendered and racialized (Sosnowski, 2002). Based on this literature, students' perception of climate may be different for women of color who perceive their peers as being treated and treating others differently or witnessing peers behaving offensively.

Faculty-Student Interaction and Student Support Services

Similar to developmental theories, the cultural-ecological model (Ogbu, 1981) and the campus climate framework (Hurtado et al., 1998, 1999; Milem Chang, & antonio, 2005) emphasize the importance of supportive environments. The models posit that behaviors and perceptions come from social interactions, experiences, and external influences. In addition, there is an underlying assumption that student outcomes are determined by available resources as well as a supportive environment. Therefore, how well students do in engineering programs might be affected by interactions with faculty, using support services like tutoring, as well as perception of climate in their academic programs.

Interactions between faculty and students have been studied often and have shown many positive effects. Faculty members are one of college students' most important influences (Cole & Griffin, 2013; Crisp & Cruz, 2009; Pascarella & Terenzini, 2005). In addition to long-term metrics, such as persistence, grade-point average, or graduate study plans (Anaya & Cole, 2001; Hathaway, Nagda, & Gregerman, 2002; Pascarella & Terenzini, 2005), faculty also aided students' integration and adjustment as well as their degree of learning (Lundberg & Schreiner, 2004; Schwitzer, Griffin, Ancis, & Thomas, 1999). Most faculty-student interactions took place inside the classroom and focused on academic issues such as learning course material (Johnson, 2007). However, the frequency and nature of the interaction may matter to certain outcomes as well and thus, different types of interaction should be examined (Cotten & Wilson, 2006; Cox, McIntosh, Terenzini, Reason, & Quaye, 2010; Lundberg & Schreiner, 2004).

Within STEM, and specifically engineering fields, interactions between faculty and students have had mixed outcomes. Faculty-student interaction and support is shown to have many positive influences for engineering students (Brown, 2002). For example, Shain (2002) showed that Black engineering students who stayed in the field had strong relationships with faculty members. Several studies have also found positive impacts for students particularly women of color in STEM in general or engineering specifically (Ellington, 2006). The perception and nature of students' interactions with faculty, however, may differ for engineering student groups, and yield more negative results for certain groups, for example, women. Female engineering students felt less respected and had more reports of being sexually discriminated against than male students (Hughes, 2000). Women saw their gender, race, or both demographic traits as impediments in interacting with faculty if they did not believe faculty members perceived them as being earnest students (Carlone & Johnson, 2007; Seymour & Hewitt, 1997). As a result of not establishing good relationships, women students may also become discouraged (Espinosa, 2011). Furthermore, Kim and Sax (2007) found both differences in frequency and differences in the effect of faculty-student interaction by gender and race. Women of color may have different experiences than White women and those experiences may lead to different outcomes in engineering programs. Furthermore, interactions with peers and faculty members may not only form students' perceptions of climate in the program, but also their career projections.

Another resource and source of support in the structure of a university's academic plan are academic support services. Universities and engineering programs have provided a variety of academic support services to increase performance and persistence, often particularly for women and racial minority students. Studies indicate that all types of students, regardless of their academic preparedness, benefit from the use of academic support services (Amenkhienan & Kogan, 2004; Keenan & Gabovitch, 1995; Moore, 2006). The importance of student support services, however, might differ by students' individual characteristics, especially race. Studies report that African American students usually call attention to their lack of math preparation for collegiate mathematics (McGee & Martin, 2011), which is a major part of the engineering curriculum, Thus, African American students might perceive the role of learning and tutoring centers more importantly and use the services more often than other student groups. Students' perception of the climate and resources are reflected in the ways in which they report differential treatment, offensive behaviors, perceptions of their career projections, interaction with faculty, and importance of support services.

Framework

This research is guided, primarily, by critical race theory and intersectionality and focuses on examining the perceptions, interactions, and structures faced by engineering women students. As such, it aims to understand salient multiple identities and how they relate to student experiences within an academic environment such as engineering classrooms (Bell, 1992; Delgado & Stefancic, 2001; Ladson-Billings, 1998). Furthermore, it addresses the three issues spoken of by Cole (2009) which advise researchers to consider who is included in the categories examined, the role inequality play, and the patterns of similarities. Thus, this study explored intersections of race and gender, students' environmental factors, and considered similarities across groups in order to gain a better understanding of the experiences of women of color in engineering.

In examining the contextual factors within the academic environment, the Campus Racial Climate framework was utilized. Campus climate frameworks (Hurtado et al., 1998, 1999; Milem Chang, & antonio, 2005) demonstrate the importance of the institutional context (Cabrera, Nora, Terenzini, Pascarella, & Hagedorn, 1999; Hurtado & Carter, 1997) and have been applied to assessing climate not only for race, but other aspects of diversity as well. Of particular focus was the perceptions, interactions, and organizational structures within the student's' environment. Students' perception of the psychological climate, (represented by the treatment shown by peers and instructors) has been shown to have key consequences on student outcomes (Hurtado et al., 1999). In addition, the behavioral dimension of climate, the interactions between individuals, as seen in faculty-student interactions, is important to students' performance, confidence, and identity (Chen et al., 2008; Hughes, 2000; Carlone & Johnson, 2007).

Finally, organizational structures of supports, such as access to resources (Ro, Terenzini, & Yin, 2013), also influence individual's experiences and outcomes.

Therefore, this study examines whether specific racial groups of engineering women students differ in their perceptions of climate, their interactions with faculty, and in their perceptions of the importance of learning and tutoring center services. Specifically, the research questions guiding our study are: Are there significant differences among women (i.e., White, Asian, African American, and Latina women students) in engineering in relation to perceptions of treatment by peers and instructors, career projections, faculty-student interaction, and importance of learning and tutoring center services?

METHOD

Design, Population, and Samples

This study was part of a larger, cross-sectional investigation of curricular, instructional, organizational practices and policies, and undergraduate engineering students' educational experiences and learning. The overall goal of the study was to examine the curricular, instructional, and organizational features that support high-quality learning and diversification in engineering programs. The dataset is suitable for this study because it includes various questions on students' self-reported perceptions of climate, interaction with faculty members, and view on the importance of support services. These factors have been underscored by engineering undergraduate programs, ABET accreditation, and engineering workforce to improve women and racial minorities' pathway.

Data were collected from 32 four-year colleges that are representative of all four-year U.S. engineering schools offering two or more ABET-accredited programs in seven engineering disciplines: biomedical/bioengineering, chemical, civil, electrical, general, industrial, and mechanical. In the aggregate, these disciplines accounted for over 70 percent of all baccalaureate engineering degrees awarded in 2007. The stratified sample design of institutions was also representative on three levels of highest degree offered (bachelor's, master's, or doctorate), and two levels of type of control (public or private).

The student population for the study was defined as all sophomore, junior, and senior students (2nd, 3rd, and 4th year) in one of the seven targeted engineering disciplines. The sample does not include first-years because some engineering schools do not allow students to declare a major in a specific engineering discipline until their sophomore year. Chi-square Goodness-of-Fit tests indicate that respondents were marginally unrepresentative of the overall population of engineering students at the 32 institutions (institution population-sample differences ranged from 1 to 11 percentage points). Consequently, individual weights were created to adjust for any campus-specific response bias based on respondents' gender, race/ethnicity, class year, and engineering discipline, as well as for different response rates across institutions. An overall weight was calculated (by multiplying the gender by race/ethnicity by class-year by discipline weight by institutional response weight) and applied to all student respondents to produce a sample that can be considered representative of the population of engineering students (as specified) on each campus. Appendix A summarizes the demographic characteristics of the student population and the sample of respondents.

Missing data were imputed using the Expectation-Maximization (EM) algorithm of the Statistical Package for the Social Sciences (SPSS) software (v.18). This procedure is recommended by Dempster, Laird, and Rubin (1977) and by Graham (2009). Although EM algorithms is perhaps the most commonly used in the educational literature (Cox, McIntosh, Reason, & Terenzin, 2014), EM yields standard errors that are artificially small, threatening the validity of subsequent hypothesis testing (Graham, 2009; Von Hippel, 2004). The increased likelihood of Type I errors, led to an adoption of more conservative critical p-values (i.e., .01 instead of .05) (Cox et al., 2014). After cleaning the data, analyses used the weighted responses of 850 women students in 31 colleges of engineering during the 2009 spring and summer terms (one institution sampled was unable to provide student contact information in a timely fashion). The weighted sample includes 470 White women, 93 Black women, 114 Latina women, and 173 Asian women.

Measures

Instrument Development

A team of education and engineering researchers collaborated on instrument development, beginning with an extensive literature review on key topics related to key learning outcomes identified by the National Academy of Engineering's (2004)*Engineer of 2020* report. In addition to providing conceptual guidance for survey development, findings from this literature review generated a bank of potential survey items related to engineering students' college experiences and learning outcomes. In cases where available scales had acceptable psychometric properties, items were adopted or minimally revised. The team also conducted interviews and focus groups with engineering administrators, faculty members, students, and alumni at the following five campuses to develop new survey items and ensure appropriate coverage of key topics: Penn State–University Park, Penn State– Altoona, City College of New York, Borough of Manhattan Community College, and Hostos Community College. Drafts of potential survey items were reviewed by engineering faculty and administrators to evaluate and refine the survey, and the instrument was pilot tested with students at Penn State–University Park and Penn State–Altoona (n = 482) for newly developed items.

Dependent Variables

Student perception of climate is one of the criterion measures in this study, represented by 9 items. Although the research team developed three scales for these items, this study used each item because some items only focus on gender and others on race/ethnicity. Three gender-focused items, which we call gender treatment, are (1) Male students treat other male students better than female students; (2) When working in groups, male students treat other male students better than female students; and (3) Instructors treat male students better than female students. The race-focused items (i.e., race treatment) include a similar set of items: (1) White students treat other White students better than non-White students; (2) when working in groups, White students treat other White students better than non-White students; and (3) Instructors treat White students better than non-White students. Four additional individual items were evaluated regarding peers' offensive behaviors based on gender and race/ethnicity and career projection in engineering fields based on gender and race/ethnicity.

Three different types of student-faculty interaction were also included: 1) discuss academic or course-related matters; 2) ask about careers or get professional advice; and 3) talk informally. Finally, to measure the importance of support services for students' learning and tutoring, a single item was used.

Control Variables

Controls were in place to minimize the potential confounding effects of selected pre-college student characteristics, such as parents' education, transfer status, and high school achievement (as reflected in students' total SAT scores and high school Grade Point Average). Also included were students' class years and academic majors based on the relatively higher percentage of women students in some academic programs, such as Chemical Engineering, General Engineering, Bio-medical/Bio-engineering, and Industrial Engineering. The other category contains Civil Engineering, Electrical Engineering, and Mechanical Engineering. While this paper focuses on individual student-level analysis, three institutional-level control variables: undergraduate enrollment size, highest-degree awarded, and Historically Black College and University (HBCU) and Hispanic Serving Institutions (HSI) institution status were included. Appendix B provides operational information on all variables' contents and metrics.

The main demographic variables in this study were race/ethnicity of women students. Four variables represented race/ethnicity of women students: African American ("Black"), Hispanic/Latino/a ("Latina"), Asian American ("Asian"), and Caucasian/ White ("White").

Analytical Methodology

Ordinary Least Squares regression method was employed to examine racial differences among engineering women students in their perceptions of climate, interaction with faculty, and perceptions of importance of learning and tutoring services. The reference group of the race variable was White women students. Thus, the interpretation of each regression model is, for example, compared to White women, Black women students reported more or less that male students treat other male peers better than female students, controlling for other individual students' and institutional characteristics.

Results

Perception of the Environment

Students' perception of their engineering programs' environment was operationalized through four aspects: perceptions of gender treatment in their classroom (Table 1); perceptions of race treatment in their classroom (Table 2); peers' offensive behaviors (Table 3); and career projection based on gender and race (Table 4). Perceptions of gender treatment was based on three items related to how often male students and instructors treated male students better than female students. Perceptions of race treatment were from a similar set of three items.

Differential treatment by gender was seen in students' responses in Table 1. Compared to White women, Black women students, reported that male students treated other male students better than female students when they worked in groups. How male students treated other male students in general or how instructors treated male students was not viewed statistically differently between Black and White women. Asian and Latina women reported fewer frequencies than White women with respect to differential gender treatment. Compared to White women, Asian and Latina women reported that they were less likely to experience that male students treated other male students better than female students in general or in their group work. The Asian and Latina women students also had less report of instructors treating male students better.

Table 1. Gender treatment

How often in your courses:	Male students treat other male students better than female students			When working in groups, male students treat other male students better than female students			Instructors treat male students better than female students		
	Beta	SE		Beta	SE		Beta	SE	
African American female	0.259	0.116		0.407	0.124	**	0.130	0.102	
Asian American female	-0.340	0.073	***	-0.318	0.078	***	-0.052	0.064	
Latina female	-0.302	0.096	**	-0.379	0.103	***	-0.422	0.084	***
Parental education (BA or advanced)	-0.191	0.068	**	-0.091	0.073		-0.191	0.059	**
High school GPA	-0.036	0.081		0.005	0.087		0.199	0.071	**
SAT composite score	-0.011	0.003	***	-0.013	0.004	***	-0.014	0.003	***
First-time college student	-0.196	0.089		-0.227	0.095		-0.244	0.078	**
Major in BE, IE, and GE	0.006	0.058		-0.124	0.062		-0.107	0.051	
Junior	0.114	0.073		0.112	0.079		0.205	0.064	**
Senior	0.128	0.068		0.019	0.073		0.136	0.060	
Intercept	3.072	0.206	***	3.286	0.220	***	2.727	0.180	***
R^2		0.095	***		0.101	***		0.098	***

$**p < .01$ $***p < .001$

Table 2. Race treatment

How often in your courses:	White students treat other White students better than non-White students			When working in groups, White students treat other White students better than non-White students			Instructors treat White students better than non-White students		
	Beta	SE		Beta	SE		Beta	SE	
African American female	0.528	0.107	***	0.642	0.104	***	0.259	0.077	***
Asian American female	0.020	0.067		-0.026	0.065		0.066	0.048	
Latina female	-0.088	0.088		-0.012	0.086		-0.102	0.063	
Parental education (BA or advanced)	-0.220	0.063	***	-0.210	0.061	***	-0.059	0.045	
High school GPA	-0.009	0.075		0.055	0.073		0.059	0.054	
SAT composite score	-0.002	0.003		-0.001	0.003		-0.007	0.002	**
First-time college student	-0.332	0.082	***	-0.345	0.079	***	-0.278	0.059	***
Major in BE, IE, and GE	-0.055	0.053		-0.148	0.052	**	-0.094	0.038	
Junior	0.053	0.068		0.083	0.066		0.204	0.049	***
Senior	0.007	0.063		0.041	0.061		0.126	0.045	**
Intercept	2.294	0.190	***	2.158	0.183	***	1.929	0.136	***
R^2		0.082	***		0.103	***		0.098	***

$**p < .01$ $***p < .001$

Table 3. Using offensive words of peers based on gender or race

Do you agree or disagree:	**Some engineering students use offensive words, etc. directed at students due to *gender***			**Some engineering students use offensive words, etc. directed at students due to *race/ethnicity***		
	Beta	SE		Beta	SE	
African American female	0.461	0.118	***	0.400	0.111	***
Asian American female	-0.231	0.075	**	-0.207	0.070	**
Latina female	-0.233	0.098		-0.115	0.092	
Parental education (BA or advanced)	-0.217	0.069	**	-0.221	0.065	***
High school GPA	0.080	0.083		-0.025	0.078	
SAT composite score	-0.003	0.003		-0.005	0.003	
First-time college student	-0.387	0.091	***	-0.254	0.085	**
Major in BE, IE, and GE	-0.126	0.059		-0.003	0.056	
Junior	0.084	0.075		0.072	0.071	
Senior	0.031	0.070		0.072	0.066	
Intercept	2.438	0.210	***	2.364	0.198	***
R^2		0.083	***		0.077	***

**p < .01 ***p < .001

Table 4. Career projection perception based on gender or race

Do you agree or disagree:	**My *gender* will negatively influence my engineering career.**			**My *race/ethnicity* will negatively influence my engineering career**		
	Beta	SE		Beta	SE	
African American female	0.091	0.109		0.343	0.098	***
Asian American female	-0.102	0.069		0.071	0.061	
Latina female	-0.264	0.090	**	0.074	0.081	
Parental education (BA or advanced)	-0.153	0.064		-0.172	0.057	**
High school GPA	-0.070	0.077		-0.113	0.068	
SAT composite score	-0.004	0.003		-0.011	0.003	***
First-time college student	0.034	0.084		-0.025	0.075	
Major in BE, IE, and GE	-0.178	0.054	**	-0.077	0.049	
Junior	0.026	0.069		0.013	0.062	
Senior	0.145	0.064		-0.009	0.057	
Intercept	2.345	0.193	***	2.379	0.173	***
R^2		0.041	***		0.087	***

**p < .01 ***p < .001

Table 2 indicates that compared to White women, Black women seem to experience and perceive more differences in treatment based on race. In comparison to White women, Black women reported more experiences that White students treated other White students better than non-White students in general and in workgroups. Similar findings hold true for perceptions about how instructors treated White and non-White students. There was no difference between Asian, Latina, and White women in experiences with differential race treatment.

Similar results regarding student reports of peers' offensive behaviors were found (Table 3). Black women reported that they experienced the use of offensive words, which were directed at students due to their gender and race/ethnicity, more than White women. On the other hand, Asian women students reported fewer offensive behaviors by their peers than White women students.

Career projections also differentiated based on gender and race/ethnicity (Table 4). Compared to White women, Black women reported a more negative career projection for their career based on their race/ethnicity. Compared to White women, Latina students reported a less negative projection for their career based on their gender. In short, the overall findings suggest that Black women engineering students experience and perceive more differential treatment because of their race/ethnicity than White women students. Compared to Asian and Latina women, White women report more differential treatment based on gender in their classroom and college, as well as career projection.

Faculty and Student Interaction

Table 5 presents the number of faculty-student interaction by different racial group of women students. Compared to White women students, Black women students are more likely to interact with faculty members while Asian women students are less likely to interact. Black women students talk informally

Table 5. Student and Faculty Interaction

Number of discussions with faculty:	Discuss academic or course-related matters			Ask about careers or get professional advice			Talk informally		
	Beta	SE		Beta	SE		Beta	SE	
African American female	2.204	1.230		0.400	0.548		4.552	1.164	***
Asian American female	-3.151	0.774	***	0.336	0.345		-1.021	0.732	
Latina female	-0.923	1.016		0.633	0.453		-0.643	0.961	
Parental education (BA or advanced)	0.430	0.719		0.021	0.320		-0.081	0.680	
High school GPA	0.774	0.864		0.249	0.385		2.527	0.817	**
SAT composite score	-0.080	0.035		-0.050	0.016	**	-0.115	0.034	***
First-time college student	-0.029	0.943		0.732	0.420		0.186	0.892	
Major in BE, IE, and GE	0.821	0.614		0.389	0.273		0.599	0.581	
Junior	2.470	0.779	**	0.089	0.347		0.888	0.737	
Senior	2.563	0.723	***	0.867	0.322	**	2.130	0.684	**
Intercept	12.543	2.179	***	4.619	0.971	***	8.313	2.062	***
R^2		0.075	***		0.070	***		0.111	***

**p < .01 ***p < .001

Table 6. The Importance of Learning/Tutoring Center Services

How important to your academic success in engineering are the services of a learning/tutoring center at your college?			
	Beta	SE	
African American female	0.927	0.142	***
Asian American female	0.271	0.089	**
Latina female	0.120	0.117	
Parental education (BA or advanced)	-0.178	0.083	
High school GPA	-0.049	0.100	
SAT composite score	-0.018	0.004	***
First-time college student	-0.068	0.109	
Major in BE, IE, and GE	-0.077	0.071	
Junior	-0.306	0.090	***
Senior	-0.491	0.083	***
Intercept	4.456	0.251	***
R^2		0.171	***

**p < .01 ***p < .001

with professor more than White women students. Black women students are also more likely to have discussions with faculty members for academic and course-related issues than White women students. Asian women students are less likely to talk with professors. There is no difference in student-faculty interaction between White women and Latina women.

Importance of Learning/Tutoring Center Services

Black and Asian women students reported that the services of a learning/tutoring center at their college are important to their academic success in engineering more than White women students (Table 6). Among Black and Asian women students, the coefficient (effect) of the importance of the services is almost three times bigger in Black women (0.927) than in Asian women students (0.271). Learning/tutoring center services thus are of very high importance to some students, particularly Black and Asian women. We did not find significant differences between Latina and White women.

DISCUSSION AND RECOMMENDATIONS

This study examined several factors which represent aspects of climate such as the psychological dimension of climate (how people feel treated, career projections), the behavioral dimension of climate (experiencing offensive behaviors, interacting with faculty) and the structural dimensions of climate (supportive resources). These multiple aspects of climate are of concern because climate can influence women of color students' academic success and career choices in engineering fields (Baker, Tancred, & Whitesides, 2002). This study found that women of color experienced these aspects of climate differently from their White female counterparts. Overall, the findings confirm the need for researchers to examine fields individually, investigate experiences as well as outcomes, and distinguish groups further. Aligned with our framework, the differences by groups, contextual factors, and structural issues to consider in the future are discussed below.

This study, like others (Ong, 2005; Rankin & Reason, 2005; Varma, Prasad, & Kapur, 2006), provides evidence that women of color experience the climate differently based on their race. Black women had a more disparate experience than other group members. Black women's peer interactions were less favorable that White women when examined in a general classroom context, within class working groups, and through offensive behaviors from their peers. In regards to faculty-student interaction, Black women were more likely to interact with faculty members informally and because of course work. Black women students also reported that instructors treated them differently because of their race, but not because of their gender. The tutoring/learning center was of most importance to Black women. Finally, Black women had more negative career projections than White women because of their race. Unlike, Ogbu's findings, the Black women students in this study did not disengage from faculty or devalue support centers.

The other women of color had different experiences than Black women. Asian and Latina women did not report different peer interactions with students based on gender or race. There is no difference in career projection between Asian and White women. Latina women had less negative career projection than White women. Asian women interacted with faculty less and viewed tutoring/learning centers as more important than White students. Compared to White women, Latina women did not interact differently with faculty members or feel treated differently. They also did not view tutoring/learning centers differently. These variations by group show that women of color experiences are different from majority groups and different from each other.

Variations may exist between students' experiences because of the saliency of their identities or the lived experiences of the participants. Cole (2009) notes that inequality plays a role in how different people experience the same situations. The students in this study may report differential treatment because their experiences with past social inequities allow them to notice displays of unequal treatment more readily. By the same reasoning, students may also be more keenly aware of the differential treatment towards their membership groups, but not to others. For example, Black women were the only female group to experience differential treatment based on race and they noted this in all areas- in general, in work groups, and amongst instructors. They also reported differential gender treatment in work groups but not by instructors. White women were more likely to report differential treatment by gender in all areas as compared to Asian and Latina women, however they did not note that White students treated non-White students differently. The findings from Black and White women students support both Crenshaw's (1989) claim that Black women face a double-discrimination as well as the explanation that one's identities influence perceptions and experiences. Neither the double-discrimination nor the identity-experience link, however, held true for other women of color.

In identifying the experiences by subgroup among women students, focus must shift from only examining individuals or how individuals perceive climate issues, to the organizational/structural systems in place which create an environment which facilitates these patterns of disparate experiences. Engineering programs should consider changes within its organizational/structural dimensions of climate that would be more welcoming to women of color. For example, programs must evaluate the ways in which they define meritocracy, how they consider disadvantaged backgrounds, and how they address vestiges of racism and sexism (Ong et al., 2011). Action in this area is imperative because past studies indicate that unsupportive environments, discouragement, offensive actions, and lack of resources affect students' satisfaction and in turn their departure. Engineering programs need to proactively address the climate for diversity by conducting climate assessments and responding to concerns, educating community members on expectations of inclusion and equal treatment, being advocates for students who have experienced differential treatment and offensive behaviors, promoting a culture of faculty involvement, and providing resources for students such as tutoring and other support services.

Creating an environment that has resources that are important to students is essential to their success. After studying both poor and middle class African-Americans, Ogbu, like other educational researchers, recommended that programs include academic counseling, study skills, student dialogues about institutional and cultural pressures, teacher workshops on raising expectations and building trust, parental involvement, and having like role models (Foley, 2005). It is important to include support services and resources for students because they may have significant beneficial effects on how students view themselves in their environments and how they view and navigate the campus climate (Ong, 2002). In addressing these issues and providing these services, engineering programs should consider that they may have to reach out to students in different ways and that students may value different opportunities. While some may seek to gain acceptance by interacting more with faculty and support services (Ong, 2002), others may not feel safe interacting or asking for help but this resource might still be very important to their experience. For example, Black and Asian women students may benefit greatly from support services. Additionally, Asian women students might not interact frequently with faculty, especially if they do not actively approach faculty members, but this does not mean that they would not benefit from interaction. Black students with strong faculty-student relationships were more likely to persist in the engineering field (Shain, 2002). This finding may hold true for other specific subgroups as well. Thus, faculty members and administrators should encourage women of color to take advantages of the resources available to them from faculty and support services.

FUTURE RESEARCH DIRECTIONS

This study highlights the importance of studying the experiences of engineering women of color. Women of color are uniquely affected by their academic environment. This research lends support to Carlone and Johnson's (2007) conceptual model which claims that although recognition in the community is needed for the science identity development of women of color, it is not often gained. Women of color report inequitable treatment in many aspects of their college career including the classroom, work groups, and interacting with faculty. It would be beneficial to understand which aspects of the environment are more likely to influence college outcomes and career projections for these students. In addition, while differential relationship among women students were found, it was not possible to examine further distinctions regarding students' analysis of inequity based on race versus gender. Empirical research investigating this issue within specific racial groups in a qualitative manner would be an insightful follow-up study. In particular, our results repeatedly suggest that priority could be given to explore differences between Black and White women students in engineering programs. Looking at specific sub-fields, such as chemical engineering or civil engineering where a different level of gender compositional diversity exists, would also help to make distinct implications from the findings.

Considering the past exclusion of women of color from both the research design and the analysis of studies (Clewell & Ginorio, 1996), much more information should be gathered about this population in a variety of contexts. Additional studies are needed on other factors that are related to the climate. Furthermore, a longitudinal study design would be useful in assessing whether the climate, student-faculty interaction, or usage of services influence women of color students' persistence, degree completion, and career choice in STEM disciplines. National agencies, such as the National Science Foundation, would do well to develop nationally-representative data having a large sample size of women of color. This would enable more empirical quantitative research to be conducted.

CONCLUSION

With the low number of women of color in engineering programs and even lower numbers in the workforce, it is time to address the double bind these women may face as they navigate inequitable treatment in their classrooms from peers and instructors. As critical race theory suggest, experiences are differential and women of color must navigate differently and inequitable. Findings reiterate continued disparities among women of color and show that these disparities are evident in their experiences through interactions and opportunities to engage as well as in their career projections. Considering the intersectional identities of students enabled access to information which would have remained missing had the groups been examined only as women or only as minorities. This study demonstrates the need for administrators and faculty members to create engineering programs which have a better environment for women students and women of color in particular. Engineering programs, which have a few women and minorities, should examine their classrooms through the campus climate framework to assess dimensions of their climate such as their history of inclusion, perceptions of treatment, interactions, and structures of support.

REFERENCES

Amenkhienan, C. A., & Kogan, L. R. (2004). Engineering students' perceptions of academic activities and support services: Factors that influence their academic performance. *College Student Journal*, *38*(4), 523–540.

Anaya, G., & Cole, D. G. (2001). Latina/o student achievement: Exploring the influence of student-faculty interactions on college grades. *Journal of College Student Development*, *42*(1), 611–622.

Baker, S., Tancred, P., & Whitesides, S. (2002). Gender and graduate school: Engineering students confront life after the B. Eng. *The Journal of Engineering Education*, *91*(1), 41–47. doi:10.1002/j.2168-9830.2002.tb00671.x

Bell, D. (1992). *Faces at the bottom of the well*. New York: Basic Books.

Bonilla-Silva, E. (2006). *Racism without racists: color-blind racism and the persistence of racial inequality in the United States*. Lanham, MD: Rowman & Littlefield.

Brown, R. D., Clarke, B., Gortmaker, V., & Robinson-Keilig, R. (2004). Assessing the campus climate for gay, lesbian, bisexual, and transgender (GLBT) students using a multiple perspectives approach. *Journal of College Student Development*, *45*(1), 8–26. doi:10.1353/csd.2004.0003

Brown, S. W. (2002). Hispanic students majoring in science or engineering: What happened in their educational journeys? *Journal of Women and Minorities in Science and Engineering*, *8*(2), 123–148. doi:10.1615/JWomenMinorScienEng.v8.i2.20

Burge, S. W. (2011). Academic programs: Undergraduate, graduate, and professional. In B. J. Bank (Ed.), *Gender and higher education* (pp. 171–181). Baltimore, MD: The Johns Hopkins University Press.

Cabrera, A. F., Nora, A., Terenzini, P. T., Pascarella, E., & Hagedorn, L. S. (1999). Campus racial climate and the adjustment of students to college: A comparison between White students and African-American students. *The Journal of Higher Education*, *70*(2), 134–160. doi:10.2307/2649125

Carlone, H. B., & Johnson, A. (2007). Understanding the science experiences of successful women of color: Science identity as an analytic lens. *Journal of Research in Science Teaching*, *44*(8), 1187–1218. doi:10.1002/tea.20237

Chen, H. L., Lattuca, L. R., & Hamilton, E. R. (2008). Conceptualizing engagement: Contributions of faculty to student engagement in engineering. *The Journal of Engineering Education*, *97*(3), 339–353. doi:10.1002/j.2168-9830.2008.tb00983.x

Clewell, B. C., & Ginorio, A. B. (1996). Examining women's progress in the sciences from the perspective of diversity. In C.-S. Davis, A. B. Ginorio, C. S. Hollenshead, B. B. Lazarus, & P. M. Rayman (Eds.), *The equity equation: Fostering the advancement of women in the sciences, mathematics, and engineering* (pp. 163–231). San Francisco: Jossey-Bass.

Colbeck, C. L., Cabrera, A. F., & Terenzini, P. T. (2001). Learning professional confidence: Linking teaching practices, students' self-perceptions, and gender. *The Review of Higher Education*, *24*(2), 173–191. doi:10.1353/rhe.2000.0028

Cole, D. (2008). Constructive criticism: The role of student-faculty interactions on African American and Hispanic students' educational gains. *Journal of College Student Development*, *49*(6), 587–605. doi:10.1353/csd.0.0040

Cole, D., & Griffin, K. A. (2013). Advancing the study of student-faculty interaction: A focus on diverse students and faculty. In *Higher education: Handbook of theory and research* (pp. 561–611). Springer Netherlands. doi:10.1007/978-94-007-5836-0_12

Cole, E. R. (2009). Intersectionality and research in psychology. *The American Psychologist*, *64*(3), 170–180. doi:10.1037/a0014564 PMID:19348518

Cotten, S. R., & Wilson, B. (2006). Student–faculty interactions: Dynamics and determinants. *Higher Education*, *51*(4), 487–519. doi:10.1007/s10734-004-1705-4

Cox, B. E., McIntosh, K., Reason, R. D., & Terenzini, P. T. (2014). Working with missing data in higher education research: A primer and real-world example. The Review of Higher Education, 37(3), 377-402. doi:10.1353/rhe.2014.0026 doi:10.1007/s11162-010-9178-z

Crenshaw, K. W. (1989). Demarginalizing the intersection of race and sex: A Black feminist critique of antidiscrimination doctrine, feminist theory, and antiracist politics. *University of Chicago Legal Forum*, *140*, 139–167.

Crenshaw, K. W. (1994). Mapping the margins: Intersectionality, identity politics, and violence against women of color. In M. A. Fineman & R. Mykitiuk (Eds.), *The public nature of private violence* (pp. 93–118). New York: Routledge.

Crenshaw, K. W. (2014). The structural and political dimensions of intersectional oppression. In P. R. Grzanka (Ed.), *Intersectionality: A Foundations and Frontiers Reader* (pp. 17–22). Boulder, CO: Westview Press.

Crisp, C., & Cruz, I. (2009). Mentoring college students: A critical review of the literature between 1990 and 2007. *Research in Higher Education*, *50*(6), 525–545. doi:10.1007/s11162-009-9130-2

Delgado, R., & Stefancic, J. (2001). *Critical race theory: an introduction*. New York: NYU Press.

Dempster, A. P., Laird, N. M., & Rubin, D. B. (1977). Maximum likelihood from incomplete data via the EM algorithm. *Journal of the Royal Statistical Society. Series A (General)*, *38*(1), 1–38.

Ellington, R. (2006). *Having their say: Eight high-achieving African-American undergraduate mathematics majors discuss their success and persistence in mathematics*. Doctoral dissertation. Retrieved from ProQuest Dissertations and Theses database, Publication No. 305304604.

Espinosa, L. L. (2011). Pipelines and pathways: Women of color in undergraduate STEM majors and the college experiences that contribute to persistence. *Harvard Educational Review*, *81*(2), 209–241. doi:10.17763/haer.81.2.92315ww157656k3u

Ferrini-Mundy, J., & Güçler, B. (2009). Discipline-based efforts to enhance undergraduate STEM education. *New Directions for Teaching and Learning*, *2009*(117), 55–67. doi:10.1002/tl.344

Foley, D. (1990). *Learning capitalist culture: deep in the heart of Tejas*. Philadelphia, PA: University of Pennsylvania Press.

Foley, D. (2005). Elusive prey: John Ogbu and the search for a grand theory of academic disengagement. *International Journal of Qualitative Studies in Education*, *18*(5), 643–657. doi:10.1080/09518390500224986

Goodman, I. F., Cunningham, C. M., Lachapelle, C., Thompson, M., Bittinger, K., Brennan, R. T., & Delci, M. (2002). *The Women's Experiences in College Engineering (WECE) Project final report* (pp. 1–264). Cambridge, MA: Goodman Research Group.

Graham, J. W. (2009). Missing data analysis: Making it work in the real world. *Annual Review of Psychology*, *60*(1), 549–576. doi:10.1146/annurev.psych.58.110405.085530 PMID:18652544

Grzanka. (Ed.). (n.d.). *Intersectionality: A Foundations and Frontiers Reader*. Boulder, CO: Westview Press.

Haines, V. A., Wallace, J. E., & Cannon, M. E. (2001). Exploring the gender gap in engineering: A re-specification and test of the hypothesis of cumulative advantages and disadvantages. *The Journal of Engineering Education*, *90*(4), 677–684. doi:10.1002/j.2168-9830.2001.tb00659.x

Hamann, E. T. (2004). Lessons from the interpretation/misinterpretation of John Ogbu's scholarship. *Intercultural Education*, *15*(4), 399–412. doi:10.1080/1467598042000313421

Hathaway, R. S., Nagda, B. A., & Gregerman, S. R. (2002). The relationship of undergraduate research participation to graduate and professional education pursuit: An empirical study. *Journal of College Student Development*, *43*(5), 614–631.

Huang, G., Taddese, N., & Walter, E. (2000). *Entry and persistence of Women and Minorities in ollege science and engineering education (NCES Rep. No. 2000-601).* Washington, DC: U.S. Government Printing Office.

Hughes, W. (2000). Perceived gender interaction and course confidence among undergraduate science, mathematics, and technology majors. *Journal of Women and Minorities in Science and Engineering*, *6*(2), 166–167. doi:10.1615/JWomenMinorScienEng.v6.i2.40

Hunter, A. B., Laursen, S. L., & Seymour, E. (2007). Becoming a scientist: The role of undergraduate research in students' cognitive, personal, and professional development. *Science Education*, *91*(1), 36–74. doi:10.1002/sce.20173

Hurtado, S., & Carter, D. F. (1997). Effects of college transition and perceptions of the campus racial climate on Latino college students' sense of belonging. *Sociology of Education*, *70*(4), 324–345. doi:10.2307/2673270

Hurtado, S., Milem, J. F., Clayton-Pedersen, A., & Allen, W. R. (1998). Enhancing campus climates for racial/ethnic diversity: Educational policy and practice. *The Review of Higher Education*, *21*(3), 279–302. doi:10.1353/rhe.1998.0003

Hurtado, S., Milem, J. F., Clayton-Pedersen, A. R., & Allen, W. R. (1999). Enacting diverse learning environments: Improving the campus climate for racial/ethnic diversity. ERIC Clearinghouse on Higher Education.

Johnson, A. (2007). Unintended consequences: How science professors discourage women of color. *Science Education*, *91*(5), 805–821. doi:10.1002/sce.20208

Johnson, A. (2010). Privilege, power, difference, and us. In M. S. Kimmel & A. L. Ferber (Eds.), *Privilege, A Reader* (2nd ed.; pp. 77–86). Boulder, CO: Westview Press.

Johnson, D. (2012). Campus Racial Climate Perceptions and Overall Sense of Belonging Among Racially Diverse Women in STEM Majors. *Journal of College Student Development*, *53*(2), 336–346. doi:10.1353/csd.2012.0028

Justin-Johnson, C. (2004). *Good fit or chilly climate: An exploration of the persistence experiences of African-American women graduates of predominantly White college science programs.* Doctoral dissertation. Retrieved from ProQuest Dissertations and Theses database, Publication No. 305172242.

Keenan, K., & Gabovitch, R. (1995). *Evaluating the impact of a freshman seminar program on student development and retention (Rep. No. ED395614).* ERIC Document Reproduction Service.

Kim, Y. K., & Sax, L. J. (2007). *Different Patterns of Student-Faculty Interaction in Research Universities: An Analysis by Student Gender, Race, SES, and First-Generation Status.* A Student Experience in the Research University (SERU) Project Research Paper. Research & Occasional Paper Series: CSHE. 10.07. Center for Studies in Higher Education.

Kristof, A. L. (1996). Person-organization fit: An integrative review of its conceptualizations, measurement, and implications. *Personnel Psychology*, *49*(1), 1–49. doi:10.1111/j.1744-6570.1996.tb01790.x

Ladson-Billings, G. (1998). Just what is critical race theory and what's it doing in a nice field like education? *International Journal of Qualitative Studies in Education, 11*(1), 7–24. doi:10.1080/095183998236863

Lattuca, L. R., Terenzini, P. T., Harper, B. J., & Yin, A. C. (2010). Academic environments in detail: Holland's theory at the subdiscipline level. *Research in Higher Education, 51*(1), 21–39. doi:10.1007/s11162-009-9144-9

Lord, S. M., Camacho, M. M., Layton, R. A., Long, R. A., Ohland, M. W., & Washburn, M. H. (2009). Who's persisting in engineering? A comparative analysis of female and male Asian, Black, Hispanic, Native American, and White students. *Journal of Women and Minorities in Science and Engineering, 15*(2), 167–190. doi:10.1615/JWomenMinorScienEng.v15.i2.40

Lord, S. M., Layton, R., & Ohland, M. W. (2011). Trajectories of electrical engineering and computer engineering students by race and gender. Education. *IEEE Transactions on, 54*(4), 610–618.

Lundberg, C. A., & Schreiner, L. A. (2004). Quality and frequency of faculty-student interaction as predictors of learning: An analysis of student race/ethnicity. *Journal of College Student Development, 45*(5), 549–565. doi:10.1353/csd.2004.0061

Malcom, S. M., Hall, P. Q., & Brown, J. W. (1976). *The double bind: The price of being a minority woman in science (No. 76-R-3)*. Washington, DC: American Association for the Advancement of Science.

Malicky, D. (2003). A literature review on the underrepresentation of women in undergraduate engineering: Ability, self-efficacy, and the "chilly climate". In *Proceedings of the 2003 American Society for Engineering Education Annual Conference & Exposition*. Nashville, TN: American Society for Engineering Education.

McGee, E. O., & Martin, D. B. (2011). "You would not believe what I have to go through to prove my intellectual value!" Stereotype management among academically successful Black mathematics and engineering students. *American Educational Research Journal, 48*(6), 1347–1389. doi:10.3102/0002831211423972

McQuillan, P. (1998). *Educational Opportunity in an Urban American High School: A Cultural Analysis*. Albany, NY: State University of New York Press.

Metz, S. S. (2011). Engineering. In B. J. Bank (Ed.), *Gender and higher education* (pp. 196–202). Baltimore, MD: The Johns Hopkins University Press.

Milem, J. (2003). The educational benefits of diversity: Evidence from multiple sectors. In M. Chang, D. Witt, J. Jones, & K. Hakuta (Eds.), *Compelling interest: Examining the evidence on racial dynamics in higher education* (pp. 126–169). Stanford, CA: Stanford University Press.

Milem, J. F., Chang, M. J., & antonio, A. L. (2005). *Making diversity work on campus: A research-based perspective*. Washington, DC: Association American Colleges and Universities.

Moore, J. L. III. (2006). A qualitative investigation of African American males' career trajectory in engineering: Implications for teachers, school counselors, and parents. *Teachers College Record, 108*(2), 246–266. doi:10.1111/j.1467-9620.2006.00653.x

National Academies. (2010). *Expanding underrepresented minority participation: America's science and technology talent at the crossroads*. Washington, DC: National Academies Press.

National Academy of Engineering. (2004). *The engineer of 2020: Visions of engineering in the new century*. Washington, D.C.: National Academies Press.

National Center for Education Statistics. (2011). *Digest of Education Statistics: 2011. Table 301. Bachelor's degrees conferred by degree-granting institutions, by sex, race/ethnicity, and field of study: 2009-10*. Author.

National Science Board. (2006). Science and Engineering Indicators 2006. National Science Foundation.

National Science Foundation. (2015). *Women, Minorities, and Persons with Disabilities in Science and Engineering: 2015*. Arlington, VA: National Science Foundation.

Ogbu, J. (1974). *The Next Generation: An Ethnography of Education in an Urban Neighborhood*. New York: Academic Press.

Ogbu, J. (1981). School ethnography: A multilevel approach. *Anthropology & Education Quarterly*, *12*(10), 3–29. doi:10.1525/aeq.1981.12.1.05x1281g

Ogbu, J. (2003). *Black American Students in an Affluent Suburb: A Study of Academic Disengagement*. Mahwah, NJ: Lawrence Erlbaum Associates.

Ohland, M. W., Brawner, C. E., Camacho, M. M., Layton, R. A., Long, R. A., Lord, S. M., & Washburn, M. H. (2011). Race, gender, and measures of success in engineering education. *The Journal of Engineering Education*, *100*(2), 225–252. doi:10.1002/j.2168-9830.2011.tb00012.x

Ong, M. (2002). *Against the current: Women of color succeeding in physics*. Doctoral dissertation. Retrieved from ProQuest Dissertations and Theses database, Publication No. 304803810.

Ong, M. (2005). Body projects of young women of color in physics: Intersections of gender, race, and science. *Social Problems*, *52*(4), 593–617. doi:10.1525/sp.2005.52.4.593

Ong, M., Wright, C., Espinosa, L., & Orfield, G. (2011). Inside the double bind: A synthesis of empirical research on women of color in science, technology, engineering, and mathematics. *Harvard Educational Review*, *81*(2), 172–208. doi:10.17763/haer.81.2.t022245n7x4752v2

Pascarella, E. T., & Terenzini, P. T. (2005). *How college affects students: A third decade of research* (Vol. 2). San Francisco: Jossey-Bass.

Persaud, A., & Salter, D. W. (2003). Understanding women's perceptions of classroom 'fit' and participation as interactions between psychological and environmental types. *Journal of Classroom Interaction*, *38*(2), 1–10.

Rankin, S. R., & Reason, R. D. (2005). Differing perceptions: How students of color and White students perceive campus climate for underrepresented groups. *Journal of College Student Development*, *46*(1), 43–61. doi:10.1353/csd.2005.0008

Ro, H. K., & Loya, K. I. (2015). The effect of gender and race intersectionality on student learning outcomes in engineering. *The Review of Higher Education*, *38*(3), 359–396. doi:10.1353/rhe.2015.0014

Ro, H. K., Terenzini, P. T., & Yin, A. C. (2013). Between-college effects on students reconsidered. *Research in Higher Education*, *54*(3), 253–282. doi:10.1007/s11162-012-9269-0

Sax, L. J. (2008). *The gender gap in college: Maximizing the developmental potential of women and men*. San Francisco, CA: Jossey-Bass.

Schwitzer, A. M., Griffin, O. T., Ancis, J. R., & Thomas, C. R. (1999). Social adjustment experiences of African American college students. *Journal of Counseling and Development*, *77*(2), 189–197. doi:10.1002/j.1556-6676.1999.tb02439.x

Seymour, E., & Hewitt, N. M. (1997). *Talking About Leaving: Why Undergraduates Leave the Sciences*. Boulder, CO: Westview Press.

Shain, C. H. (2002). *Revisiting the problem of engineering school persistence in African-American women students*. Doctoral dissertation. Retrieved from ProQuest Dissertations and Theses database, Publication No. 304798888.

Solorzano, D., Ceja, M., & Yosso, T. (2000). Critical race theory, racial microaggressions, and campus racial climate: The experiences of African American college students. *The Journal of Negro Education*, 60–73.

Sosnowski, N. H. (2002). *Women of color staking a claim for cyber domain: Unpacking the racial/ gender gap in science, mathematics, engineering and technology (SMET)*. Doctoral dissertation. Retrieved from ProQuest Dissertations and Theses database, Publication No. 275796259.

Steele, J., James, J. B., & Barnett, R. C. (2002). Learning in a man's world: Examining the perceptions of undergraduate women in male-dominated academic areas. *Psychology of Women Quarterly*, *26*(1), 46–50. doi:10.1111/1471-6402.00042

Terenzini, P. T., & Reason, R. D. (2005, November). *Parsing the first year of college: A conceptual framework for studying college impacts*. Annual meeting of the Association for the Study of Higher Education, Philadelphia, PA. Education, Philadelphia, PA.

Terenzini, P. T., & Reason, R. D. (2010, June). *Toward a more comprehensive understanding of college effects on student learning*. Paper presented at the Annual Conference of the Consortium of Higher Education Researchers (CHER), Oslo, Norway.

Varma, R., & Hahn, H. (2007). Gender differences in students' experiences in computing education in the United States. *International Journal of Engineering Education*, *23*(2), 361–367.

Varma, R., Prasad, A., & Kapur, D. (2006). Confronting the "socialization" barrier: Crossethnic differences in undergraduate women's preference for IT education. In J. M. Cohoon & W. Aspray (Eds.), *Women and information technology: Research on underrepresentation* (pp. 301–322). Cambridge, MA: MIT Press. doi:10.7551/mitpress/9780262033459.003.0010

Vogt, K. E. (2005). *Asian American women in science, engineering, and mathematics: Background contextual and college environment influences on self-efficacy and academic achievement*. Doctoral dissertation. Retrieved from ProQuest Dissertations and Theses database, Publication No. 304996659.

Von Hippel, P. T. (2009). How to impute interactions, squares, and other transformed variables. *Sociological Methodology*, *39*(1), 1–27. doi:10.1111/j.1467-9531.2009.01215.x

Whitt, E. J., Edison, M. I., Pascarella, E. T., Terenzini, P. T., & Nora, A. (1998). *Influences on Students' Openness to Diversity and Challenge in the Second and Third Years of College*. Paper presented at the Annual Meeting of the Association for the Study of Higher Education.

Willis, P. (1977). *Learning to Labor: How Working Class Kids Get Working Class Jobs*. New York: Columbia University Press.

Willis, P. (1981). *Learning to labor: how working class kids get working class jobs*. New York: Teachers College Press.

Worthington, R. L., Navarro, R. L., Loewy, M., & Hart, J. (2008). Color-blind racial attitudes, social dominance orientation, racial-ethnic group membership and college students' perceptions of campus climate. *Journal of Diversity in Higher Education*, *1*(1), 8–19. doi:10.1037/1938-8926.1.1.8

APPENDIX A

Table 7. Characteristics of the population of 2008 engineering students, survey respondents, and their institutions

Characteristic	288-Institution Population [a] (N = 136,761)	32-Institution Sample [a] (n = 32,565)	Respondents (n = 5,249 [b])
Individual			
Discipline			
Biomedical	6.5%	6.5%	8.7%
Chemical	10.4	10.4	14.4
Civil	19.5	16.0	17.3
Electrical	21.8	21.4	17.5
Industrial	6.1	6.0	4.3
Mechanical	32.1	27.8	29.1
General	3.6	11.9	8.1
	100.0%	100.0%	100.0%
Gender			
Men	81.5%	80.7%	71.9%
Women	18.5	19.3	28.1
	100.0%	100.0%	100.0%
Race/Ethnicity			
Black	5.2%	5.9%	2.8%
Asian	12.1	12.3	8.1
Latino/a	6.5	6.1	5.8
Native American	.6	.6	.2
Other	6.1	7.2	5.9
Foreign	5.9	7.1	12.5
White	63.5	60.7	64.8
	100.0%	100.0%	100.0%
Level			
Sophomore	6.1%	27.9%	17.9%
Junior	39.0	29.0	33.9
Senior	54.9	43.1	48.2
	100.0%	100.0%	100.0%
Institution			
Institution Type [c]			
Doctoral	66.3%	61.3%	88.0%
Master's	26.1	19.4	9.5
Baccalaureate	7.6	19.4	2.5

continued on following page

Table 7. Continued

	288-Institution Population [a]	32-Institution Sample [a]	Respondents
Characteristic	**(N = 136,761)**	**(n = 32,565)**	**(n = 5,249 [b])**
	100.0%	100.0%	100.0%
Control			
Public	66.7%	61.3%	73.8%
Private	33.3	38.7	26.2
	100.0%	100.0%	100.0%

[a]Source: American Society of Engineering Education.
[b]The number of the respondents is larger than the final sample that we studied.
[c]Based on highest engineering degree offered.

APPENDIX B

Table 8. Table of measures

Variables	Coding
Student characteristics	
White woman	1=yes, 0=no
Black woman	1=yes, 0=no
Latina woman	1=yes, 0=no
Asian woman	1=yes, 0=no
Mother's education	1= College graduate and above, 0= Some college experiences or below
Father's education	1= College graduate and above, 0= Some college experiences or below
SAT composite score	Continuous; SAT composite score (the sum of math, critical reading, and writing divided by three) adjusted from one- to ten-point intervals
High school GPA at least A-	1=yes, 0=no
Class year	Categorical: sophomore (reference group), junior or senior
First-time college student	1=yes, 0=no (No: students transfer from 2-year or 4-year institutions)
Academic majors	1= Chemical engineering, Industrial Engineering, General Engineering, or Bio-medical or Bio-engineering, 0=other majors
Institutional characteristics	
Institution size	1= large (over 10,000); 0= small (undergraduate enrolment 1,000-3,000) or medium (3,000-10,000)
Highest degree awarded	1= doctorate; 0= bachelor's or master's
HBCU/HIS	1=HBCU or HSI; 0=other institutions
Outcome variables	

continued on following page

Table 8. Continued

Variables	Coding
Climate items	Nine individual items are used. In your engineering courses, how often do: (1) Male students treat other male students better than female students; (2) White students treat other White students better than non-White students; (3) When working in groups, male students treat other male students better than female students. (4) When working in groups, White students treat other White students better than non-White students. (5) Instructors treat male students better than female students. Each item is ordinal, from 1=never to 5=very often. Do you agree or disagree with the following: (6) Some engineering students use offensive words, behaviors, or gestures directed at students because of their gender; (7) Some engineering students use offensive words, behaviors, or gestures directed at students because of their race/ethnicity (8) My gender will negatively influence my engineering career; (9) My race/ethnicity will negatively influence my engineering career. A Likert scale is used from 1=strongly disagree to 5=Strongly agree.
Student and faculty interaction items	Three individual items are used. During the past six months, about how many times did you meet outside of class with a faculty member to (count only conversations of 10 minutes or more): (1) Discuss academic or course-related matters; (2) Ask about careers or get professional advice; (3) Talk informally.
Importance of learning/tutoring center services	A single item is used. How important to your academic success in engineering are the services of a learning/tutoring center at your college? Each item is ordinal, from 1=not at all important to 5=Extremely important.

Chapter 3
Barriers to Success:
Disadvantages of Gender Normative Language to Women in STEM

Melissa M. Haswell
Davenport University, USA

ABSTRACT

Cultural ideals of gender normativity creates stereotypes that lead to the identification of specific occupations as being male-oriented or female-oriented, further perpetuating institutional sexism. Research shows that stereotypical beliefs are pervasive and exist in all professions, including higher education. Women academics, especially in the STEM fields, are still present in low numbers and often have to sacrifice marriage and family for their careers due to the overarching masculine organizational structure that forms the basis of most higher education institutions. This chapter will discuss the history of women in academia, gender equity in higher education, and the consequences of gender normative language on women in academia.

INTRODUCTION

Oppression or exploitation of a person based on their biological sex forms the basis of sexual discrimination (Benokraitis, 1997). Sexual discrimination is pervasive because it invades all aspects of society due to its culturally internalized and socially organized roots (Benokraitis, 1997; Prewitt-Freilino, Caswell, & Laakso, 2012). The Gender stereotypes preservation within a culture occurs through the consistent use of gender normative language that stemmed from proscribed cultural characteristics and roles (Prewitt-Freilino, Caswell, & Laakso, 2012). Further, the overarching cultural norms projected by society become part of the foundation of the hierarchal, organizational structure of institutions including institutions of higher education. Institutional sexism further perpetuates gender role stereotypes of masculinity and femininity because academia has deemed certain disciplines to be either male-oriented or female-oriented (Heilman, Wallen, Fuchs, & Tamkins, 2004). Further, in the workplace, role stereotyping leads to negative expectations, or the belief, that a woman cannot perform the same occupational role as well as a man because according to cultural norms, a woman cannot, or should not, possess the qualities necessary to

DOI: 10.4018/978-1-5225-0174-9.ch003

be successful (Heilman, et. al., 2004). Women in academia, especially those at high-ranking research institutions, are still present in low numbers despite the fact that the number of women entering higher education as students and as faculty has increased dramatically since the passage of laws, such as Title IX in 1972, that prohibit sex discrimination in educational programs (American Association of University Professors, 2006). This chapter focuses on the cultural influence that gender normative language has in perpetuating the stereotypical views and hierarchal structure of higher education institutions to reinforce the discrimination of female faculty members, particularly in the STEM (Science, Technology, Engineering, and Mathematics) disciplines. Therefore, the objectives of this chapter include the following:

1. To outline the hierarchal, masculine organizational structure that forms the basis of the culture at the majority of higher education institutions in the United States, and which creates an inherently unequal workplace for women in academia.
2. To bring to light how the use of gender normative language derived from cultural norms influences the hiring and publication rate of women in academia.

BACKGROUND

A Tangled Web: Gender Normativity, Language, and Microaggressions

A continuum that ranges from blatant discrimination to covert discrimination best represents the deeply embedded cultural roots of sexism (Benokraitis, 1997). Both blatant sexism and covert sexism involve unequal treatment or harassment of women with the main difference being that blatant sexism is explicit whereas covert sexism is ambiguous (Benokraitis & Feagin, 1995; Benokraitis, 1997). Although both types are intentional, covert sexism is often malicious and involves direct sabotage to undermine the position of a woman in the workplace, and fits the description of old-fashioned sexism because it endorses and perpetuates traditional gender roles and norms (Benokraitis, 1997). The constructs behind cultural gender normativity creates a foundation for discriminatory practices because categorization by sex promotes both descriptive and prescriptive stereotypical connotations of gender which are then promoted as normal societal beliefs and behaviors (Swim, Mallett, & Stangor, 2004; Bobbit-Zeher, 2011). This practice reinforces institutional sexism because the gender normative beliefs of a culture become the established norms through policies and practices of an institution (Benokraitis & Feagin, 1995; Benokraitis, 1997). Thus, discrimination against people on the sole basis of their sex or identified gender becomes a normal practice (Benokraitis & Feagin, 1995; Benokraitis, 1997). Further, in the middle of this spectrum lies subtle sexism, which is less discernible because it consists of an accumulation of experiences related to the various institutional structures or cultural norms that society internalizes and accepts as normal behavior (Benokraitis & Feagin, 1995). Subtle sexism, also known currently as modern sexism, neosexism, and everyday sexism, encompasses the deeply embedded gender-normative structure that exists in contemporary society that hides among the advances in equality that women have made in the last century and gives society the impression that sexism in no longer prevalent (Benokratis, 1997; Swim, et al., 2004). However, research presented by Basford, Offermann, and Behrend (2014) suggests that instead of highly visible, blatant sexism, contemporary workers experience subtle forms of sexism in the form of gender microaggressions, which mainly affect women.

Microaggressions, as defined by Sue (2010) include, "commonplace verbal, behavioral, or environmental indignities, whether intentional or unintentional, that communicate hostile, derogatory, or negative slights and insults to people of color" (p. 29). The development of the construct of microaggressions initially provided a way for examining racial discrimination (Sue, 2010.) However, its relevance and subsequent application to gender discrimination provides a way to understand the paradigm shift from overt sexism to covert sexism (Basford, et al., 2014). There are three subcategories of microaggressions, microinsults, microassaults, and microinvalidation (Sue, 2010). Insensitive or demeaning behaviors, which imply that a person belongs to a lesser group, that he or she is less intelligent, or the assumption that a person possesses gender-specific characteristics, are *microinsults* (Sue, 2010). For example, assuming that women are less intelligent than men is a common microinsult. Whereas, *microassaults* are deliberate, purposeful attacks perpetrated through both verbal and nonverbal attacks meant to cause psychological harm to the recipient (Sue, 2010). An example of gender-specific microassaults include nonconsensual sex, staring at the breasts of a woman while speaking to her, or the purposeful display of pornography in the workplace (Sue, 2010). Finally, *microinvalidations*, like microinsults, are often unconscious manifestations that "exclude, negate, or nullify the psychological thoughts, feelings, or experiential reality" of another person (Sue, 2010, p. 29). An example of a microinsult related to gender would be denial of the existence of sexism in general, or the belief that being male enhances the chances of success in the workplace (Sue, 2010). The aggregation of workplace microaggressions creates a hostile working environment that reinforces sexual objectification of women, the use of sexist language, identification of women as second-class citizens, assumption of the inferiority of women, and denial of both institutional and individual sexism (Sue, 2010). Unfortunately, due to the insidious and invisible nature of the majority of microaggressions both genders often lack the ability to identify them, although the identification of a microaggression is often easier for women than men (Basford, et al., 2014).

Historical gender role assignments translate into descriptive and prescriptive stereotypes that are now part of contemporary societal norms because of their incorporation into the linguistic habits of contemporary cultures (Parks & Roberton, 2004). These deep-seated stereotypes have metamorphosed from overt acts of sexism to microaggressions that predominately consist of acts related to both verbal and nonverbal language abuse and labeled as linguistic sexism (Stahlberg, et al., 2007). Culture and language share an intimate relationship identified through the study of linguistic structure and its relationship with cognitive development, which brings together cognitive psychology and anthropology to examine the reciprocal relationship between languages and societal constructs (Whorf, 1941). Language has the capacity to become a reflection of cultural beliefs and the use of the constructed language, including body language and written text, become habits of a culture and forms the basis of Whorf's principle of linguistic relativity, also known as the Sapir-Whorf Hypothesis (Whorf, 1941). Whorf posited that the lived experience develops through the interplay of everyday language and culture and thus influences the way societal concepts are categorized (Whorf, 1941). Therefore, the Sapir-Whorf Hypothesis has been one of the major theoretical constructs applied to the study of the influence of sexist language on society. The reciprocal relationship between language and culture generates sexist language that perpetuates the favored status of the dominant sex, which in most cultures tends to be males (Parks & Roberton, 2004; Prewitt-Feilino, et al., 2012). Contemporary language carries subtle sexist meaning related to its cultural roots because sex is one of the social categories that is integrated into the lexicon of all studied languages around the world (Stahlberg, Braun, Irmen, & Sczesny, 2007; Prewitt-Feilino, et al., 2012). Gender-normative characteristics attributed to, and expected of, women often penalizes women who are successful in occupational roles traditionally assigned to men (Heilman, Wallen, Fuchs, & Tamkins, 2004).

The use of these descriptive "female" terms is a form of discrimination known as radiant devaluation (Benokraitis, 1997). On the surface, it appears that a positive, gender normative characteristic regarding the perceived characteristics or ability of a woman is complementary (Benokraitis, 1997). However, the use of gender descriptive terms to evaluate the skills of a woman is a form of intellectual harassment because it attributes stereotypical characteristics to women and then uses these characteristics as a weapon to prevent advancement (Benokraitis, 1997; Bobbit-Zeher, 2011). For example, the descriptive belief that women *will be* nurturing leads to a prescriptive belief that they *should be* nurturing (Bobbit-Zeher, 2011). Thus, words such as nurturing take on a female connotation when used in everyday language that can significantly affect the perception of a woman (Briere & Lanktree, 1983). In addition, successful women who do not fit the assigned gender normative characteristic may end of carrying such labels as "bitch" or "dragon lady," which not only creates a negative working environment, but often prevents women from achieving higher status positions because they do not fit the traditional gender ideology (Heilman, et al., 2004, p. 426; Bobbitt-Zeher, 2011).

The extent to which a society incorporates sex or gender into its language varies depending on the stereotypes and gender schema of the respective culture (Stahlberg, et al., 2007). Three forms of expression of sex in the lexicon of language exist in the linguistic study of culture and language. These include gendered, natural gender, or genderless languages (Stahlberg, et al., 2007). Gendered languages, such as French, German, and Spanish, use sex as a grammatical category; therefore, reference to biological sex is frequent throughout a typical conversation (Stahlberg, et al., 2007). In gendered languages, animate nouns have either masculine or feminine associations, and have a corresponding personal pronoun (Stahlberg, et al., 2007). For example, in Spanish, if a female cat is a *gata*, while a male cat is a *gato* (Stahlberg, et al., 2007). Thus, animate nouns ending in –a denote female/feminine and nouns ending in –o denote male/masculine. In contrast, natural gender languages, such as English, use pronouns, such as he/him or she/her, to denote male/female, but there is no gender association for nouns (Stahlberg, et al., 2007). Finally, genderless languages, such as Finnish and Turkish, lack grammatical nouns as well as masculine/feminine pronouns (Stahlberg, et al., 2007). Genderless languages have lexical expressions to signify distinctions between male and female only when necessary and instead use one pronoun that represents both males and females the majority of the time (Stahlberg, et al., 2007). In summary, the three types of gender expression found in languages from around the world exist on a continuum with grammatical gender using the most references to sex and genderless languages using the least (Stahlberg, et al., 2007).

One of the commonalities that grammatical gender, natural gender, and genderless languages share is the use of masculine generics (Stahlberg, et al., 2007). Masculine generics "are used sex-specifically in reference to male persons and generically in reference to mixed groups and to people whose sex is unknown or irrelevant" (Stahlberg, et al., 2007, p. 169). The use of masculine generics in written text or everyday language may affect women negatively by presenting masculinity as the norm (Benokraitis, 1997; Stahlberg, et al., 2007). Common examples of masculine generics include calling all people who fight fires, firemen, regardless of gender; or referring to all professors as he or him; using the phrases such as "our forefathers"; or referring to human rights as the rights of man (Stahlberg, et al., 2007). Constant exposure to masculine generics implies that being male is the norm and may affect the ability of young girls to envision themselves in various academic disciplines or occupations in the future (Stahlberg, et al., 2007). Thus, the use of masculine generic language may generate long-lasting cognitive effects that influence the decision-making process of both women and men (Stahlberg, et al., 2007). In fact, several key studies suggest that English-speaking participants, regardless of their identified gender, increasingly envision women in traditionally ascribed male roles when gender-fair versions of the word replace masculine generics (Stahlberg, et al., 2007). For instance, the use of firefighter versus fireman creates a different image in the mind of the listener and implies that a firefighter could be a man or a woman. Not surprisingly, the use of masculine generics has been at the heart of the feminist debate surrounding the use gender-fair language in contemporary society (Stahlberg, et al., 2007). Feminists posit that the derivation of masculine generic use originates from an androcentric worldview that has a hidden agenda to prevent women from becoming a dominant, visible part of society (Stahlberg, et al., 2007). For example, a recent study by Prewitt-Freilino, Caswell, and Laakso (2012) suggests that gendered language use correlates to an increased level of gender inequality, especially in the workplace, when compared to natural gender or genderless language countries. Interestingly, countries with natural gender languages, which range in the middle of the continuum for gender reference, actually show a higher political empowerment rating than the grammatical gender and genderless countries. Thus, genderless languages, which seem gender-fair on the surface, still allow for gender bias depending on the audience (Prewitt-Freilino, et al., 2012). The bias exhibited by genderless language countries may exist because of the androcentric roots of these languages and the ability of the language to make, and keep, women invisible (Prewitt-Freilino, et al., 2012). Consequently, natural gender language countries, which are in the middle of the continuum of gender reference, actually seem to be better at improving the status of women in society (Prewitt-Freilino, et al., 2012).

In summary, the perpetuation of traditional gender roles through nonverbal and verbal language contributes to microaggressions that prevent women from succeeding in male-dominated occupations (Burgess & Borgida, 1999). Instead of the overt sexual discrimination of the past, contemporary women experience a series of gender microaggressions or microinequities that result in negative work outcomes such as poor work performance, as well as physical and psychological health issues (Basford, Offerman, & Behrend, 2014). Therefore, this chapter explores historical use of gender normative language and the ramifications of its use for women in the STEM fields of academia.

Historical Roots: Gender Equity in Academia

Institutions of higher education are not immune to sexism because inequalities stemming from the historical roots of the United States form the basis of institutional structure, which began as a way to promote "traditional" cultural values that oppress women (Benokraitis, 1997; Rury, 2009). The development of the modern school system in the United States grew from a need for educating the future workforce and the newly arrived immigrants during the Industrial Revolution in the early 1800s (Rury, 2009). Emergent ideology derived from the dominant religious and political values of the time, which were the American Protestant religion and a white Eurocentric view of the world of domination and conformity (Rury, 2009). Thus, the development of the educational system was essentially a way to impart the virtues and values of the dominant societal groups, which tended to be white males (Rury, 2009). Since "traditional values" formed the basis of the educational system, the exclusion of women in all forms of education was common until the beginning of the Revolutionary War during the 19th century (Rury, 2009). However, by the end of the 19th century women outnumbered men in high school and gained a foothold in higher education (Rury, 2009). The increased numbers of women in the education system is due in part to Horace Mann, a proponent of free public education who also believed in the education of women (Rury, 2009). Mann did not promote equal education; he promoted the education of women because he felt that due to their ascribed intrinsic nature to be maternal and affectionate, that women were better suited to teaching (Rury, 2009). The education of women grew in the 19th century largely due to the notion of training young women for the republican motherhood; thus, education was a way to train future mothers under the guise of domestic feminism (Rury, 2009). Domestic feminism was encouraged as a way to advance the education of women; however, its main premise was to provide educated companions for men and to enhance their skills as mothers (Rury, 2009). Essentially, the transformation of the education of women was to create cultural capital through the transference of "skills and knowledge consistent with social class differences in family roles and expectations" to maintain traditional gender normative traits and skills in modern society (Rury, 2009, p. 103). At this point in history, white women had achieved educational equity, but they had not achieved social equality.

Over time, women began to challenge traditional gender roles during a paradigm shift in the later 19th century stemming from the women's rights movement (Rury, 2009). This shift was largely due to women gaining access to higher education, which eventually led to women joining the ranks of professions with a higher status, such as a physician or lawyer (Rury, 2009). However, access to full-time employment for women academics in higher education institutions did not come to fruition until approximately 1969 (Gappa, Austin, & Trice, 2007). In contemporary society, women in academia are generally "clustered at the lowest faculty ranks, receive lower salaries, are more likely to occupy temporary or part-time positions, and advance in rank or are tenured less frequently or more slowly" than men (Benokraitis, 1997, p. 54-55). Thus, occupational equity for women in academia is not yet a reality. In the workplace and in the legal world the definition of equity includes "equal pay, equal access to opportunities to enter an occupation and to advance in it, and to freedom from harassment" (Bailyn, 2003, p. 139). If a woman is successful in a male-dominated role, her competence falls under scrutiny because she has gone against the gender-normative, prescribed role for females (Heilman, et al., 2004). The notion that women are less competent has historically led to unequal pay for equal work, and has thus created the "gender gap," as well as hostility and social isolation for working women (Koppelman, 2014, p. 250; Heilman, et al., 2004). In fact, statistics for higher education show that men are more likely to be full professors (36 percent versus 18 percent), which creates an additional pay gap between men and women as the average salary would be $73,433 for men versus $57,699 for women (Gappa, et al., 2007). This basic form of inequity is the dominant form acknowledged by higher education administrators and easily demonstrated via the low number of women in the faculty pool, especially in the sciences, as well as the documented lower pay for the same professorial rank, low representation in high-ranking leadership positions (Bailyn, 2003; Marchant, Bhattacharya, & Carnes, 2007). Therefore, women have equal opportunity (*equality*) to have an academic career; however, the discrepancy in pay and ability to achieve academic rank provide evidence to a lack of *equity* (Bailyn, 2003). The "model of the ideal academic" is not gender-neutral, which puts women at a disadvantage from the very beginning of their academic careers, especially if they have children, want to have children, or are the caregiver for another family member (Bailyn, 2003, p. 139). This is not an equitable situation because the constraints of fulfilling the heavy academic load of teaching, research, service, becoming an expert in your field, writing grants, and meeting the required deadlines to obtain tenure lead many women to sacrifice either their career or their family (Bailyn, 2003). Although women have the opportunity to succeed, they often do not reap the full benefits of success achieved by their male colleagues. This is due, in part, to the continued use of gender normative language to perpetuate gender schemas that allow the underlying male dominance to be the underpinning culture force that drive institutional and departmental policy (Benokraitis, 1997).

THE INFLUENCE OF A GENDERED HIGHER EDUCATION SYSTEM ON WOMEN IN STEM

Women have been present in the sciences and have contributed to major scientific discoveries, yet their representation in the STEM disciplines is still lacking (Settles, Cortina, Steward, & Malley, 2007). The embedded effects of gender normative language on both societal and institutional cultures may be one of the leading factors that affect retention of women in these fields. The topics outlined in the following sections provide a detailed understanding of how gender normative language has become a part of the "normal" structure of academic communications and policy, which limits the achievement of women in STEM disciplines.

Influence of Gender Normative Language on Female Students in STEM

Faculty members act as role models to students; however, many inadvertently send the wrong messages to students through both conscious and unconscious actions that reduces female persistence in STEM disciplines (Morrell & Parker, 2013). The unintentional communication of gender normative micromessages, such as "facial expressions, gestures, words, or tone of voice" in combination with the language patterns of a culture provide validity to gender stereotypes and become instances of covert sexism (Morrell & Parker, 2013). Examples of unintentional communications include the use of masculine generic language or male dominant language patterns while lecturing in the classroom, speaking to students, or with colleagues; verbally discouraging or questioning the choice of female students from participating in STEM courses or activities; and authoritative communication practices that prevent women from participating in meetings and collaborating with peers.

Reinforcement of stereotypes may begin as early as adolescence when female students are discouraged from participating in STEM-related courses or activies by teachers or counselors in their high school (Grossman & Porche, 2014). A qualitative study conducted with both male and female students in ninth and tenth grade revealed that 29% of the female students interviewed indicated they had experienced gender-based microaggressions (Grossman & Porche, 2014). Some of the female adolescents reported that teachers questioned their academic ability in STEM-related classes and some described how teachers told them they did not fit "the traditional images of scientists" (Grossman & Porche, 2014, p. 717). Consistent exposure to gender-normative microaggressions leads to an aggregation of microinequities that reinforces negative connotations about a person's identity and concept of self (Morrell & Parker, 2013). In contrast, faculty members at all levels can counteract negative micromessages by using microaffirmations, such as inclusive language and incorporation of lessons that include historical female role who are often not included in normative textbooks (Morrell & Parker, 2013). In addition to direct comments regarding female ability, students also experience less direct gender normative stereotyping from faculty members (Morrell & Parker, 2013). A common example of this type of microaggression occurs when faculty members assign a historically gendered task, such as note taking, to a female student while directing a male student to conduct the experiment or lead the group activity (Morrell & Parker, 2013). The relegation of female students to historically gendered tasks at the direction of their instructor is a subtle way of telling her that she is not competent enough to conduct the experiment and that she should act more like a girl (Morrell & Parker, 2013). Further, student exposure to the continuous use of masculine generics, such as he or him, to describe all people in a STEM discipline, significantly affects how students rate the attractiveness of a STEM-related career is to women (Briere & Lanktree,

1983). Therefore, young students are very astute at identifying inequalities or inequities, which leads to internalization and reinforcement of gender normative stereotypes when given no other alternative (Morrell & Parker, 2013). In contrast, pilot programs in which teachers are trained using a professional development model aimed at reducing gender microaggressions in STEM fields show a higher test scores and pass rates in STEM classes for both boys and girls than students at the same school whose teachers did not receive the training (Morrell & Parker, 2013). Therefore, the education of faculty regarding gender normative language use and its direct effect on student success is essential for retaining female students in STEM.

In addition, research suggests that the observation of communications that occur between faculty members in the academic departments of higher education institutions also play a crucial role in the development of student identity (Curtis, 2013). Students who witness a verbal exchange between a male and female faculty member often unknowingly observe internalized or institutionalized dominance, which cognitively reinforces their childhood exposure to gender normative language patterns (Benokraitis, 1997). From a young age, children learn to use specific language patterns to dominate or withdraw from a conversation (Benokraitis, 1997). Since male adults are promoted as being the dominate sex through cultural cues, males are more likely to dominate a conversation by interrupting females or talking over them (Benokraitis, 1997). Thus, dominating a conversation becomes an internalized behavior that adult males subconsciously display leading to a verbal microassault towards the women they are conversing with (Benokraitis, 1997). In fact, the literature confirms that both male colleagues and male students interrupt and challenge the knowledge of female professors during a lecture at significantly higher rates than male professors (Benokraitis, 1997). Male faculty members who observe a microassault on a female faculty member without intervening cognitively reinforce male dominance (Benokraitis, 1997).

Finally, women not deterred from STEM disciplines during their undergraduate years and who subsequently choose to pursue a graduate degree often experience subtle verbal and nonverbal language cues that restrict their ability to fully participate in the academic community of their discipline (Fox, 2001; Ferreira, 2003). At the departmental level, female graduate students in both biology and chemistry, reported being told that they need to be more aggressive and that teamwork or collaboration should be limited (Ferreira, 2003). Thus, there is reinforcement of gender-normative stereotypes through language that conveys to females that they need to be more masculine (increase aggressiveness) and less feminine (reduce collaboration). In addition, female doctoral students described subtle language differences that restricted the development of a mentor-mentee or collegial relationship with their academic advisor (Fox, 2001). Instead, female students were more likely to describe this relationship as a faculty-student relationship, whereas, male students were more likely to describe their advisor as a mentor or colleague (Fox, 2001).

In summary, women have historically been present in the sciences and have contributed to major scientific discoveries. However, the women who have endured in the scientific disciplines are subject to the aggregated results of the gender normative microaggressions previously described. The experiences of Rosalind Franklin provide a contextual example of the profound effects that gender normative language has on the career of a woman trying to forge a career in a STEM discipline. At the age of 15, Rosalind wanted to be a scientist and eventually earned a doctorate in physical chemistry from Cambridge University in 1945, which was rare for a woman at the time (U.S. National Library of Medicine, 2014). In 1950, Rosalind began studying the structure of DNA using x-ray diffraction at King's College in London at the suggestion of the assistant lab chief, Maurice Wilkins (U.S. National Library of Medicine, 2014). Unfortunately, there was an initial miscommunication regarding the role that Rosalind was to occupy in the laboratory, which led to a strained working relationship between Wilkins and Franklin (U.S. National Library of Medicine, 2014). To exacerbate the situation, Rosalind faced harassment by male researchers in the lab, including James Watson, because they knew that she was on to something with her research (Watson, 1969). Watson and her other male colleagues frequently provoked her by calling her "Rosy," which was demeaning, as well as constantly questioning her research capabilities and harassing her to the point that their discussion would result in a shouting match (Watson, 1969). Consequently, the male researchers in the lab felt she was secretive and disagreeable because she eventually gave up trying to collaborate with them because of the demeaning manor in which she was treated (Watson, 1969). Franklin played a significant role in the discovery of the helical structure of DNA. However, no official credit is on record for her work because Wilkins stole her data and showed it to Watson and Crick, two other scientists working on the theoretical model of DNA (Watson, 1969; U.S. National Library of Medicine, 2014). The male researchers in question deemed this necessary because they assumed that she did not know how to interpret the data correctly and they wanted to be the first to publication (Watson, 1969; U.S. National Library of Medicine, 2014). Watson and Crick were the first to publish the data and eventually received a Nobel Prize for their work. Watson even went as far as to belittle Franklin in his biography about their experience, in the book entitled the *Double Helix,* although he did provide a half-hearted apology in the epilogue. Eventually, Crick and Wilkins admitted Franklin's role; however, by that time she had died from ovarian cancer (Watson, 1969; U.S. National Library of Medicine, 2014). Nobel Prizes are not posthumously given, thus Rosalind never received official credit for identifying the helical structure of DNA and her contribution is generally not included in normative textbook descriptions of the discovery of DNA (U.S. National Library of Medicine, 2014).

The case of Rosalind Franklin provides a heartbreaking historical example of the discrimination and harassment she endured. However, she is only one of dozens of women scientists who struggled in a historically male dominated field. Other examples include Jocelyn Bell Burnell whose male supervisor won the 1974 Nobel Prize in Physics for her discovery of pulsars, or Esther Lederberg whose husband used her invention of replicate plating of bacteria to win a Nobel Prize in 1958 (Lee, 2013). Unfortunately, similar practices occur in contemporary scientific research communities; however, the instead of blatant acts have become more subtle. This is in line with the transition from blatant sexism to more subtle forms of sexism previously described. The subtle forms sexism experienced currently experienced by women include the use of gender normative language and the patronization of women via gatekeeping practices used in the male-dominated scientific community to maintain the social oppression of women.

Gatekeeping: Use of Gender Normative Language to Reinforce Gender Schema

The underrepresentation of women in both tenure-track faculty positions and academic administrative positions in higher education is still a reality in contemporary society, especially in the STEM disciplines (Gappa, et al, 2007; Grossman & Porche, 2014). Title VII of the Civil Rights Act (1964) outlawed discrimination against a person based on sex; nevertheless, the discrimination, or at least discouragement of women to apply for certain jobs, continues to socially oppress women due to lingering perceptions that question the capability of women to perform in science, engineering, and math fields (Bem & Bem, 1983; Grossman & Porche, 2014). Several gate-keeping practices related to gender-biased language used by males in positions of power or privilege limit the advancement of women in higher education (Trix & Psenka, 2003; Riegle-Crumb & King, 2010). Examples of these gate-keeping practices are found in written documents such as performance evaluations of women faculty, letters of recommendation for female faculty members, posted job descriptions for faculty and leadership positions, and departmental policy, such as tenure criteria (Trix & Psenka, 2003; Marchant, et al., 2007; Marschke, Laursen, Nielson, & Dunn-Rankin, 2007). For example, characteristics ascribed to successful scientists such as a high level of assertiveness and competitiveness are also characteristics generally associated with the Western view of masculinity (Bailyn, 2003). Further, Marchant, et al. (2007) found that Western culture influences both men and women to "hold the bias that high authority figures should be men" (p. 998). Interestingly, gender biased language, such as that previously described, is evident in the wording of academic job description and tenure requirements (Marchant, et al., 2007). For instance, job descriptions that contain male-gendered attributes, ultimately limits the attractiveness of a position to female applicants (Marchant, et al., 2007). Job recruiting documents for tenure-track faculty positions from several leading medical schools indicated that the use of the word "leader" (male-gendered) seems to automatically activate culturally entrenched assumptions that the position should be filled by a man (Marchant, et al., 2007). Typically, women scientists are more likely to excel in characteristics such as curiosity and persistence, but a woman who displays assertiveness, or do not present a nurturing and communal demeanor, is often viewed negatively because their personality does not match the associated assumptions of the prescriptive gender-normative characteristics (Bailyn, 2003; Heilman, et. al., 2004). Therefore, women exhibiting feminine characteristics during a job interview or employee evaluation experience denial of employment or promotion in rank because they do not fit the prescribed gender role (Burgess & Borgida, 1999; Bobbitt-Zeher, 2011). This is a form of prescriptive stereotyping known as "walking a tight rope" because women are not allowed to appear too feminine or too masculine or their competence is question Williams, Phillips, & Hall, 2014, p. 3). In addition, women often experience a phenomenon known as "prove-it-again," which forces women to provide proof beyond that expected of their male counterparts to prove that they are competent (Williams, et al., 2014, p. 3). This is disturbing because women hold 29 percent of tenure-track research positions in the biomedical sciences at major universities and research institutions (Marchant, et al., 2007; Martinez, et al., 2007). Further, only 19 percent actually hold tenured senior investigator positions at these universities even though 45 percent of the postdoctoral fellows at the same institutions are women (Martinez, et al., 2007). This seemingly innocuous use of gender normative language may be holding women back from tenure-track faculty or high-level administration positions.

Although medical education has been a leader in gender equity, women are still lagging behind in academic medical leadership positions (Marchant, et al., 2007). White males who tend to be socially conservative have traditionally dominated the field of medicine in the United States (Trix & Psenka, 2003). Therefore, the department heads at medical colleges often consist entirely of white, older males who, perhaps unconsciously, practice a form of subtle sexism (Trix & Psenka, 2003). Department heads act as gatekeepers for medical education because they frequently write letters of recommendation for students and faculty applying for academic promotion or a faculty position upon completion of residency (Trix & Psenka, 2003). Trix and Psenka (2003) revealed systematic differences in the language used to describe female candidates versus male candidates. The descriptions of female candidates contained gender-normative language that created negative inferences about the candidate (Trix & Psenka, 2003). For example, a qualifier denoted that the candidate was a "lady physician," and as having "challenging personalities"; however, similar language was not used for male candidates (Trix & Psenka, 2003; p. 202-203). In addition, female applicants had a twice as many letters of recommendation that contained negative language, which would raise doubt in the mind of a potential employer than their male counterparts (Trix & Psenka, 2003).

Finally, the language of tenure criteria or policy may act as another gatekeeping practice. Marchant et al. (2007) examined the tenure criteria for 24 high-ranking medical schools for male gendered words such as leader, analytical, competitive, independent, defends, and leadership (Marchant et al., 2007). At least one male gendered term appeared in all but two schools with a total of 183 occurrences total (Marchant et al., 2007). Whereas, female-gendered words such as sensitive and understanding appeared three times for all 24 schools (Marchant et al., 2007). Interestingly, schools that had a higher average number of the word "leader" in their tenure documents employed fewer female faculty members (Marchant et al., 2007). Thus, gendered terms in tenure documents may play a greater role in deterring women from schools that employ this language.

Gatekeeping: Effects of Gender Normative Language on the Academic Publication Process

Inequities in the academic publication process, especially in the sciences have also been identified (Budden, Tregenza, Aarssen, Koricheva, Leimu, & Lortie, 2007). Early studies conducted in Sweden regarding the peer-review process in the sciences, revealed that female publication applicants received significantly lower scores than male applicants, and were deemed to be scientifically incompetent (Wenneras & Wold, 1997). Tenure-track academics at the majority of colleges and universities require research and subsequent publications as the gold standard for academic achievement. For example, two female researchers, one from the University of Sussex in the United Kingdom, and the other a researcher from Australian National University, submitted a paper to the Journal PLOS ONE (Else, 2015). An anonymous peer reviewer from the journal rejected their article and suggested in the comments that collaboration with a male biologist would strengthen the quality of their manuscript (Else, 2015). Further, the PLOS ONE reviewer voiced several opinions that suggested in a patronizing tone that women were incompetent scientists due to the slight physiological differences between men and women (Else, 2015).

In addition, Budden, et al. (2007) have shown that the peer-reviewed publication process is not without bias due to the association of a gender with a particular name. A double-blind review process, in which the identity of both the author and the reviewer is unknown, is not common practice, especially in scientific journals (Budden, et al., 2007). A small study conducted at the offices of the journal, *Behavioral Ecology*, compared the difference in the number of papers published by female first-authored papers before and after they converted to a double-blind review process (Budden, et al., 2007). In this study, Budden, et al. (2007) found a significant increase (7.9%) in the number of female first-authored studies after the policy was instituted (Budden, et al., 2007). Finally, Budden, et al. (2007) indicated that lack of fair assessment and the ability to publish papers by women researchers could be a barrier to the obtainment of tenure and other professional advances for female faculty members.

SOLUTIONS AND RECOMMENDATIONS

One solution to counteract the perpetuation of gender schema with gender normative language is to change the language used within the higher education culture. Gender normative language is a habitual part of everyday culture and members of contemporary may not recognize the subtle sexism and cumulative microaggressions that result from its use (Swim, et al., 2004). However, training to recognize gender normative language and microaggressions often results in a purposeful use of gender-neutral language regardless of gender role beliefs (Swim, et al., 2004; Morrell & Parker, 2013). Further, efforts for policy change regarding gender bias issues require clear and consistent attention and commitment throughout the recruiting, hiring, tenure, and promotions processes (Williams, et al., 2014). Written institutional and departmental policies, tenure criteria, or job descriptions that contain gender normative language require revision using gender-neutral language. Best practices may include the following (Williams, et al., 2014):

1. Use gender-neutral terms when drafting the advertisement. Terms associated with masculinity, such as "competitive" and "assertive" tend to deter women applicants.
2. Conduct blind screening of resumes by removing all indications of gender of the applicant.
3. Use an evidence-based hiring protocol that requires training for gender bias recognition throughout the interview and campus visit process.
4. Creating policies that allow for work-life balance for both genders.

Another possible solution involves the use of Metrics-Based Bias Interrupters, which is a new approach to organizational change that specifically addresses gender bias (Williams, et al., 2014). This approach emphasizes the real-time interruption of subtle bias and consists of an iterative four-step process (Williams, et al., 2014). The first step of this process begins with an assessment of the hiring and promotion processes of the institution through interviews or focus groups (Williams, et al., 2014). Identification of a bias requires the implementation of a strategically determined bias interrupter along with a monitoring process to determine if the interrupter is successful (Williams, et al., 2014). An example of an identified bias may include surveying departmental faculty to determine if distribution of task assignments is equal between male and female faculty members (Williams, et al., 2014). Williams, et al. (2014) emphasizes that monitoring and follow-up are necessary to ensure that improvement of the indicated bias has occurred.

Finally, it is highly probable that the tenured faculty members conducting research in their academic department are men, while the women are contingent faculty members (Curtis, 2013). Further, there are also fewer women in leadership roles in the STEM discipline (McCullough, 2011). This sends a micromessage to female students that women are not as qualified in this field as men (Curtis, 2013). This is especially troublesome to for female graduate students who look to faculty members and administrators as mentors and role models (Curtis, 2013). In addition, if a female graduate student at high-ranking research institution is lucky enough to have a female faculty member as a mentor, that faculty member may be a role model that is difficult to obtain for young women that hope to have a family (Morrell & Parker, 2013). Full-time faculty members at high-ranking research institutions are less likely to be mothers due to the difficulty in maintaining an active research laboratory with family life (Curtis, 2013). In fact, the literature indicates that few high-level women researchers are able to combine families with their careers (Curtis, 2013). This can be discouraging to aspiring female researchers when they feel that they have to choose between a research career and the possibility of having a family. The Association of Women in Science (2014) outlined five family-friendly practices to include in tenure criteria that all universities should emphasize to increase the number of women faculty members. These include:

1. Establishing a dual career policy that allows hiring of a faculty member and their spouse at the same institution.
2. Allow for the postponement of tenure for faculty members who wish to have children (both men and women) and wish to provide childcare or elder care.

3. Part-time tenure tracks with the ability to return in order to accommodate family needs.
4. An increase in paid leaves for a faculty member that needs to provide parental or elder care. 5.Onsite, affordable childcare.

Perhaps if more institutions focused employment descriptions and requirements with the importance of both family and the academic career in mind, then there would be more women in high-level positions in academia to as mentors for female students. In addition,

FUTURE RESEARCH DIRECTIONS

Three main directions for additional research stem from this dialogue. First, research that examines the relationship between leadership related to the effects of societal constraints and women in STEM disciplines is minimal (McCullough, 2011). The literature suggests that the path to leadership for women in STEM is no longer a glass ceiling, but is instead a labyrinth (McCullough, 2011). Several questions that require examination with regard to leadership in STEM include:

1. Does gender normative language and communication differences between men and women continue to play a major role?
2. Does the communication style of a particular form of transformational leadership, which is a leadership style used more frequently by women, influence advancement?

Second, further examination of departmental culture in the STEM disciplines is required to enhance the understanding of the interactions between male and female faculty members. Recent research examining the cultural influences of a biology department on scholarly identity and role prioritization during a transformational change involving faculty composition and curriculum modernization suggests that female gendered characteristics may play a role in the success of a department (Haswell, 2015). This study elucidated several key female-gendered characteristics such as the use of collaboration and a transformational leadership style within the department that may share a relationship with the use of gender-neutral language in departmental policy (Haswell, 2015). Thus, further examination of the role of that gender-neutral language has on promoting transformational changes within a department is needed.

Finally, additional research regarding the gendered racism experienced by women of color in academia is required (Pittman, 2010: Williams, et al., 2014). Gendered racism involves the intersection of race and gender; however, the literature describing this phenomenon with regard to gender normative language is limited. Further, much of the language used in higher education policy writing stems from white, Eurocentric male normative terminology (Pittman, 2010). Many of the gender bias experiences related to gender normative language described by women of color are similar to that of white women; however, the effect may be quite different between black and white women (Pittman, 2010). Pittman (2010) acknowledges the distinct experiences of women of color that can be described as racism or sexism, but some of these experiences cannot be reduced to either alone and appear to be quite systematic in nature. For example, women of color have an acute awareness of the challenges they expect to face in the classroom either as a student or as a faculty member (Pittman, 2010). The majority of the challenges stem from white male students who use their position of privilege to undermine them by challenging their competency which indicates that this behavior is not racism or sexism exclusively, but an indirect combination of both (Pittman, 2010). Underrepresentation of women of color in the STEM fields and in higher education in general creates an academic world that lacks diversity (Chronicle Almanac, 2014). Therefore, it is imperative that we continue to study gendered racism in the context of higher education and the STEM disciplines to enhance the pool of women scientists with better representation of all racial and ethnic groups.

CONCLUSION

The progression of gender representation and equity for women in higher education has made little headway since the 1970s (Marschke, et al, 2007). "Excellence cannot be achieved without equity" and defining gender equity based almost exclusively on the pay gap ignores the fact that the higher education system is structured around the life experiences of men and the established definition of masculinity as "normal" by Western culture (Shakeshaft 1986, p. 499Bailyn, 2003). Although women may be self-selecting out of high-level academic research positions, they are not leaving academia altogether (Curtis, 2013). The number of women employed in higher education grew by 266 percent from 1976-2009 (Curtis, 2013). However, many of them have taken contingent faculty positions or have opted to teach at the community college level instead because these positions are non-research intensive (Curtis, 2013). These lower paying contingent positions have been on the rise at most universities exacerbate the gender gap in pay as more women are employed as contingent faculty members (Curtis, 2013). In addition, more women are opting to teach at community colleges where the ratio between male and female faculty members is 44 percent and 55 percent respectively (Rifkin, 2014).

The continuous exposure of female students to gender normative microaggressions from the beginning of their secondary school experience through their postsecondary education discourages young women from considering the STEM disciplines as a career. Therefore, acknowledgement of the gender-normative structure of higher education is imperative because it will prompt a paradigm shift to create gender-neutral working environments. Many institutions, including the Massachusetts Institute of Technology (MIT), have already reexamined the commonly held gender-normative assumptions and practices that suppress the academic advancement of women (Bailyn, 2003). Female faculty members and administrators at MIT have outlined the difficulties that women face at high-level research institutions and are using this knowledge to promote retention of women (Bailyn, 2003). Academic administrators, especially in the STEM disciplines must realize that academic diversity is a key factor for conducting the well-rounded research investigations and collaborative projects (Leggon, 2006; Association of Women in Science, 2014). Without change in the current hierarchal, gender-normative schema of the STEM disciplines, the best and brightest women are becoming increasingly dispirited and continue to choose other academic disciplines (Leggon, 2006; Association of Women in Science, 2014). Instead, STEM disciplines to need eliminate the ideology of the lone, white male researcher and embrace equity by including the feminine aspects of each discipline into everyday language and institutional policy (McCullough, 2011).

REFERENCES

American Association of University Professors. (2006). *AAUP faculty gender equity indicators 2006*. Retrieved from http://www.aaup.org/NR/rdonlyres/63396944-44BE-4ABA-9815-5792D93856F1/0/AAUPGenderEquityIndicators2006.pdf

American Association of University Women. (2013). *Improve girls' and women's opportunities in science, technology, engineering, and math*. Retrieved from http://www.aauw.org/files/2013/02/position-on-STEM-education-111.pdf

Association of Women in Science. (2014). *Institutions without family-friendly practices risk losing their best and brightest*. Retrieved from http://awisblog.wordpress.com/2014/06/18/institutions-without-family-friendly-practices-risk-losing-their-best-and-brightest/

Bailyn, L. (2003). Academic careers and gender equity: Lessons learned from MIT. *Gender, Work and Organization*, *10*(2), 137–153. doi:10.1111/1468-0432.00008

Basford, T. E., Offermann, L. R., & Behrend, T. S. (2014). Do you see what I see? Perptions of gender microaggressions in the workplace. *Psychology of Women Quarterly*, *38*(3), 340–349. doi:10.1177/0361684313511420

Bell, N. (2008). *Graduate enrollment and degrees: 1997 to 2007*. Washington, DC: Council of Graduate Schools.

Benatar, D. (2003). The second sexism. *Social Theory and Practice*, *29*(2), 177–210. doi:10.5840/soctheorpract200329213

Benokraitis, N. V. (1997). *Subtle sexism: Current practice and prospects for change*. Thousand Oaks, CA: SAGE Publications, Inc.

Benokraitis, N. V., & Feagin, J. R. (1995). *Modern sexism: Blantant, subtle, and covert discrimination* (2nd ed.). Englewood Cliffs, NJ: Prentice Hall.

Bobbitt-Zeher, D. (2011). Gender discrimination at work: Connecting gender stereotypes, institutional policies, and gender composition of workplace. *Gender & Society*, *25*(6), 764–786. doi:10.1177/0891243211424741

Budden, A. E., Tregenza, T., Aarssen, L. W., Koricheva, J., Leimu, R., & Lortie, C. J. (2007). Double-blind reviews favours increased representation of female authors. *Trends in Ecology & Evolution*, *23*(1), 4–6. doi:10.1016/j.tree.2007.07.008 PMID:17963996

Burgess, D., & Borgida, E. (1999). Who women are, who women should be: Descriptive and prescriptive gender stereotyping in sex discrimination. *Psychology, Public Policy, and Law*, *5*(3), 665–692. doi:10.1037/1076-8971.5.3.665

Chronicle of Higher Education Almanac. (2014). *Diversity*. Retrieved from http://chronicle.com/article/Almanac-2014-Diversity/148173/?sn

Civil Rights Act of 1964 § 7, 42 U.S.C. § 2000e et seq (1964).

Curtis, J. W. (2013). What roles are we modeling? Gendered academic employment and its consequences. *Peer Review, 15*(3). Retrieved from http://www.aacu.org/peerreview/pr-su13/Curtis.cfm

Else, H. (2015). 'Sexist' peer review causes storm online. *Times Higher Education.* Retrieved from https://www.timeshighereducation.co.uk/news/sexist-peer-review-causes-storm-online/2020001.article

Gappa, J. M., Austin, A. E., & Trice, A. G. (2007). *Rethinking faculty work: Higher education's strategic imperative*. San Francisco, CA: John Wiley & Sons, Inc.

Grossman, J. M., & Porche, M. V. (2014). Perceived gender and racial/ethnic barriers to STEM success. *Urban Education*, *49*(6), 698–727. doi:10.1177/0042085913481364

Haswell, M. M. (2014). *Exploring the influence of departmental culture on perceptions of scholarship and role prioritization of biology professors.* (Doctoral dissertation), Department of Educational Leadership, Central Michigan University, Mt. Pleasant, MI.

Hedgepeth, A. (2013). *Balancing work and life: Family-friendly workplace policies*. Retrieved from http://www.aauw.org/files/2013/02/position-on-work-life-balance-112.pdf

Heilman, M. E., Wallen, A. S., Fuchs, D., & Tamkins, M. M. (2004). Penalties for success: Reactions to women who succeed at male gender-typed tasks. *The Journal of Applied Psychology*, *89*(3), 416–427. doi:10.1037/0021-9010.89.3.416 PMID:15161402

Koppelman, K. L. (2014). *Understanding human differences: Multicultural education for a diverse America* (4th ed.). Boston, MA: Pearson Education, Inc.

Lee, J. J. (2013). 6 women scientists who were snubbed due to sexism. *National Geographic*. Retrieved from http://news.nationalgeographic.com/news/2013/13/130519-women-scientists-overlooked-dna-history-science/

Leggon, C. B. (2006). Women in science: Racial and ethnic differences and the differences they make. *The Journal of Technology Transfer*, *31*(3), 325–333. doi:10.1007/s10961-006-7204-2

Marchant, A., Bhattacharya, A., & Carnes, M. (2007). Can the language of tenure criteria influence women's academic advancement? *Journal of Women's Health*, *16*(7), 998–1003. doi:10.1089/jwh.2007.0348 PMID:17903076

Martinez, E. D., Botos, J., Dohoney, K. M., Geiman, T. M., Kolla, S. S., Olivera, A., & Cohen-Fix, O. et al. (2007). Falling off the academic bandwagon: Women are more likely to quit at the postdoc to principal investigator transition. *European Molecular Biology Organization Reports*, *8*(11), 977–981. PMID:17972894

McCullough, L. (2011). Women's leadership in science, technology, engineering, and mathematics: Barriers to participation. *Forum on Public Policy: A Journal of the Oxford Round Table*, *2011*(2). Retrieved from http://forumonpublicpolicy.com/vol2011.no2/papers2011vol2.html

Morrell, C., & Parker, C. (2013) Adjusting micromessages to improve equity in STEM. *Diversity and Democracy*, *16*(2). Retrieved from http://www.aacu.org/diversitydemocracy/vol16no2/morrell_parker.cfm

Parks, J. B., & Roberton, M. A. (2004). Attitudes toward women mediate the gender effect on attitudes toward sexist language. *Psychology of Women Quarterly*, *28*(3), 233–239. doi:10.1111/j.1471-6402.2004.00140.x

Pittman, C. T. (2010). Race and gender oppression in the classroom: The experience of women faculty of color with white male students. *Teaching Sociology*, *38*(3), 183–196. doi:10.1177/0092055X10370120

Pollack, E. (2013, October 3). Why are there still so few women in science? *The New York Times*. Retrieved from http://www.nytimes.com/2013/10/06/magazine/why-are-there-still-so-few-women-in-science.html?_r=0&pagewanted=print

Prewitt-Freilino, J. L., Caswell, T. A., & Laakso, E. K. (2012). The gendering of language: A comparison of gender, equality in countries with gendered, natural gender, and genderless languages. *Sex Roles*, *66*(3-4), 268–281. doi:10.1007/s11199-011-0083-5

Riegle-Crumb, C., & King, B. (2010). Questioning male advantage in STEM: Examining disparities in college majors by gender and race/ethnicity. *Educational Researcher*, *39*(9), 656–664. doi:10.3102/0013189X10391657

Rifkin, T. (2014). *Public community college faculty*. Retrieved from http://www.aacc.nche.edu/Resources/aaccprograms/pastprojects/Pages/publicccfaculty.aspx

Rose, A. S., & Hill, C. (2013). *Women in community colleges: Access to success*. Retrieved from http://www.aauw.org/files/2013/05/women-in-community-colleges.pdf

Rury, J. L. (2009). Education and social change (3rd ed.). New York, NY: Routledge.

Settles, I. H., Cortina, L. M., Steward, A. J., & Malley, J. (2007). Voice matters: Buffering the impact of negative climate for women in science. *Psychology of Women Quarterly*, *31*(3), 270–281. doi:10.1111/j.1471-6402.2007.00370.x

Shakeshaft, C. (1986). A gender at risk. *Phi Delta Kappan*, *67*(7), 499–503.

Stahlberg, D., Braun, F., Irmen, L., & Sczesny, S. (2007). Representation of the sexes in language. In K. Fiedler (Ed.), *Social communication* (pp. 163–187). New York, NY: Pscyhology.

Sue, D. W. (2010). *Microaggressions in everyday life: Race, gender, and sexual orientation*. Hoboken, NJ: John Wiley & Sons.

Swim, J. K., Mallett, R., & Stangor, C. (2004). Understanding subtle sexism: Detection and use of sexist language. *Sex Roles*, *41*(3/4), 117–127. doi:10.1023/B:SERS.0000037757.73192.06

Trix, F., & Psenka, C. (2003). Exploring the color of glass: Letters of recommendation for female and male medical faculty. *Discourse & Society*, *14*(2), 191–220. doi:10.1177/0957926503014002277

U.S. National Library of Medicine. (2014). *National Library of Medicine Profiles in Science Rosalind Franklin paper*. Retrieved from http://profiles.nlm.nih.gov/ps/retrieve/Narrative/KR/p-nid/183

Watson, J. D. (1969). *The double helix*. New York, NY: The Penguin Group.

Wenneras, C., & Wold, A. (1997). Nepotism and sexism in peer-review. *Nature*, *387*(6631), 341–343. doi:10.1038/387341a0 PMID:9163412

Whorf, B. L. (1941). The relation of habitual thought and behavior to language. In J. B. Carroll (Ed.), *Language, thought, and reality: Selected writings of Benjamin Lee Whorf* (pp. 134–159). Cambridge, MA: MIT Press.

Williams, J. C., Phillips, K. W., & Hall, E. V. (2014). *Double jeopardy? Gender bias against women of color in science*. Retrieved from http://www.uchastings.edu/news/articles/2015/01/double-jeopardy-report.pdf

KEY TERMS AND DEFINITIONS

Blatant Sexism: Explicit, obvious, unequal treatment of the opposite sex.

Covert Sexism: Purposefully hidden, intentional unequal treatment of the opposite sex.

Gender Normative Language: A lexicon that assigns gender to specific words; promotes gender schemas.

Microaggression: Commonplace verbal, behavioral, or environmental indignities, whether intentional or unintentional, that communicate hostile, derogatory, or negative slights and insults to people of color. (Sue, 2010, p. 29)

Modern Sexism: Sexist beliefs that indirectly condone unequal treatment of a member of the opposite sex. (Benokraitis & Feagin, 1995)

Subtle Sexism: Sexism that is "just below the surface" and considered customary. (Benokraitis, 1997, p. xi)

Chapter 4
Overcoming the Myriad of Obstacles:
The Ongoing Journey as a Female African American Physicist

Helen Cassandra Jackson
Wright State University, USA

ABSTRACT

This chapter documents the experiences of the ongoing journey of an African American female physicist. They correspond to those in documented studies of other African Americans and females in both the specific field of physics as well as the broader area encompassing Science, Technology, Engineering, and Mathematics (STEM). While there are some anomalies, when scaled with the norm of these groups, there is a thread of consistencies in the obstructions and difficulties that seem to be unique to mostly African Americans and on a smaller scale to White females. The intent of this writing is to shine a light on the status of affairs particularly in the scientific Ph.D. community, an area that many have felt was immune to the difficulties faced by African Americans on the lower end of society. It is evident that our society is neither "post-racial" nor "post-sexist", even on the higher intellectual turf.

INTRODUCTION: THE TROUBLING DATA

After climbing a great hill, one only finds that there are many more hills to climb. I have taken a moment here to rest, to steal a view of the glorious vista that surrounds me, to look back on the distance I have come. But I can rest only for a moment, for with freedom comes responsibilities, and I dare not linger, for my long walk is not yet ended. (Mandela, 1975)

History is decorated with African American inventors and scientists who have made world changing contributions despite the obstacles that surrounded them (List of African American Scientists and Inventors, 2015). Despite a rich history as contributors in all arenas, many decades after Jim Crow has formally ended, African Americans are poorer and have harder lives than their counterpoints in White society.

DOI: 10.4018/978-1-5225-0174-9.ch004

They still face racism at every turn. Their women also face sexism. Echoed in Mandela's quote above, and as recently stated in his speech on the 50th celebration of Bloody Sunday, President Barack Obama clarified the fact that the Civil Rights movement did not end decades ago but is an ongoing movement, a work still in progress. (Obama, Selma, 2015)

It is a constant rebuttal to discriminatory practices that the adequate intelligence, skills or motivation is lacking for advancement. Nonetheless, for the underrepresented, while moving forward with all the required facilities such as brain power, ambition, and diligence to accomplish and succeed being in place, their crossing the bridge to success is still not a given. Too often, Blacks in the higher level technical arena have to double or triple prove themselves over and above what is required of Whites. This is a systemic problem for Blacks and women and is discussed at length in the report "Double Jeopardy" (Williams, Phillips, & Hall, 2014). Additionally, amongst a significant segment of the scientific community, African American females still are stereo-typed and perceived as outside of "the best and brightest", which the technical field pursues. The same words can come out of a White male and an African American female; however, the former is accepted as credible and brilliant and the latter is subject to many levels of, sometimes never ending scrutiny. In the term coined as "Prove it Again" in the above mentioned report (Williams, Phillips, & Hall, 2014), the resulting unfair requirement stems from the perception of what is and is not credible and scientifically brilliant by the status quo scientific community. One's perception of scientific/technical worth has been documented in studies, such as" Merit alone is not enough" (Eichler, 2012), and further discussed throughout this chapter. So for the underrepresented in these situations, acceptance of credibility is not based on merit but on personal preference and bias, which is highly un-scientific and illogical. Unfortunately discriminatory thinking is hardwired into the minds of a too large segment of society. This is however, racist, period.

If success was based on merit alone, the status of STEM for African Americans and women would be very different. In the same article by the previously referenced Eichler (2012), it was acknowledged that merit alone will not put one ahead. Not everyone can get through or around "the good ole boys" system. In some of these relevant scenarios, power is in numbers, and the underrepresented are just outnumbered. As for government instituted remedies, too often but not always, Affirmative Action efforts are staging shows set up to appease. They do a poor job of resolving the issues of fairness for all the underrepresented within a STEM group. In fact, due to sometimes insincere implementation, too often these programs birth another set of problems.

While it may be well known that racial tension and injustices against African Americans has escalated in some arenas since the election of our first Black President, it may not be known how those in STEM fields have been affected in parallel but on a different plane. In particular, those seeking Ph.D.'s in the hard physical science areas, which have always been White male dominated, are suffering despite their intellect and the quality of the metrics they possess for success.

The current US President, at the time of this writing, has put forth more STEM related initiatives than any US President known to date. In light of that and to the contrary, there has and is a huge disparity that exists in these fields such that African Americans in particular are still disproportionately underrepresented.

The reasons for this disparity are complex, but it is not because there is less capability or brain power in African Americans. After making it through the many levels of difficulties that African Americans face such as inequality in K-12 education and a greater percentage of poverty and single parent families; there is still a tremendous racial and, if female, gender firewall to surmount. All of this is encountered once one has been proven just as technically competent as White male counterparts.

All the outside world sees is the small numbers in a time that so many are saying this is a "post-racial" society. And it is assumed that certainly "supposedly" intelligent scientists would not be bigots in their interactions with African American females. History begs to differ. Most of Academia cooperated with Jim Crow in its discriminatory practices. Many African Americans still today come from a blood line that was not allowed to attend public education until Brown vs. the Board of Education was enacted fully. The schools available were whatever the southern descendants of slaves could build or the Catholic Charities provided, prior to this.

BACKGROUND

Here, the background of this chapter's author is summarized and tied to the STEM data. The author's pathway is very important, as still, very few African Americans in particular have evolved far enough since Jim Crow has ended to have a family line of professionals, as compared to other ethnicities. Many are still the "first" to finish college and even the "first" African American in their institution, if at the Ph.D. level.

1. Surviving Childhood

In describing one's pathway, the description cannot start in the middle as one is merging into a robust STEM environment. The life and path prior to entrance into the STEM environment is just as important as the details of the professional endeavor. Like all groups, women and African American women in particular, come from a variety of backgrounds entering into STEM. Some come from strong educated families, while others come from dysfunctional single parent or extremely economically deprived families. Having some model of professionalism or educational achievement is certainly a healthy place to spring from, but the author along with many others have treaded into territory where it is urgent to develop quickly the necessary missing skills.

Highlights of this journey is also shared because socioeconomically, the author was an unlikely candidate for being a Ph.D. physicist. So many women of color have a more painful and dramatic biography. This is then compounded with the additional discrimination to be endured in the STEM environments. There can be multiple modes of discrimination on the job as well as in academia in the Ph.D. process that must be overcome. The female scientists can include those who are White, Black, Asian, Indian, Hispanic and Middle Eastern. While women of any color can experience sexism, African American women are in the subgroup that must endure the majority of the racism enacted via stereotypes among the pool of female scientists.

The author's parents were born and raised in Mississippi and South Carolina, strongholds of Jim Crow and oppression. Life as a young girl was lived first in Southern Alabama and South Carolina under the iron curtain of southern racism, until later moving north to Washington D.C. There were multiple guardians, mostly uneducated grandparents. The late teenage years were primarily spent living alone, seeing the domestic worker grandmother-guardian only a few hours on the weekend.

An idea was put into the author in the midst of this, perhaps by none other than Providence, the calling and desire to be a physicist. Experimenting in dark rat infested attics, making homemade cameras, telescopes and electronic gadgets, the world of knowledge and science was an escape from the drug and crime ridden world outside the crumbling doors of the places called home.

2. Maternity and Professionalism

Fast-forwarding to many years later, after having escaped a physically abusive first marriage and making a way through the first stages of the required education as a physicist, one of the first grand epiphanies revealed was the fact that the door being knocked on was that of an elite club that the civil rights movement seemingly did not penetrate This was the royal priesthood of scientists dominated by males, mostly White but specifically anything but Black and female.

After a break to focus on rearing her children, the return to the pursuit of becoming a physicist was like being astro-projected through a time capsule. Many adults, particularly women and in a variety of disciplines, break from their career aspirations temporarily to care for young children. Temporary escape from the "Maternal Wall" (Correll, Barnard & Park 2007; Cuddy, Fisk & Glick, 2004), which is a form of gender discrimination based on motherhood, was accomplished. To the author, it is not necessary to apologize for temporarily evading the additional set of STEM workplace hurdles experienced by motherhood. While on an extended break, the decision was to embark on doing social and educational repair and upgrade, through the mechanism of home school education. This paid off well academically as far as its intended goals were documented in two publications. (R. Moore, 1988; B. Hunter, 2006).

The details of these years will be in a subsequent publication. Contrary to what is thought, this form of education is not necessarily for extremists living on the fringes of society but is somewhat popular with the parents in the scientific community. This, along with keeping to the subsequently discussed cultural and religious mandates of "good motherhood", drove this decision. However, re-entry into the world of physics over a decade later was like Rip Van Winkle awaking to a new, changed world that defined science and technology. Nonetheless, the entrance into graduate school was not simply as that of a housewife, but as former CEO of a company based on the author's patented, trade named, trademarked product and a string of software engineering, technical, consulting and adjunct teaching positions. The once husband and partner in science and invention, (Justia, 2015) passed away in 2007 after experiencing many heart wrenching problems and a prolonged illness. Promotion to family matriarch was not by choice but seemingly inevitable due to being part of a family structure still wounded and experiencing the residual effects of being rooted in Jim Crow America.

3. How Culture (Gender Roles and Stereotypes) can Effect Black Female Professionals

One's journeys and pathways are strongly affected by the cultural norms in which individuals find themselves embedded. The same cultural/religious aspects of African American female life which makes her as strong as steel, and often unnecessarily strong, can be roadblocks to success. In the lower echelons, even when the benefits of the structure of a marriage existed, it was often founded in a subculture where a woman's career aspirations were by requirement sanctioned and permitted by the husband. Too often abuse was so normal it was too often accepted. These cultural mutants did and do exist as unwritten norms. Now with children to rear alone, being a single parent mother in the African American subculture had come to be another norm.

The Disproportionately Heavy Yoke

At the time of remarriage to the author's husband, who was at the time an engineer/inventor at NASA (where the author was also employed), the early leg of the journey took a pivotal turn. The benefits of the Civil Rights movement could be felt at this point by many at NASA. As stated, with the remarriage, culture met up with science, but still the traditional roles a woman held trumped career aspirations. .While it has been proven that it can be done successfully for women with children, the needed support system for the author was lacking.

The traditional role that many women play can be a challenging juggle with a demanding STEM career. To add insult to injury some women, in particular African American women, have been culturally cast into the role of care giver. They care for the generation before, below and every stray family member who has minimally achieved in life. Aspects of this is detailed in the book, "Too Heavy a Yoke" (Barnes, 2014). Because generationally many still have family rooted from times when success and opportunity for African Americans only minimally existed, there is a "savior-like" profile attached to an even barely successful STEM family member. Certainly not all African American female professionals are tied to this "yoke". It seemingly can be a success deal breaker to cope with this set of demands compounded with the extra burden of proving oneself as both a female and African American in a technically challenging field. Just as support is needed in the career arena, so it is also needed in her role in the family structure, however broad this family structure may be.

THE MYRIAD OF OBSTACLES

Both sexism and racism discriminate, and are among a collage of areas of discrimination. Racism, it has been concluded by the author and her African American female contemporaries, is the most dominant. It stands out in bright flashing neon lights over sexism. Racist behavior that damages is not always done by individuals who purposely hate. So it should be useful to analyze racism, as the academic superiors and employers have a smorgasbord of discriminatory behaviors from which to choose. An African American female can suffer from a multiplicity of these harmful biases that are described as intersectional. In a study done on the science of racism (Gelman, 2012), it was shown that the way people categorize individuals is hardwired in their brains. Race is the trait that leaves the most dominant impression on the brain, according to such scientific studies. Gelman's research points out that race, above all other areas that can trigger discriminatory responses, is most strongly and incorrectly "essentialized". This essentialized type of categorization induces an automatic rapid fire association in the brain that often leads to very derogatory assumptions about that individual of a given race and value judges them based on pre-wired groupings based on their race. If an individual is intersectionally underrepresented, such as an African American female, it is compounded. If he/she wants to be an athlete that may be good. However, if they want to forge forward to become a respectable scientist, he/she will have to overcome the effects from stereotypes that want to degrade his/her intellectual capability.

So in the generations that have evolved since "quality" or at least marginally similar education became available, it should be known as reports in this area are made, that the small numbers (e.g., less than 1% representation by African Americans in Physics) are not due to laziness, insufficient motivation, or lower intellect. Figure 1 plots the percentages of underrepresented minorities (African American, Hispanic and Native American males and females) in STEM fields by degree level (APS & IPEDS). The causes of

Figure 1. Percentage of degrees in STEM fields earned by underrepresented minorities (APS and IPEDS, 2008)

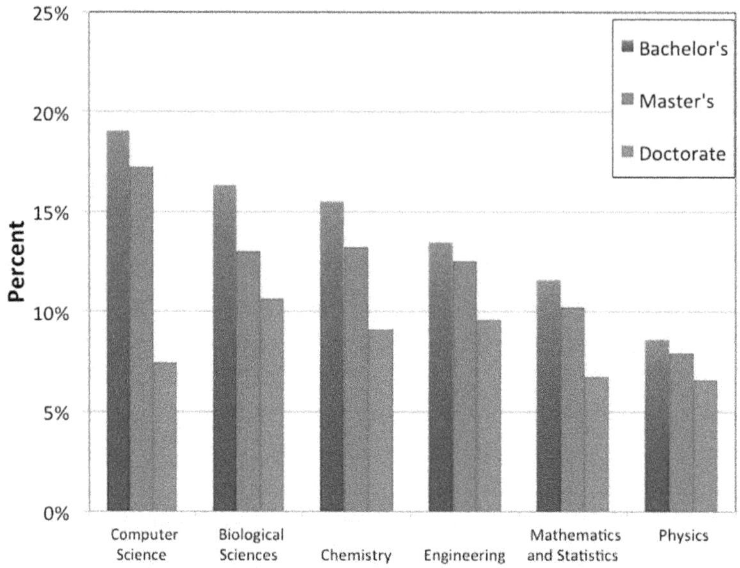

the low numbers are recorded in two reports (Jackson, APS, 2015) and another report (Journal of Blacks in Higher Education, 2006). While in the latter it was said an all-time high had been reached compared to previous decades, it is not obvious that the increasing numbers are countered by obstructionism. As documented in many reports to date, discriminatory practices are still keeping the numbers lower.

Ironically, in contrast to the 2006 report, a later report by Associated Press writer Jesse Washington (Washington, 2011) indicates that the number of Blacks going into Math and Science careers was shown to be decreasing relative to previous years. This is not surprising, for in many jurisdictions across the U.S., Affirmative Action is dead. It was also stated that Black people make up 2% of all PH.D.'s but 12 percent of the population. (National Center for Education Statistics, 2015). As stated above, in the field of physics, less than 1% of the Ph.D. degrees were awarded African Americans in physics. In the combined areas of all of the physical sciences, which would include areas like chemistry and geology, only 2% of the degrees were awarded to African Americans.

To further subtract from the numbers, once a STEM career is achieved by an African American, there is no guarantee for remaining there. Regarding retention, in another article published in 2014 by the American Research Institute(AIR, 2014) and as shown in Figure 2, women and Blacks are most likely to leave STEM fields (Turk-Bicakci, 2014). The findings were: "Women are more likely than men to leave STEM (19 percent versus 16 percent). Blacks are more likely than other races and ethnic groups to leave (21 percent compared with 17 percent of whites, 14 percent of Asians and 14 percent of Hispanics)."

1. Stereotypes and Roadblocks

In retrospect, before learning much of anything, and upon re-entering the academic and workplace environment, the first certainty discovered by the author was that amongst the indefinites was a definite. This definite was that of being heavily stereotyped and presumed incompetent. Incidentally, "Presumed

Figure 2. Retention and departure in STEM based on gender and race (AIR, 2014)

Incompetent" is the title and subject of a book edited by Gabriella Gutierrez y Muhs et al. (Gutierrez y Muhs, Flores Neiman, Gonzales, & Harris 2012) that documents the experiences of women (particular of color) in the academic world and speaks to this topic. Years before this came to be documented and affirmed as an issue, it was the elephant in the room upon walking through the doors that became open. It seemed the only way to overcome this obstacle was to work twice, even thrice, as hard and overly prove that for which White males where given the benefit of the doubt. "Prove it again" requirements is a common thread experienced in particular by African American female scientists, because prewired perception betrays the mind of their scientific peers that their research is legitimate. So be it, for now it had to be this way. It was then determined by these scientists to do what was needed to achieve the sought after goals.

Once the author was back on the path to completing the graduate physics education program and working as a research physicist, the feeling was that of being a data point in the database of stereotyping statistics. On social statistical data and statistical correlation, the current era is plagued by stereotyping, despite the fact that as a nation there have been some efforts to exit from this. One side of a given argument grabs data and uses it to justify their biases or preferences. Some major fallacies of these statistical analyses of social data need to be examined.

a. Statistics and Discrimination

All statistical models can be understood to be averaged simplifications. Proceeding from this concept, correlation does not equal cause (Correlation, Wikipedia 2015). Because there is a correlation between African Americans and their minimal presence in STEM fields does not implicate an organic causal connection based on their aptitude.

There are many hidden and complex variables not involved when characterizing social data. Characterizing individuals based on group level data, can lead to ecological fallacies, which in turn lead to stereotyping that is damaging and sometimes dangerous. The rapid fire association mentioned above that categorizes people can be very diminutive. Each individual should be subject to distinct individual analysis as opposed to being clumped into an aggregate group that is often labeled with negative stereotypes. This hurts those scientists aspiring to enter STEM fields because such behavior can lead to unfair elimination, if not initially, later down the road or whenever harmful assumptions are made based on such modus operandi.

Related to this is the characterization of statistical outliers. As a scientist, the author has found valid data and often discovery lies in outliers, so whether or not to dismiss or include them needs evaluation. To say an intellectually outstanding African American female is a statistical outlier can be interpreted as grossly offensive, regardless of the available data.

An example of incorrectly analyzing social data would be to say that significantly more people are killed by auto accidents than plane crashes or troops in combat. If plane crash deaths are considered outliers with respect to the total number of deaths, incorrect analysis of this could lead to fallacious conclusions. One airplane crash is profoundly serious and needs much attention. One loss of life by a troop is serious and warrants much evaluation and attention as to why, even though such death could be said to be outliers in the total of non-disease induced deaths of the nation's total population.

This current time period is supposedly an enlightened high tech era. One size fits all oversimplifications are unacceptable. More accurate modelling in the realm of social behavior is needed to avoid stereotyping. Often with certain news channels, statistics appear to be used and evaluated based on pre-existing nonscientific biases and to subsequently legitimize a perhaps false or weak concept.

And to those who don't hold themselves accountable to scientific thought, for example in some political or religious realms, the generalizations and subsequent damage are still unacceptable. History has proven how tragic this can be.

b. Racial Slurs and The Misuse of Statistics Involved in Stereotyping

To further establish these claims, the issues faced by some women and many African Americans are neither irrelevant outliers nor statistical oversimplifications. Stories have been told by far too many who have had to endure the racially derogatory name calling by those in University and laboratory environments. It is not anomalous. After having to double or even triple prove oneself above and beyond White counterparts, (because African American females are too often don't get the benefit of the doubt), they are agonized to still be ostracized unfairly. It was said to a female African American Ph.D. physics student, for example, "You better not mess up like Obama". The statement was loaded with unhealthy stereotypes that she reminded them of Obama (who they considered a mess up), and additionally that they did not like President Obama. The author was told that people like "her kind" had a poor work ethic (AKA African Americans are lazy Blacks. This was despite the fact that there were many years of documented success, which resulted from putting in consistent 16 hour days. These are cruel examples of stereotyping based on misuse of statistical data coming from biased individuals. A more disrespectful example was the author being compared to a cockroach, which is more beneficial being smashed, since it is generally an unwelcomed pest.

c. Consequences-Long Delays

For the author in this same setting, upon finally defending and passing the Ph.D. defense (after years of "prove-it-again" delays), there was this an awe as to how someone "in this (undervalued) category" could assemble and analyze all that technical information. The dissertation research was extraordinarily long, one of the longest in the school's history. However, the "prove it again" requirement (Williams, Phillips, & Hall, 2014) for which African American females must go through was a main reason for the extra lengthiness that resulted in the amassing of significantly larger than normal data sets. Out of caution, the invitation to attend the Ph.D. defense in virtual mode was extended to credible scientists from around the country and globe. This was done in spite of the vow by a key Ph.D. committee member that the author would never be passed, even if the outside scientific community was so impressed and awarded her the Nobel Prize for this research. With the more objective assessment from the rest of the international scientific community, the hope was that the aggregate bias would be less intense. Such assumptions were correct.

d. Stereotyping Can Lead to Lack of Support

After all, why should someone who is innately not up to par be supported? Stereotyping can be exhibited both openly and passive aggressively, but it all translates into "you don't belong here", or "you will never make it". It can be discouraging enough for some in the underrepresented category to walk away from STEM before they get started and migrate to a more welcoming environment, where such an individual is not just minimally tolerated. (Bayer Corp., 2010)

e. Some Roadblocks are Career Deal Breakers

Scientists need to publish their research findings and get their name out. In most environments for the typical researcher there is minimal obstruction to this. However, with a combined atypical high security academic and workplace environment, darts from both had to be dodged. The chain of approval was outrageously lengthy to get a paper approved and out to a journal. Some of the technical approval staff of that chain did not have a clue about the technical depth of what was written for submission. Novel material is something in particular one wants to publish, and this was an even bigger problem. Even without the extra barrier created by stereotyping, some supposedly seasoned authorities are reluctant to approve material that does not align with the current norm or what they are capable of understanding. With the stereotyping that can occur especially with an African American female, novel and significant findings can be marginalized, trashed or called error, simply because of from whom it is coming. In the previously mentioned report (Williams, Phillips, & Hall, 2014)) as well as from personal observation, it has been documented that accepting exceptional findings from a minority female is downplayed or dismissed. Nuclear

f. The Road to a Roadblock

Reflecting on one particular incident, where the author was very excited about the long toiled for research findings and wanted to publish the results, the "powers to be" of the research group said no way would they approve publishing this material because it would be an embarrassment to the organization. A few years later a prestigious researcher from a prestigious institution in Boston came and presented their findings, which concurred exactly with the author's previously rejected results. This researcher from Boston got to be credited as the first to come out with this. There is no doubt that had it been originally handed to them by a White male, the author's group would not have quickly rejected the research findings. In turn, this organization would have had this "first to discover" credit and benefited by utilizing the results to enhance the technologies being developed.

g. General Consequences of Stereotyping

Because of stereotyping, for women and underrepresented minorities, merit alone will not enable one's advancement and success. In the previously mentioned report by Eichler (2012), which addressed meritocracy, the presumption that the underrepresented would be further ahead if they tried harder was debunked. Or the notion that "such individuals" are not further ahead because of lesser qualifications, which overlooks the discriminatory stereotyping roadblock, was also dismantled.

h. Multiplicity of Effects

Corollary to the first section above where the science of racism was introduced, researchers such as Leonard Mlodinow (Mlodinow, 2014) have identified the power that subconscious prejudices are at play that marginalize in gender stereotyping. Just as this study showed that for a man and woman delivering the same scientific script, it was determined by the student audience that the man was of superior qualifications; how much more of a diminution an African American woman will suffer, as she is cast down intersectionally.

2. Lack of Support

Referring back to the author's personal account, in the world of someone growing up in housing projects or other domains in the range of poverty, compounded with a dismantled family, the structures were not in place to support the goals for being a physicist. Some, even in STEM, were such oddities who slipped through the cracks. There were no parents cheering, because the family of origin was in shambles and no one understood what a physicist was. With a main guardian, being the classic "ghetto grandmother" figure, who was first a cotton picker and then a maid, the ability to understand what the nerdy granddaughter was aspiring to be was out of reach.

a. **Support for the STEM Novice:** For many underrepresented, once navigating through an intense battlefield of obstacles to make it into the STEM environment, the next level of difficulties has to be tackled. Today's science is highly collaborative. It is complex and intertwined. Initial guidance and support is a critical necessity. It was also found that, for African American females in particular as opposed to those of other categories, support is something for which one had to beg. Some initial reactions to walking into a graduate quantum mechanics class or a high energy radiation test lab were: "You're in the wrong place. Can I direct you to where you are trying to go?" In this situation the assigned support was just as traumatized for the "intrusion" into the elite club. The relayed feelings were that it would just be a matter of time before this unfit intruder would be eliminated, so why waste valuable time in support efforts?

b. **Lack of Support can be a Byproduct of Stereotyping:** Lack of support is a major roadblock to overcome, which was addressed in a previous section. Academia is the first landing ground for those aspiring to a STEM career. When transferring from a Ph.D. graduate institution in 2008, one of the items that pierced was the very frustrating pockets of negative racial bias. Upon transferring eventually to two other schools, it was concluded that this was not an isolated southern university problem but a nationwide problem of institutional racism by members of society that are not expected to have this bias. As will be subsequently emphasized, an important clarification needs to be made. Some of the biggest and most important advocates for the underrepresented have been White males. Ironically, discriminatory behavior is also enacted by other minorities and females in particular. STEM discriminatory behavior is still disproportionately troubling because it only takes one or two people situated in strategic decision making positions in the path of a prospective STEM worker to ruin one's efforts or career. There is still enough of a remnant of such individuals who feel it is their obligation to preserve the exclusive and elite nature of, for examples, the Ph.D. physicists. So many are still unnecessarily and unfairly thrown into battle for what could otherwise be a smoother ride.

c. **The Men's Club:** First of all, the hard physical sciences at the Ph.D. level has traditionally been a men's club. So tier one for acceptance into that is getting past the gender barrier, if female. Secondly and most strongly, in keeping up with all that is being discoursed on racism, it requires some special rites of initiation for Blacks. The mandate is to be razor sharp and leave no stone unturned because even if tolerated by one's present contemporaries, on the other side of the next door the situation could be worse. One thing that many African American parents nail into their children is "you cannot be wrong, you must be perfect." It turns out that "perfect" is relative and often is still not enough.

d. **From Murky Waters for the Ultra-Filtrates to Tokenism:** Upon making it through the initial filters of the elimination process, it is time to go to work. Once situated in the workplace, life and work can range from near perfect to a nightmare. In some of these environments, many of the diversity initiatives hurt and victimize rather than provide a pathway. Because they are implemented by random middle management and not the creators of the diversity initiatives. They can be a conglomerated mess of just putting up a poster child. In the era of the 1980's to 2000, in the recent aftermath of the Civil Rights laws, diversity and Affirmative Action was better celebrated instead of greeted with drudgery. Its efforts now are minimally quantitative and not qualitative, so "the box can be checked" to please higher ups in human resources who always seem more enthused about diversifying their workplace than the middle management with whom the individual has to interact. Once the "box is checked", other underrepresented individuals (in particular African Americans,) enter into these domains at their own risk, since they are no longer needed. A term "tokenism" has been coined and is demonstrated by the arbitrary "checking of the box" routine. If not the "selected" one for display, one's future could be grim. A talented and diverse workforce that produces excellence is far from the goal in "forced diversity", which hinges on tokenism. Tokenism is bad because it pits minority against minority. On the flip side, it in fact appears to be a threatening situation if the middle management for the underrepresented STEM individual lacks the level of education and character needed to not suffer intimidation. Aforementioned, female management and professional colleagues have been shown to discriminate just as must as their male counterpoints. So therefore the solution is not in arbitrarily hiring more female management- it is in hiring fair, honest and intelligent staff who see the value of a diverse workforce.

e. **The Quota Game:** Once the minimal "quota" is met, the possibility of a strategic elimination process is initiated by those who feel threated. Because of the privileges and the associated level of comfort that are traditionally given to the dominant groups, which in some organizations include White females, a technically well groomed minority poses a threat. In such situations, there is no exploration for utilization of the talent available, but just a "dog and pony show" based on token diversity, to showcase to upper management. There is also vicious turf protecting tactics, as often the level of seniority enjoyed by the intimidated manager could not be matched outside of the organization, should they have to leave or be reassigned.

f. **Basic Facts:** For a worker who is undervalued, finding support may be difficult. In comparing women with African Americans, statistical data findings are similar, but more exaggerated with regard to the lack of support African American women have in the STEM workplace. As for careers, among the findings in the report by the American Institute for Research (AIR), researchers Tanenbaum and Upton (AIR, 2014) found that (1) Men were more likely than women to secure a position upon earning their STEM Ph.D.'s., but among those with secured positions, women were more likely than men to begin their careers in academia, (2) At the same time, males were significantly more likely than females to secure the more prestigious or difficult-to-obtain academic position: faculty at a research university, (3) Being married and having children suggests a disadvantage in securing a position at a research institution—for both men and women." In fact, having children is a strike against women in STEM (Tanenbaum, 2014), but so is being childless.

g. **The Consistency of Lack of Support:** The lack of support has been a consistent marker once past the master's degree. Relating back to the authors personal journey, which was often very difficult and bumpy, Fisk University and NASA where positive experiences in terms of support. The first was in the category of a Historically Black College and /or University (HBCU) and the second was one of NASA's east coast centers with a very rich history for diversity.

In pursuing a post master's degree and starting to tread into the higher ground, which was more hostile for African Americans, this lack of support occurred for the author while navigating the critical milestones. Black females are the least supported. Very importantly and as previously emphasized, when there was enough support for a successful outcome, this was often due to assistance from a White male. There is just a need for more such individuals. Since the process to be successful can be described as a political jungle with untreaded territory for the underrepresented STEM student or worker, support is not optional. It can be a task identifying and maintaining such support, when it is not built into a program. It is a test above and beyond one's technical ability. It is an endurance test of constantly reprocessing the rejection and marginalization, while maintaining belief one's abilities and seeking one's goals.

3. The Issue of Geography

For the author, a leg of the journey involved changing Ph.D. institutions and relocating to the mid-West military research lab, so that both employment and completion of her studies would be allowed. Now there were two wars to fight; the issues already in Academia and now to add to this a rigid STEM workplace. In the physical sciences in the mid-West and no doubt everywhere, it's a very White male dominated environment. This move marked the crossover from shallow to deep waters.

a. Geography is Critical but Not Constrained

Racism is nationally systemic, but it is easier to make it across shallow waters than deep. It is now and was in the earlier days of the Civil Rights movement. More of the east coast institutions appear to be better at retaining and graduating African Americans in STEM. While such success occurs in institutions nationwide, there seems to be a strong link to how progressive an area is in terms of civil rights and how many STEM as well as other African American high level professionals are produced.

For the author, it was previously erroneously thought that discriminatory problems existed because of living in the south. Many are not schooled about the Midwest, the "new" south; it is dotted with the former strongholds of "Sundown towns" (Loewen, 2006). Upon arriving in the area, it was observed that, apart from a handpicked few, Affirmative Action seemed to be something for White females and foreigners. In at least two institutions it was observed that the democratic process did not exist if you were African American. This is a process for which traditional academia is touted. For those from more progressive settings, perhaps these experiences may be categorized as far outside the norm. It was compounded and initially surreal, as entrance into the mangled world of drill sergeant types waiting to rip the unsuspecting victim down to their inner core while at the same time holding face, functioning in a well outfitted scientific environment. By special algorithmically determined criteria, graduate students loaded with debt translated into nothing honorable. Predictions declared one as such a security threat and may never be able to move forward regardless of one's technical capabilities. Negative categorization was multi-faceted. Once again considered double jeopardy from another angle, as a financially strapped graduate student and the single (widowed) mother of a seriously ill daughter, the immediate label was "dead upon arrival". It took innocence and audacity to enter into this guarded, rigid world.

b. Survey of the Land

Taking an excerpt from a blog post contribution (Word Press, 2014):

After 6 years in the Midwest, moving from Nashville, I feel so emotionally and spiritually drained. I feel as though I have landed in alien hostile territory. In the South the lines were clearly drawn, but up here there is a big theatre were the lines between reality and falsity were blurred; you can be subtly destroyed and no one notices. One can be left beaten and bleeding on the side of the road and it is not noticed. A big portion of African Americans seem to be at a joyfully content despite the lack of achievement, socially lobotomized, functionally rendered ineffective and in a state of domicile contentedness, as beneficial educational programs are gutted, districts gerrymandered (79% of them)(Cooper, 2015), and the unarmed gunned down. When poked, both sides like to roll over and play dead rather than invoke a response.

Additionally, there is a trend to bring on board non-native born individuals of color in lieu of native born African Americans. While it is to be applauded the acceptance of the intellectual capability of any person and any people of color and of African descent, it is suspected that the "powers to be" in many institutions are more comfortable in choosing someone not rooted in Jim Crow and American slavery. Such individuals also appear to be less damaged, as many of us who are deeply rooted here are still in recovery mode. Also this does not correct the disparity that African Americans face.

c. Nationally Systemic

In all fairness, geography is not necessarily a pre-test because some experiences are ubiquitous geographically, even if less tolerated in some areas. While there is a forward movement in parts of this overall mechanism of change, the operation of the vehicle of change seems to be currently subject to the forces of a "magnetic" reversal. That is, in some places, the race relations and the associated civil rights seem to have reverted back 40 years. Whether or not this is a temporary transient will have to be seen.

4. Specific Experiences

The author gives some specific experiences that line up with those of other aspiring or established African American STEM individuals, which pinpoints that the difficulties supersede the technical capabilities of the individual.

In progressing into both the workplace and the Ph.D. completion path, the most unexpected continues to happen as guidance and support is sought. While perhaps it is not unusual to have a drill sergeant type advisor, since at the Ph.D. level one is dealing with a professional adult student, lines can be crossed in interactions that diminish the individual's dignity with derogatory racial or gender insinuations.

a. **Marginalization 101:** One such example involves a conversation with the Ph.D. advisor about more good data results being now available. The response, in front of other people, was: "Your data is crap, and it's about time you started doing some good work". This data, incidentally, had already been sanctioned by the organizations Chief Scientist as a novel discovery.
b. **Disregard and Disrespect:** Another example of a typical recurring play out of the micro-aggressive and demeaning treatment was the impromptu cancelling of scheduled appointments. These appointments were the only way advising and counseling meetings could occur. The scheduled arrangement would be trumped by any student or person deemed more important than the author, which was effectively anyone. Since at that time, as the only African American Ph.D. student in the department for most of the time there, this behavior looked suspect. This appointment cancelling occurred far too regularly. It often occurred after waiting by the advisor's office for a lengthy amount of time for a scheduled appointment, to be bumped by another student randomly showing up (who would of course be a White male). With the requirement to give regular updates to sponsors, the delays this caused always cast a negative shadow upon the author. As opposed to almost 95% of the Engineering Physics department student body who was not employed, scheduling appointments meant taking off from work.
c. **Delays and Setbacks:** The heart and soul of an experimental physicist is doing experiments. The wait in line to use the equipment critical to performing and completing one's research (in this case an accelerator owned jointly by the research lab and the Ph.D. student's institution) was another delay that often would lead to 6 months or more each year, due to being given last priority over the other students and researchers. It is not usual for researchers to have to wait in line to do experiments on equipment that does not have duplicates. But usually when the one who is scheduled gets to the front of the line, going forward is allowed. It became obvious that this line was "status" coded. Perhaps it was mingled with some other "coding" that resulted in disregard based on certain other categories. Excuses for constantly being thrown back to the end of the line were given that did not apologize

for this behavior, which included: the students that bumped the author were working with a time frame for completion. This is mentioned because this process went on for five years. The author was thrown back down into the bottom of the queue each quarter, four times a year. Explanations given to supporting sponsors were not acceptable to them and written off as foot dragging. Since apparently all the sponsoring management had ever experienced in such a scenario were White students and employees, the notion that discriminatory behavior by these residing smart engineers and scientists was out ruled.

d. **There is More:** There was a laundry list of other items experienced. Most of the items, upon talking with fellow African American physics students in hostile institutions, were similar. Among some of the other things, one thing that most other graduate students can take for granted is having a desk at which to sit. Things like tuition assistance (a given for most graduate physics students because of their high demand) and the use of laboratories owned by the school for research were rights that were denied.

e. **Personal Cost and Resolve:** While able to work around most of these situations, at a cost of personal time and money, the most damming and hurtful experiences was the verbal abuse. Getting into the specific episodes of marginalization, an example of the verbal abuse from an advisor is given. Words hurt, and "the power of life and death is in the tongue". The greatest damage done during slavery and Jim Crow was not was done to the body, but to the mind. It took very strong mental acts of reconstruction to reject what was regularly being delivered and reverse into the brain a different message of capability. This loud verbal downplaying was projected outward to those in authority, who naturally believe the more established individual. There is then the task of further proving oneself against the word of the respected person. As stated in the above section, the author came into this graduate and research environment with a very good professional and technical resume, but this fact had no redemptive power to rescue her from these onslaughts.

So jargon that should not be directed at a professional, with derogatory comments implying ignorance and incompetence, were the darts too often to shield .Unprofessional and "mean" language thrown at individuals is very abusive and aggressive, yet the author knows of several who have had to tolerate in STEM environments. It is tantamount to a form of "bullying", and crosses the line but unfortunately cannot be proven unless overheard by someone who is willing to confirm it. A colleague from an organization on the east coast in horror described this as a form of torture. Places where things like this happen, typically have no venue for complaint processing or these incidents could not occur in the first place. While most cases of verbal abuse are much milder in the overall STEM community, there are personal accounts from a number of others, from environments where there is no accountability. Intellectual marginalization, endured by females and in particular African American females, most frequently is hurtful. Individuals must be very passionate about their studies and research to have made it to third base. The question is asked why such behavior is tolerated by the underrepresented. For these scientists, the only option seems to be continued risky institution hopping in search of a fair and just environment. That may or may not occur and adds up to years of one's life. In the absence of a fair democratic process and for a student facing outside responsibilities, the only option is sometimes just bite the tongue and bear it. While it has been observed that the situations tend not to be as extreme for White females, their intellectual marginalization by the males only club was observed like flashing neon lights. Positive change is still a work in progress.

SOLUTIONS AND RECOMMENDATIONS

1. The Need for Role Models

Role models can guide, inspire and mentor. It is somewhat natural to want a role model that looks like oneself, but that does not have to be the case. Decades ago one had to be flexible with that requirement if minority and/or female. As a young girl, physicist Marie Curie was a choice, because of being a Nuclear Physicist, wife and mother. The author's decision had been made by the age of twelve to become a Nuclear Physicist. Early in the author's career at the bachelor's degree, level there was a mentor relationship while at NASA with the mathematician Dr. Valerie L. Thomas. Dr. Shirley Jackson, who is a renowned Particle Physicist, was later revealed as someone to emulate.

When studying role models, young girls can look at their lives and see what was overcame and what was achieved. Like many recent and somewhat crippled descendants of Jim Crow, there was a frustration. Some are born into and surrounded by much more hostile territory and huge walls to tear down before making it to first base. It is not about being dealt a bad hand, just a different hand. There are multiple pathways to success. The author hopes that sharing the challenges and bumpy road will particularly encourage those from nontraditional backgrounds. With a level playing field that is established by removing unfair roadblocks, accomplishing one's dreams are possible, if given the opportunity.

As a resident of the inner city slums, by junior high school, the author was speaking French, Italian, Portuguese, Russian and Swahili. IQ tests a few years later were found to be 143. While a junior in high school, the SAT was taken and she earned a perfect score.

The problem was, as a young girl, nobody cared about all of that. Caught in a vicious fight for survival, choices were made to work maid jobs, five and dime store jobs, or whatever was available to maintain minimal subsistence. Fighting off thugs on the way home was not by choice. So in this world, nobody cared about being a high achiever. However, there were the author's dreams that were never going to be satisfied until the halls of academia could be graced by her efforts in quantum mechanics, electrodynamics and nuclear interactions.

Today, there is in fact many African Americans in STEM that can be displayed and emulated as role models. Their names need to be as popular as Black entertainers and athletes. Popular culture, while improving in this area, at the present is doing a poor job of promoting them. There are many females in STEM to serve as role models for females as opposed to decades ago. The author was fortunate at the very end of the Ph.D. process to get a brilliant female Physicist as the lead advisor. This professor was in fact the only female professor in the very long graduate school experience, spanning three schools.

2. Remedies

Since this chapter is mainly addressing the issues of professional and Ph.D. levels in STEM, the solutions to apply to those in that phase of their career will be suggested.

a. **Create a Database of Supporters to Choose From:** As part of a remedy for the intermediate tier, which is finishing graduate school, institutions favorable (having strong support and mentoring systems) to African Americans are being identified and are key in increasing the numbers of Masters and Ph.D. graduates, particularly for areas like physics.

As a case in point for the correction of a disparity and its remedy, in 2007, Vanderbilt University had not graduated an African American Ph.D. physicist in roughly 35 years. Nearby Historically Black Institution (an HBCU) Fisk University partnered with Vanderbilt University and decided to form a "Bridge" Masters to Ph.D. program in an effort to rectify this (Fisk, 2015) The program identified Vanderbilt programs and professors willing to offer support and additional mentoring if needed. Regardless of race, gender or ethnicity, a percentage of the students still do not make it, for a number of reasons. However, given a support system and a non-hostile climate, students from underrepresented groups do very well.

Due to the trigger of stereotyping the then associated lack of support and hostile environments, especially encountered by African Americans, it has been observed that it is not unusual for African Americans to have enrolled in 3 or more graduate schools to complete their Ph.D. In contrast, Whites who generally find welcome arms and support are enrolled in one institution until the completion of their degree. This unstable looking school hopping is in an effort to find a supportive environment, not an easier pass. Those who lack technical confidence or sufficient motivation usually abandon the pursuit of the advanced degree early on. Some of the African American physicists (after taking as long as 12 years to complete their degrees-as opposed to Whites taking 3-6 years), have completed their degrees and gone to work in research laboratories in Europe, where they appear to be more welcome.

b. **Work on Retention:** Figure 2 is a representation of the retention in STEM based on gender and race. (AIR, 2014) The discussion in this chapter can give credence to why the numbers for workplace retention are lower for both females and African Americans.
c. **Strong Self-Esteem is a Must:** Despite the harshness of the environment one may find him/herself in, grit iron determination and self-confidence is a must. According to a quote by Gandhi:

Man often becomes what he believes himself to be. If I keep on saying to myself that I cannot do a certain thing, it is possible that I may really end by really becoming incapable of doing it. On the contrary, if I have the belief that I can do it, I shall surely acquire the capacity to do it even if I may not have it at the beginning. - Mahatma Gandhi, (Easwaran, 1997)

FUTURE RESEARCH DIRECTIONS

An excellent report referenced throughout this chapter (Williams, Phillips, & Hall, 2014) has some guidelines for moving forward. Underrepresented STEM individuals and their management and environments need to be honestly monitored. The data from such should guide future corrective efforts. General solutions have been presented but awareness of the totality of particular, unique problems of intersectional groups should be addressed. Scenarios which are camouflaged in supposedly "progressive" environments need to be honestly analyzed along with a set of metrics for progress. This book should be an eye opener. Sexism and Racism in STEM environments is real, and often situations are occurring unknowingly under the watch of very caring upper management. A one size fits all path is not suggested but, based on the information presented in this book, mandates for change should be researched and implemented.

CONCLUSION: STEM RACISM IS SYSTEMIC

We still exist in an era where systemic racism is still running rampart. An immediate solution for bypassing much grief would be that an individual thoroughly research a prospective institution for its track record on inclusiveness and diversity. If one is already situated in a less than ideal environment, bring visibility about an inhospitable situation to those who can change it, as well as join in the chorus of those seeking to rectify troubling discriminatory practices.

REFERENCES

AIR Report. (2014). *Among New STEM Ph.D.'s, Women Likelier to Join Academia, but Men Get More Faculty Jobs*. Retrieved from http://www.air.org/news/press-release/among-new-stem-ph-d-s-women-likelier-join-academia-men-get-more-faculty-jobs

APS & IPEDS. (2015). *Percentage of degrees awarded to minorities*. Retrieved from http://www.aps.org/programs/education/statistics/urm.cfm

Corp, B. Survey. (2010). *Bayer Facts on Science and Education XIV.* Retrieved from http://www.igert.org/system/content_item_assets/files/579/Bayer_Facts_of_Science_Education_Executive_Summary.pdf?1269877505

Barnes, C. W. (2014). *Too Heavy a Yoke*. Eugene, OR: Oregon Cascade Books.

Cooper, J. (2015). *Civil Rights and Equality.* Retrieved from http://cooper.house.gov

Correlation does not imply causation. (2015). Retrieved from http://en.wikipedia.org/wiki/Correlation_does_not_imply_causation

Correll, S. J., Barnard, S., & Paik, I. (2007). Getting a Job: Is there a Motherhood Penalty? *American Journal of Sociology*, *112*(5), 1297–1339. doi:10.1086/511799

Cuddy, A. J., Fisk, S. T., & Glick, P. (2004). When Professionals become mothers, warmth doesn't cut the ice. *The Journal of Social Issues*, *60*(4), 701–718. doi:10.1111/j.0022-4537.2004.00381.x

Easwaran, E. (1997). *Gandhi the Man: The Story of His Transformation* (3rd ed.). Tomales, CA: Nilgiri Press.

Eichler, L. (2012). *You can't get ahead on merit alone.* Retrieved from http://www.theglobeandmail.com/report-on-business/careers/career-advice/you-cant-get-ahead-on-merit-alone/article4226012/

Fisk Vanderbilt Bridge Program. (2010). *Get the preparation you need to earn a Ph.D.* Retrieved from http://www.vanderbilt.edu/gradschool/bridge/descript.htm

Gutierrez y Muhs, G., Flores Neiman, Y., Gonzales, C.G., & Harris, A.P. (2012). Presumed Incompetent; the intersections of race and class in academia. Boulder, CO: Utah State University Press.

Hunter, B. (2006). *Home by Choice: Raising Emotionally Secure Children in an Insecure World*, New York: Multnomah Books.

Jackson, K. (2006). *The Status of the African-American Physicist in the Department of Energy National Laboratories*. Retrieved from http://www.aps.org/publications/apsnews/200205/backpage.cfm

Justia.com. (2015). *Patents by Inventor John W. Jackson*. Retrieved from http://patents.justia.com/inventor/john-w-jackson

List of African American Scientists and Inventors. (2015). Retrieved from http://en.wikipedia.org/wiki/List_of_African-American_inventors_and_scientists

Loewen, J. W. (2006). *Sundown Towns: A Hidden Dimension of American Racism*. New York: Touchstone publishers.

Mandela, N. (1975). On perseverance. *Excerpt from letter to Mini Mandela*. Retrieved from http://qz.com/93070/the-wisdom-of-mandela-quotes-from-the-most-inspiring-leader-of-the-20th-century/

Mlodinow, L. (2014). *Overcoming Unconscious Bias*. Retrieved from http://www.thebigidea.co.uk/overcoming-unconscious_bias

Moore, R. (1988). *Home School Burnout: What It Is. What Causes It. and How to Overcome It.* Tucson, AZ: Arizona Wolgemuth & Hyatt Pub.

National Center for Education in Statistics. (2015). *Degrees conferred by sex and race*. Retrieved from https://nces.ed.gov/fastfacts/display.asp?id=72

Obama, B. (2015). *Selma: Remarks by the President at the 50th Anniversary of the Selma to Montgomery Marches*. Retrieved from https://www.whitehouse.gov/the-press-office/2015/03/07/remarks-president-50th-anniversary-selma-montgomery-marches

Tanenbaum, C. (2014). *Early Academic Career Pathways in STEM: Do Gender and Family Status Matter*? Retrieved from http://www.air.org/resource/early-academic-career-pathways-stem-do-gender-and-family-status-matter

The Journal of Blacks in Higher Education. (2006). *Doctoral Degree Awards to African Americans Reach another All-Time High.* Retrieved from http://www.jbhe.com/news_views/50_black_doctoraldegrees.html

Turk-Bicakci, L. (2014). *Women, Blacks Most Likely to Leave STEM Careers, New Research by AIR Finds.* Retrieved from http://www.air.org/news/press-release/women-blacks-most-likely-leave-stem-careers-new-research-air-finds

Washington, J. (2011). *STEM Education and Jobs: Declining Numbers of Blacks Seen in Math, Science, Jesse Washington*. Retrieved from http://www.huffingtonpost.com/2011/10/24/stem-education-and-jobs-d_n_1028998.html?view=print&comm_ref=false Ll.

Williams, J. C., Phillips, K. W., & Hall, E. W. (2014). *Double Jeopardy? Gender Bias against Women of Color in Science*. Retrieved from www. Lifeworkflow.org

Word Press. (2014). Collected Young Minds. *Journey On*. Retrieved from http://collectedyoungminds.org/art-and-culture/journey-on/

KEY TERMS AND DEFINITIONS

Essentialized: To extract the essence from.

Intersectional: Being a part of two or more distinct groups.

Jim Crow: The mechanism of segregation and discrimination against blacks, which was a trigger for the Civil Rights movement of the 1960's.

Marginalization: The act of diminishing the significance or value of.

Meritocracy: A system in which members of an elite group are rewarded and/or advanced.

Microaggression: Nonphysical derogatory and or hostile attacks.

Myriad: A very large number of items.

Tokenism: Minimal enforcement of a policy.

Ubiquitous: Being everywhere at the same time.

Ultra Filtrate: The residue or results of a very fine filtering process.

Chapter 5
Standing Alone in Computer Science Education: A Story of Black Women in Academia

Mechelle Gittens
University of the West Indies – Cave Hill, Barbados

ABSTRACT

This chapter explores issues that have influenced Black women in Canada and the Caribbean in their pursuit of advanced degrees and regular academic roles in Computer Science (CS) and related fields. The concern is with the low representation of Black women in such roles and the reasons why some women have chosen the field and others have fled from it. The issues are related to gender and race and have been gathered from the scientific literature on Computer Science Education. Black women in Canada and the Caribbean who are on paths towards and in regular academic roles in CS were surveyed about how these issues have influenced them and their peers for or against CS in general and more specifically advanced education in Computer Science, since these such choices for CS are a requirement for regular academic roles in CS. This work delivers insights into an understudied locale for an underrepresented group in a field critical to economic development in Canada and the Caribbean. Based on our findings, we make recommendations to increase the number of this underrepresented group.

INTRODUCTION

An experience is told of a faculty member who visited the library of the Canadian university where she worked, selected her research literature, and proceeded to the desk to borrow them. On presenting her identification card, the only requirement for faculty, she was pressed to provide a letter from her Department Chair to validate her credentials. Though an incident of singular bias in her environment, she was anomalously, a Black female university professor on regular academic (RA) staff in Computer Science (CS) in Canada. One may expect that the demographics of the Caribbean provide a strong population from which Black, female, RA staff may emerge, however, Caribbean female university professors in CS are also few. This library experience is a notably disconcerting example underlining the lack of black women on RA staff in CS.

DOI: 10.4018/978-1-5225-0174-9.ch005

The rarity of women and minorities as distinct groups in CS is well documented in North America(Trauth, Quesenberry, & Yeo, 2008). The underrepresentation is discussed in the 2013 Taulbee Survey (Zweben & Bizot, 2014) done by the Computing Research Association (CRA), which is an association of over 200 North American academic, governmental and industrial departments in Computer Science, Computer Engineering, and related fields. Further, Taulbee reveals that black computer scientists graduating with PhDs are only 1.5% of all CS PhDs in the United States (US) and Canada. Black female Computer Scientists with Doctoral degrees make up 3% of all females graduating with PhDs in CS (Zweben & Bizot, 2014). The void may be expected in a country where blacks make up less than 3% of the population (Statistics Canada, 2012); however, in the Caribbean where the demographics place blacks at over 80% of the population(Caribbean Community (CARICOM) Secretariat, 2011), black females (and females in general) also number few amongst PhD graduates. Though no official statistics linking gender and race to the graduates was available for the Caribbean, there is anecdotal record by administrative and academic staff that there have been at most two female PhD graduates, from the University of the West Indies at Cave Hill and at St. Augustine within the history of the CS programs at these universities.

The Research Problem

With such a low level of black female PhD graduates in Canada and the Caribbean, we formulated the following research question:

What are the issues that influence some Black women in Canada and the Caribbean to pursue regular academic roles in Computer Science and their peers to choose other paths?

It is desirable to have consistent representation of Black women in academic staff in order to build research capacity with a diversity of experiences in approaches to innovation in society (Rheingans, Brodsky, Scheibler, & Spence, 2011).

Canes also states that female role models are helpful in the support of women towards success in a given field (Canes & Rosen, 1995). This means that there must be long-term contact with Black female faculty members so that students can see them as representations of success, stability and possibility for themselves. Such faculty members must represent a history of active and successful practice in the area of CS through previous employment, ongoing research and consulting based on expertise, as well as the ability to understand the material and succeed in delivering it through strong teaching.

Approximately 20% of RA staff at Caribbean universities above are black women; but the percentages are still low compared to the overall population of black females in Caribbean countries (50%). In Canadian schools, Black women constitute at most 1.3% of all PhD graduates in CS (Zweben, 2014; Zweben & Bizot, 2014). In the course of our research only two Black women could be identified as members of CS RA staff at Canadian universities.

The numbers for Canada are based on estimates calculated as a product of the percentage of females graduated and the percentage of blacks graduated in the programs indicated. There are no direct data for black females. Tracking this information is a problem even in the US where more detailed data are kept. For example, in the study completed by the National Science Foundation (National Center for Science and Engineering Statistics, 2013), black females graduating with PhDs are unspecified. That group is reported with underrepresented minorities, who are earning CS doctorates at a rate that has plateaued well below 10% since 2000. The category of RA staff that are Black women is also unavailable as a separate category.

Objective

This chapter is an initial exploration of the issues seeding the underrepresentation of Black women in Canada and the Caribbean in regular academic roles in CS. We investigate the issues by surveying the literature to understand the primary issues for Blacks and women in CS. We also present the results of a survey of Black women in Canada and the Caribbean who are on paths towards and in regular academic roles in CS. The survey is comprised of questions about influences on themselves and their peers with regard to CS. The overall goal is to understand the differences in what influenced those who chose the field and those who did not, to make recommendations that can influence upcoming cohorts to choose CS RA roles in greater numbers.

BACKGROUND: THE ISSUES IN CANADIAN AND CARIBBEAN CONTEXTS

The focus on Canada and the Caribbean derives from both the scarcity of direct data on, and the apparent dearth of, Black women in Computer Science careers in these regions. In Canada, the population of black women, comprising approximately 1.3% of the population, has produced notable Canadians who have excelled in the Arts, Humanities and Law. Examples include Michaëlle Jean, journalist and former Governor General of Canada (2005 to 2010); Measha Brueggergosman, noted Canadian soprano; and Violet King Henry, the first black female lawyer in Canada. However, representation in CS and related fields are lacking. The same situation exists in the Caribbean.

Insights into Experiences of Black Women in Computer Science

Much of the work that has been done on the topic of Black women in Computer Science has shown that the primary concerns preventing passage into the field are social, cultural and financial (Burge & Suarez, 2005; Lopez & Schulte, 2002; Ojokoh, Adeola, Isinkaye, & Abraham, 2014). Burge (Burge & Suarez, 2005) suggests that from the social viewpoint, there are certain stereotypes that present an image of CS as the domain of white males. Black women with such perceptions of the normative population are oblivious to the reality that it is possible for them to find success in CS. Lopez and Schulte (Lopez & Schulte, 2002) agree but believe that the issues of black female representation in the field is a problem of race and not gender since there is virtually no difference in occurrences of black men to Black women in CS. As a result, the authors state that the trend is that blacks in general, choose other fields. The researchers further state that social issues such as isolation as a woman and a minority in the classroom contribute to the flight from the field. In addition, there are social concerns such as the attitudes of family and peers that discourage women from CS (Scragg & Smith, 1998).

Alumanah (Alumanah, 2005) discusses the cultural expectations for girls in Nigeria and states that girls receive contradictory messages about their roles from parents, teachers, peers and the media and the messages from these groups perpetuate gender stereotypes. Computer Science is a field that requires the ability to work independently, yet, as Alumanah notes, there is a "hidden curriculum" that imposes the expectation of conformity and discourages self-reliance. These are some of the cultural themes that enforce lack of participation in CS by Nigerian girls. Finances are also crucial to the facilitation of studies in CS. An example of the financial facts for blacks is the low accessibility to computers for women in Nigeria and the lack of infrastructure to support the computing tools needed to learn about and pursue the field (Ojokoh et al., 2014). In examining cases in the US, also in terms of culture, Burge states that the feeling of being "out of place" discourages women from the CS classroom.

These studies are in the US and in Africa, specifically Nigeria, and bear an interesting parallel to Canada and the Caribbean. In the Caribbean and in Africa, blacks constitute the majority of the population. In Canada and the US blacks are a minority. The literature therefore implies that regardless of the overall population proportion (minority of majority), the occurrence of Black women in CS is low.

The Lack of Black Women in CS in the Caribbean and Canada

There is little work examining the lack of representation of Black females in CS for the Caribbean or Canada. Though the numbers for Canada can be gleaned from work such as the Taulbee survey, the group of Black women in CS is not directly studied. This chapter is an initial offering for these regions. The opportunity to compare the trends and contribution of Black women in the Canadian and Caribbean contexts, while considering the minority versus majority status of this group as part of a region's demography, is of particular interest since the trends within each population will allow us to consider whether there is any impact of being different (by various measures within the overall population) on the group's career choices.

The importance of technology -- and, by extension, CS as a contributor to economic development -- is currently underlined by both the Canadian government and governments in the Caribbean: "Canadians understand that, if we want to create jobs and opportunity in a competitive world and address the key societal challenges that confront us in the 21st century, STI (Science, Technology and Innovation) must be an integral part of the national agenda."(Science, Technology and Innovation Council, 2013).

Also, Caribbean governments are increasingly noting the need to leverage STI to increase their capacity for research and development (Nurse, 2007). This interest is driven by a need to increase industrial production and export capabilities in order to earn foreign exchange. With these governmental emphases in mind, the contribution of all aspects of society cannot be overlooked. The inflow of black women, into CS careers, as for any other portion of the population, must be facilitated by removing deterrents. Additionally, the Caribbean region is a major contributor to immigration into Canada. Understanding the trends for the Caribbean, beyond studies of the Canadian population itself, may implicitly contribute to the increase of Black women in academia in CS.

Canada

The total population of Canadian university academic staff is 40,480. Women comprised 34.84% of that population as of 2011. If Black RA staff who are women in Canada are to come from the Canadian population, the numbers that they can be drawn from are diminutive, and are at most 1.3% black female PhD graduates within all PhD graduates; with Black females constituting at most 0.2% of the research-based Master's graduates in CS (Zweben, 2014; Zweben & Bizot, 2014) in Canada. Currently only two Black Women who are RA staff in CS in Canada have been identified.

The Caribbean

There are a number of private, public, accredited and unaccredited universities in the Caribbean. The Organization of American States lists the following universities in their Consortium of Universities. (The English-speaking Caribbean is the focus of this initial investigation.)

- University of the West Indies, Cave Hill Campus (UWICHC), Barbados
- St. George's University (SGU), Grenada
- University College of the Caribbean (UCC), Jamaica
- University of Technology (UTech), Jamaica
- University of the West Indies, Mona Campus (UWIMona), Jamaica
- University of the West Indies, St. Augustine Campus (UWISTA), Trinidad and Tobago

This OAS sample is used to examine the number of Black women on RA staff in CS at key Caribbean universities. Where information was not readily available from a university's website, we contacted the university to gain information about their staff. In a few cases there were no responses; however, we were able to gather the information shown in Table 1 about the RA staff in CS as the sample for the Universities in the Caribbean.

The average population of males to females is equal; however, within the population of CS RA staff, the percentage of Black women ranges from 9.00% to 25% in schools that offer advanced degrees. A

Table 1.

University	Caribbean Territory (Female percentage of population - (CIA, 2015))	Degrees Granted (CS)	CS RA Staff (All)	CS RA Staff (Female)	Percentage (Female) of RA Staff	CS RA Staff (Black Women)	Percentage (Black Women) of RA Staff
UWICHC	Barbados (52%)	Bachelor's, Master's, Doctorate	11	1	9.00%	1	9.00%
SGU	Grenada (50%)	Bachelor's	5	1	20.00%	-	-
UCC	Jamaica (50%)	Bachelor's, Master's	-	-	-	-	-
UTech	Jamaica (50%)	Bachelor's	10	5	50.00%	5	50.00%
UWIMona	Jamaica (50%)	Bachelor's, Master's, Doctorate	12	2	16.67%	1	8.33%
UWISTA	Trinidad and Tobago (49%)	Bachelor's, Master's, Doctorate	12	3	25.00%	3	25.00%
Totals			**50**	**12**	**24.00%**	**10**	**20.00%**

- : *Data not accessible.*

noted exception are the universities that offer only Bachelor's degrees in CS. The representation in the RA staff is not representative of the general population at the schools offering advanced degrees, raising questions about the limited correlation. It is noted that the data for blacks and women in CS is not as accessible through studies such as the Taulbee survey for North America; however, discussions with faculty/administrative staff at the UWICHC and the UWISTA reveal at most two female doctoral graduates in CS on each campus. These graduates have not pursued academia and, therefore are not direct role models for the students attending these campuses.

Summary of the Literature: The Eight Influences

Our review of the literature (as discussed below) reveals that there are generally eight categories of influences that lead to the *pursuit of* or *flight from* CS. These factors affect the likelihood of a woman continuing her studies to become an academic in CS.

1. **Biases About Self, Women, and Blacks in the Field:** These biases proceed from the existence of stereotypes, gender biases and perception of the field (Steele & Aronson, 1995). Redmond et al. (Redmond, Evans, & Sahami, 2013) found that a large fraction of women in the minority may fall prey to the "stereotype threat" (Steele & Aronson, 1995). With this phenomenon, "high-ability" women and minorities were impacted by negative stereotypes about their particular group. That is, if stereotypically their group performs poorly on a task, and they are presented with these typecasts in assessment conditions, they will generally perform worse than their norm. There is anecdotal evidence that women in CS programs often have a negative perception of their abilities in programming, and so seek careers upon graduation that limit technical tasks. This also means that they avoid further studies in CS. Some have proposed that the tasks in CS in tertiary education settings must be done individually, and researchers suggest that women prefer and thrive in collaborative environments (Berenson, Slaten, Williams, & Ho, 2004).
2. **Experiences in Education:** Including experiences with pre-tertiary educators and blocking of paths leading to CS by those in higher education determine whether Black women will be able to pursue CS (Alumanah, 2005; Mody & Brainard, 2005). Wang et al. (Wang, Hong, Ravitz, & Ivory, 2015) state that exposure to CS courses prior to the tertiary level significantly influences the choice for CS.
3. **The Demographics of the Field:** May influence those who are deciding on their academic paths. Male domination, the lack of representation of Black women in the field, and a scarcity of role models can make some feel uncomfortable (Camp, 2012; Rheingans et al., 2011). Female role models are helpful in the support of women towards success in a given field (Canes & Rosen, 1995). Young women who are engaged in computing sessions led by female faculty see an example of successful women in CS (Garnett, 1999). Female students are often attracted to classes taught by female CS faculty (Frenkel, 1990). This leads us to hypothesize that the lack of role models in CS and existing stereotypes may deter Black women from CS. Furthermore, we believe that the lack of Black women as RA staff or in top roles in industry causes doubt that such achievement is possible. This initial study is used to investigate the influence of role models on advanced-degreed Black women in CS.
4. **Internal Motivation:** May determine what Black women study and how far they go with their studies. Motivation is impacted by personal goals, values and character traits (Lopez & Schulte,

2002). Scragg et al. (Scragg & Smith, 1998) have noted that women tend to have less confidence in CS classroom situations and this influences their persistence in the field. "Learning how to deal with difficulty and failure, to go beyond what one thought was possible" is the "ordeal" approach to learning (Rheingans et al., 2011) and is somewhat common in CS-related classrooms. Female students often state that this approach is discouraging. Motivation and determination are required characteristics, and hence the style of' the CS classroom may cause flight.

5. **Attraction and Commitment to the Field:** May be influenced by how the individual is engaged within the field. Peters discusses relevance and importance of forums in creating belonging within peer groups for women in CS (Peters, 2013; Rheingans et al., 2011).
6. **Support:** Important for those interested in CS. Support may be formal or informal and may be familial, financial or otherwise (Burge & Suarez, 2005). It is crucial to longevity in the field. Financial supports such as scholarships rather than loans, and considerations for familial responsibilities are necessary for women to stay in the field. (Redmond et al., 2013) found that the support of a parent is crucial to a woman choosing to pursue a degree in CS. The role of the mother was found to be particularly necessary.
7. **Experiences in and Related to CS:** As well as an understanding of what daily life of a career in CS is like may solidify black women's interest in the field. Facilitating such experiences (as early as possible) is crucial (Alumanah, 2005; Mody & Brainard, 2005).
8. **Societal Factors:** Influence career choices and these factors are based on societal difficulties and the socioeconomic background of the individual. These factors influence black women's choices (Alumanah, 2005; Burge & Suarez, 2005; Lopez & Schulte, 2002; Ojokoh et al., 2014). For example, Lopez et al. mention that women often are focussed (by their own desires and the desires of others) on goals such as raising a family, and they perceive that studies and careers in CS do not match these goals.

Based on these eight categories of influences many women find CS unappealing. While most studies focus on women in general, many of the themes coming from the research underline and imply similarities with Black women and CS. We therefore step forward to explore the data for these eight categories for Black women.

METHODOLOGY

The focus is on the views of Black women who are RA staff, CS doctoral students, and thesis-based Master's CS students because they have achieved, or are on paths to achieving, success in CS and becoming RA CS staff. We focused on opinions of how certain issues have influenced their career choices, and may have influenced their peers' choices for or against CS.

The first part of the survey collected basic demographic information for Black women in CS:

1. Age,
2. Ethnicity,
3. Type of employment,
4. Marital status, and
5. Level of education.

The second part of the survey focused on understanding the experiences of the respondents and their peers with the issues of race and gender. Did these issues influence their choices of CS in education or a CS career? The survey collected data to answer the following concerns:

1. How did the eight influences impact the pursuit of the women's career in Computer Science and related fields?
2. How did the eight influences impact the pursuit the of women's peers' careers in Computer Science and related fields?

Procedure

The following Canadian organizations for and including women in Computer Science, distributed over all provinces of Canada were contacted with a request for participation.

1. Society for Canadian Women in Science and Technology (SCWIST)
2. The Committee on the Status of Women in Computing Research (CRA-W)
3. University of Waterloo, Cheriton School of Computer Science – Women in Computer Science
4. Canadian Celebration of Women in Computing (CAN-CWIC)
5. Canadian Coalition of Women in Engineering, Science, Trades and Technology (CCWESTT)
6. Carleton University- Women in Science and Engineering (CU-WISE)
7. Canadian Women in Technology (CanWIT)
8. The Women Unlimited Society
9. Canada's Association of I.T Professionals
10. Information Technology Association of Canada

In the Caribbean the female RA staff members at the universities listed in the OAS sample discussed in the Background section were contacted.

There was low response, even with follow-up requests. We believed this was due to the lack of representation of Black women in CS – the very issue that being studied. Due to the low response participation was solicited using snowball referrals – asking those who responded to refer other Black women in CS. The questions asked were purposefully open-ended to facilitate a narrative of the respondents' journeys.

The importance of surveying those in leadership roles in CS-related industry is also understood; however, for this initial investigation into a domain that had previously received little attention, the focus was on those who lead the way in instructing those who will take up leadership positions in the future. It is believed that low representation in these RA roles is at the root of the scarcity of representation in industry. The data cover two primary foci — demographics, and experiences with issues that create barriers. This is a collection, interpretation, and analysis of data which currently does not exist for Canada and the Caribbean region.

THE DATA

Part 1 of the survey focused on Demographics and Part 2 focused on Stories of Experiences. Answers were not required to all questions. For Part 1 there were thirteen respondents, for Part 2 – ten respondents. Three of the respondents noted too little time to answer all of the questions but agreed to answer as many as possible. The data was therefore analyzed in accordance with each part of the questionnaire, and the analysis considered the number of responses for each part.

Part 1: Respondent Demographics

There were thirteen female respondents ($\mu = 33.85$, $s = 13.35$), with a range from 22 to 56 years. Of the participants, 92.31% were black and 7.69% mixed. Caribbean respondents were 46.15% of our respondents and 53.85% were in Canada. The women from the Caribbean are from UWISTA and UWICHC. Respondents were invited from other Caribbean territories but did not receive responses from Jamaica or Grenada. The types of work engaged in by our respondents, their type of appointment, highest level of degree and marital status are shown in Figures 1, 2, 3 and 4 respectively.

Figure 1. Applicable categories of work

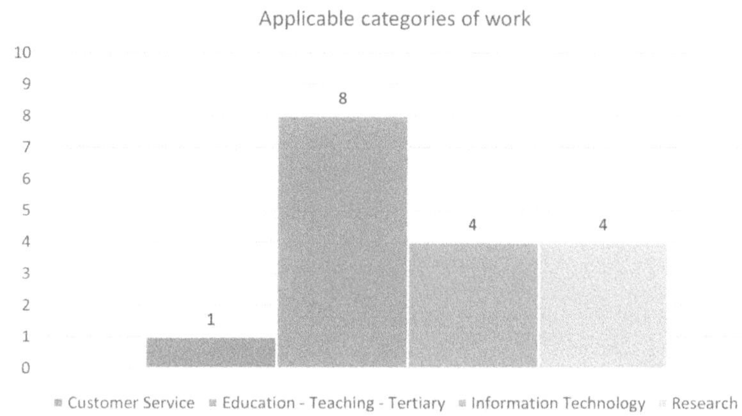

Figure 2. Type of appointment

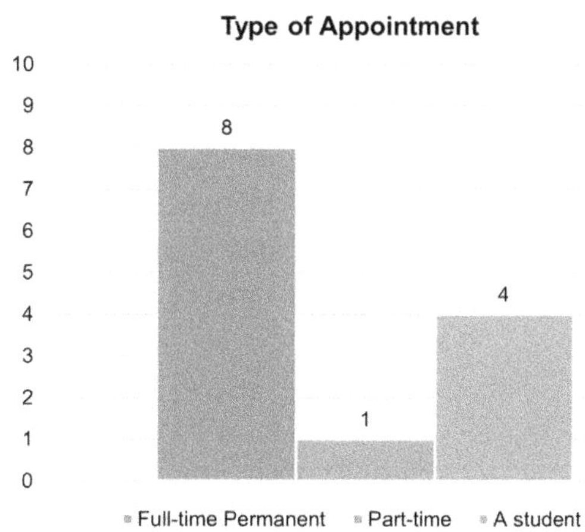

Figure 3. Highest level of degree attained

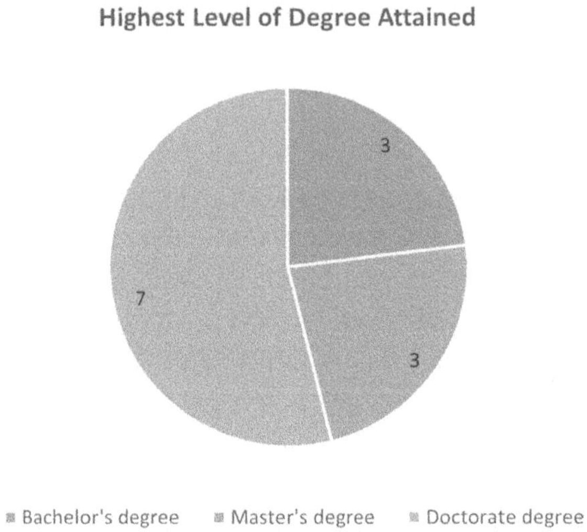

Figure 4. Marital status

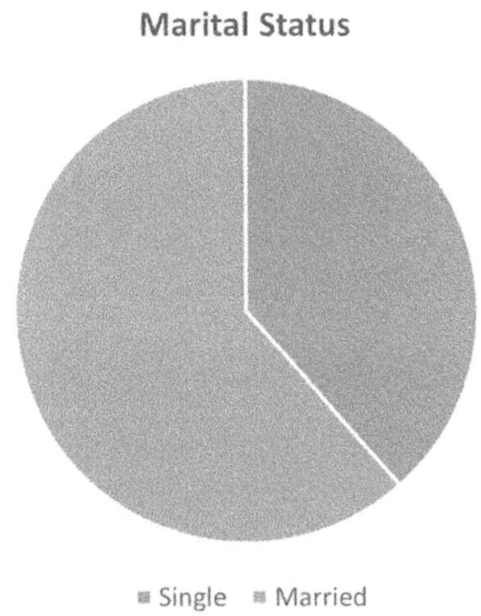

There were three doctoral students (two in Canada and one in the Caribbean) and two research-based Master's students (in the Caribbean). Eight of the participants are employed in full-time roles at tertiary level as RA staff both in Canada and the Caribbean. One participant works as part-time/temporary teaching staff and four others are students. The participants classify their work as shown in Figure 1. There is overlap in Figure 1 since the respondents chose three of the most suitable categories of work.

Part 2: The Stories

To elicit candid responses, respondents were asked to say how the eight influences discussed in the Background section affected their education and career choices using the following question:

Please briefly tell us what you think about how each factor has influenced your pursuit of a career in CS or a related field.

The participants were then asked to comment on the opinions of their peers based on their interaction with them:

Please briefly tell us what you think about how each factor listed in this section has influenced peers' perceptions of pursuing a career in CS or a related field.

For each group the following statement was added:

There are no right or wrong responses; we are merely interested in your personal opinions. Please list the thoughts **that come immediately to mind**.

The stories are presented using the following subsections:

1. A summary of the data showing the *percentage* of women that believed that the factor, e.g. *Biases Toward and About Women and Blacks in the Field* influenced them or their peers in their choice of CS or whether the factor had no influence.
2. A discussion of the women's views of *how* the factor influenced *their choice* of CS.
3. A discussion of the women's views of *how* the factor influences *their peers' choice* of CS.

P1: Biases Towards and About Women and Blacks in the Field

Summary

See Table 2.

Table 2. The percentages influenced/not influenced by biases

a. Stereotypes	**Influence**	**No Influence**
• The Women	40%	60%
• The Peers	63%	37%
b. Gender Biases	**Influence**	**No Influence**
• The Women	50%	50%
• The Peers	50%	50%
c. Perception	**Influence**	**No Influence**
• The Women	29%	71%
• The Peers	63%	37%

The Women

As shown in Table 2, 50%-71% of the respondents said that the factors related to biases towards and about women and blacks in the field have no influence on them. The respondents gave accounts of determination, confidence and parental equity in the treatment and support of their capabilities. Some were even motivated by these often negative stereotypes. One of the respondents stated that she was previously captivated by being in a male-dominated field. Another participant said that there was a perception that women could not do mathematics or computing in her country, and this sexism motivated her to disprove the detractors. The respondents are diverse in background and their rearing occurred in diverse places including Nigeria, the US, Kenya, and Barbados. Notably in the case of Canada the women are immigrants. Through this diversity common strength is observed in the face of bias.

One respondent speaks of a father who was an elementary school teacher who loved education and mathematics. He encouraged her to work at mathematics at a young age and showed her that girls could do just as well as boys. Both her mother and father prioritized their children's education, and this education both buoyed her confidence and left her unaffected by the stereotypes of girls not being good in mathematics. Another respondent moved from Africa and spent the final two years of high school in the US. In high school, she discovered (racist) perceptions that black people were not smart. The respondent learned to ignore the stereotypes since she performed well in her studies. She already had confidence and values instilled during her upbringing in Africa. With this confidence, she was able to continue to excel and focus on her goals. One more participant said that although she was aware of biases against women and blacks she also felt challenged to defy these biases. Notably, in early life, each of these women was raised as part of the racial majority and then immigrated to minority status.

Their Peers

More than half of those who responded to the questions about the issues of stereotypes, biases and perceptions believed that racial and gender stereotypes influenced their peers who often:

- Believed that CS is not for women
- Believed that the field is often associated with males and people with little social skills.
- Had the potential to excel in CS, but were hindered by their ethnicity because they were aware of their minority status in the field.
- Believed that the field is too difficult and too technical and, therefore, would not pursue it as a career

With regard to gender biases, one participant stated that some persons switched from CS to nursing because that was what was expected of them.

In the case of their peers' perceptions of the field, they stated that persons thought that even though CS offered great opportunities, there were a number of drawbacks to careers in the field as shown in Figure 5.

Figure 5.

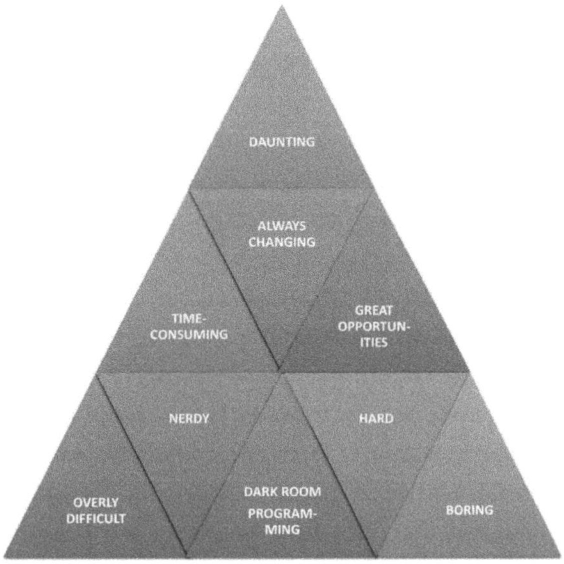

P2: Experiences in Education

Summary

See Table 3.

The Women

Here, the concern is with the attitudes in education, including the pre-tertiary educators that our participants may have encountered, and any blocking of their progress in higher education. As shown in Table 3, 80% of the respondents believed that the attitudes of pre-tertiary educators influenced them. However, no one stated that they were affected by blocking by higher education. The attitudes at the pre-tertiary level elicited many detailed and inspiring responses. The responses of those who spoke about their pre-tertiary educators speak best for themselves and are listed in Figure 6.

Table 3. The percentages influenced/not influenced by education experiences

d. Attitudes of Pre-Tertiary Educators	**Influence**	**No Influence**
• The Women	80%	20%
• The Peers	50%	50%
e. Blocking by Higher Education	**Influence**	**No Influence**
• The Women	0%	100%
• The Peers	0%	100%

Figure 6.

- *For elementary school, I went to a boys and girls school. Influence from home gave me confidence at school. The elementary school teachers were supportive. In secondary school, I went to an all-girls boarding school. In our system most children go to boarding school for secondary school, and most are separated, boys and girls. Because I was in an all-girls secondary school, there was no gender bias/stereotypes being felt. The attitudes of our secondary school teachers were very positive -- they encouraged us to work hard. Schools were ranked at the national level after national exams, so the educators wanted their schools to be first, and this meant encouraging us. But I must add that my experience should not be seen as a standard in my original country. I was just lucky. Although my school was boarding like many others, it was a top national school that attracted the best students in the country, best evaluated through competitive national exams. So I believe because I went to this school, my experience is different. The educators, as I said, wanted the school to remain tops and therefore provide a very positive attitude.*
- *I went to an all-girls school where we were lucky to be encouraged that girls can be as good as boys, so I always loved technical subjects*
- *I was encouraged to pursue a career in technology from high school years, so the other factors didn't matter.*
- *I once was slapped by a math teacher in Grade 7 because I corrected him in class when he did a math problem wrong. That was the most negative response I got. I think for most of my teachers, they were initially surprised at my ability, but at some point just accepted it.*
- *Positive attitudes.*
- *They were all supportive.*
- *At my schools, teachers seemed to believe that we could do anything we placed our minds on, and there was a focus on mathematics and the sciences. I was pretty good at these and I so pursued them. These led me to Computer Science.*
- *Encouraging. Helpful.*

Mostly positive attitudes are highlighted in the pre-tertiary educators described by these women. They sought excellence from their students, regardless of gender, gave support and cultivated belief in abilities. They also pursued strong outcomes for their schools and so students were supported, regardless of gender.

Their Peers

The general impressions of the participants were that their peers perceived no great influence of pre-tertiary educators or blocking on the choices for or against CS. One response about the Caribbean was interesting to note. It indicated that since the field requires mathematics for entry into degree programs, this can be a hindrance in the Caribbean system where generally the strong, early performers in mathematics are given the most support to reach their potential in the area. Some peers are turned off when they do not receive the necessary support to perform adequately in mathematics.

P3: The Demographics of the Field

Summary

See Table 4.

The Women

Feelings are mixed on the demographics of the field and the often discussed *male domination*, presence/absence of *role models*, and any overall *lack of representation*. Not every woman surveyed believed that she has been influenced by these issues (Table 4); however, more than half (50%-56%) do.

The proliferation of men in CS has been described as intimidating. Some struggle to fit into a male-dominated field and others use the fact to propel themselves forward. The scarcity of women – as well as being the sole black person in classes – sometimes generated survival mechanisms in the respondents like learning about football and sports statistics to fit in. This distinctiveness in the survey group was often compounded by being the only immigrant in the classroom. The approaches here are interesting and different – in some cases the women attempted to fit in, while others developed comfort in standing out.

The respondents' thoughts on *role models* revealed a distinct thread of *not* having an explicit role model. Some were propelled by wanting to do well and change the world, and well-known figures such as Mark Zuckerberg (Facebook), and Steve Jobs (Apple), inspired their education and career choices. The inspirations for them to go into CS were certainly not Black women because there were very few in the field. However, in addition to these prominent individuals, faculty members (who were most often male), research supervisors and colleagues played an important part in cultivating seeds of interest to engage and do well in CS.

The Black women responding to the survey have related different experiences because of *lack of representation*, not just as women but as Black women. The lack of representation of others like themselves

Table 4. The percentages influenced/not influenced by the CS demographic

f. Lack of Representation	**Influence**	**No Influence**
• The Women	50%	50%
• The Peers	57%	43%
g. Male Domination	**Influence**	**No Influence**
• The Women	56%	44%
• The Peers	43%	57%
h. Role Models	**Influence**	**No Influence**
• The Women	50%	50%
• The Peers	57%	43%

impacted them as university students and continues to impact them even now as professionals. They have struggled to fit into a framework that is different than who they are, and it has become an accepted fact of life for them that they are always in the minority. While always being in the minority creates an acute awareness of this fact, it is not something that has a negative impact – instead, it motivates excellence in order to set an example and create knowledge of what this specific minority can achieve. One respondent notes that in the years of undergraduate study in the Caribbean, there was reasonable representation of females in CS classes (at least 40%). In travelling to pursue graduate studies and to work in Canada, she noticed the issue of no representation. Once again this scarcity was motivation to be the example of achievement in CS.

Their Peers

The women believed that male domination in the field may have intimidated their peers. They also believe that lack of representation of persons such as themselves in CS has played a major role in deterring their peers. With meagre representation of persons similar to themselves, the peers often believe that the courses and majors are not for them. Also, since there is a large social network at the undergraduate level and in high school, they were told that taking classes like Advanced Placement Calculus, or being in Engineering or CS class that none of one's friends attend, may be influential in their peers' choices for other majors.

On the topic of role models, however, the participants appear to have experiences with peers that show that role models are influential. One participant said that "we try to follow a path that we see and, for those who cannot see a path from a role model," it is difficult to place one's cards into a given basket. Based on the more traditional careers and identifiable models for medical doctor, lawyer, high school teacher, and business person, their peers followed those paths. They also state that there are very few role models, and many of their peers believe that their male counterparts are much better at this kind of work (especially programming) than they are. The lack of role models hence affected confidence in the skill of the gender.

P4: Internal Motivation

Summary

See Table 5.

The Women

Our respondents are influenced by internal motivation (Table 5) and wanted to reach "great heights in the industry". They set high *personal goals* for themselves in education and in their careers. Some goals have been hard-coded into their plans. For example, one participant's outline for accomplishment was: *Education – Knowledge – PhD – Career*. One of our respondents had the initial personal goal of being a professor of mathematics, the subject that she describes as being her first love. She, however, made the shift to CS under the advisement of her mentor in graduate school. Other participants enjoyed technology, wanted to be engineers, or enjoyed the general experience of learning, and so were driven by their own desires to pursue a career in the field despite any social influence or external persuasion. One

Table 5. The percentages influenced/not influenced by internal motivators

i. Personal Goals	**Influence**	**No Influence**
• The Women	100%	0%
• The Peers	86%	14%
j. Values	**Influence**	**No Influence**
• The Women	75%	25%
• The Peers	50%	50%
k. Character Traits	**Influence**	**No Influence**
• The Women	100%	0%
• The Peers	57%	43%

other participant said that, in addition to wanting to be an expert in her field, she also wants to transfer knowledge and mentor others, and so these goals inspire her.

The *values* expressed here are generally common to this group and seem foundational to their attitudes and success. They are as follows:

1. You are good at what you are good at.
2. If one has talent, one must use it to be the best one can.
3. Hard work is important.
4. Discipline is paramount.
5. Women can do what they set their minds to.
6. Meritocracy is very important. Computing is matched to that value, so Computing is for me.
7. We must have passion for what we do.

The women also possess the following *character traits* that were extracted from their many answers: (i) a self-described strong personality; (ii) meticulous; (iii) inquisitive; (iv) affinity for logic; (v) perfectionist; (vi) driven; (vii) enjoyment of challenges; (viii) independent worker; (ix) enjoyment of problem solving; (x) adaptable; (xi) aspiration to achievement; and (xii) determined. Answers significantly overlapped. Interestingly, much of the identification of what distinguishes their attitude toward CS is in more cases expressed in terms of their gender rather than their ethnicity.

Their Peers

The *personal goals* of many of the peers of our group seemed limited to more traditional destinations. From the responses, it would seem that the women's peer group was focused on having a stable job and family – several years of CS education did not fit those goals. This point was also noted when the survey participants were sharing their views on their peers' *values*. One respondent noted that in the Caribbean many families value having their children earlier in life. This may not coincide with pursuing graduate studies or long hours in software development.

The perceptions of what Computer Scientists do may also have influenced oversight of the opportunities CS has to offer. For example, persons would often see computer technicians and one respondent said that it was difficult for her friends to see past that limited role. However, medical doctors and lawyers (seen as more prestigious and interesting than a technician) was a more attention-grabbing and easy-to-understand option that influenced personal goals.

Most persons were not able to comment on the *character traits* of their peers, except in one case to say that most of their Caribbean peers would not commit to the long hours required to be successful in CS.

P5: Attraction and Commitment to the Field

Summary

See Table 6.

The Women

Engagement refers to the commitment of employees to their employment and their likelihood of making a strong contribution to their organizations (Crabtree, 2013). 50% of the women did not believe that it influenced the pursuit of their studies or careers in CS (Table 6). *Professional organizations* also influenced 50% of the women in their participation in CS. However, participants stated that such organizations are important to helping women "stay on course once they are on the course." Respondents also indicated that there needs to be more outreach to women in the Caribbean to encourage them to join the field. The respondent who was most outspoken on this point did not know of any professional Caribbean organizations specifically for women in CS. There are some general CS, but not specifically for Black women. There was no mention of these concerns for Canada.

Their Peers

The only significant comment here was that for Engagement in her locale: "It is true that there is no outreach specifically geared to females in IT/CS." In all cases, respondents thought professional organizations had no influence in their peers.

P6: Support

Support

See Table 7.

Table 6. The percentages influenced/not influenced by attraction/commitment factors

l. Engagement	Influence	No Influence
• The Women	50%	50%
• The Peers	20%	80%
m. Professional Organizations	**Influence**	**No Influence**
• The Women	50%	50%
• The Peers	0%	100%

Table 7. The percentages influenced/not influenced by support factors

n. Lack of Support	**Influence**	**No Influence**
• The Women	33%	67%
• The Peers	29%	71%
o. Formal or Informal Supports	**Influence**	**No Influence**
• The Women	50%	50%
• The Peers	29%	71%
p. Familial Factors	**Influence**	**No Influence**
• The Women	63%	38%
• The Peers	50%	50%

The Women

Table 7 shows that 33% to 63% of the women noted support as influencing their choices. The supportive elements in the lives of the women were family members and educators - typically parents and husbands. There were *formal support* mechanisms in the sense of the provision of financial support for studies. The women also received *informal support* through "cheers", assistance with family responsibilities (before and after marriage), and continuous encouragement to move to the next step. Additionally, academic staff have been noted to help one participant to fit in with "the guys" and excel in programming. In the case of *lack of support*, all of the women said it did not apply to them since they had an excess of support from family, and this was what mattered to them.

Their account of the *familial factors* influencing their choice of CS include common themes of parents who valued education. One participant relates the following family history:

No one in my family is into numbers or programming, but both of my parents have doctoral degrees and all of my sisters have excelled in their chosen profession. Also, being from a Nigerian background and in a family with all girls when people used to laugh at my dad because he had no sons, was a motivating factor for all of us (four girls) to excel.

One respondent also notes that she had children later in life. This gave her the opportunity to focus on her studies and career.

Their Peers

The women in Canada noted that support is important for immigrants in particular. This would apply to *formal and informal support*. This is especially so when one has familial obligations in one's country of origin to address. This may affect their peers' decisions to follow a career in CS. In the Caribbean, one of the women noted that immediate outcomes are sometimes important, so the time it takes to understand and achieve in CS and related fields may not be supported by family and spouses. Many women are thinking about a family, so this discipline may become unattractive to them because it is unattractive to those they care about due to time requirements.

On the topic of *familial factors*, one of the women in Canada stated that her peers were often told to go into nursing because the time to achieve in that profession is shorter and the work is more conventional. The hours of work for nursing are also thought (by the culture) more conducive for married women, and so more appealing to the family. Also, a respondent from the Caribbean stated that responsibilities at home sometimes create difficulty for women in particular in spending the time required for graduate studies and to excel in CS.

P7: Experiences in Computer Science and Technology

Summary

See Table 8.

The Women

There were a number of *significant experiences leading to interest in Computer Science* for all of the women; hence 100% stated that this factor was influential as shown in Table 8. The experiences they noted were about what they enjoyed or currently enjoy about their experiences in Computing. These include:

- The immediate rewards of programming.
- The basics of computer programming learned in high school.
- The abstraction coupled with realism started my interest in high school. ‘
- A third-grade two-week mathematics assignment completed in one day for the love of numbers.
- The older sisters' sixth-grade mathematics homework done in third grade for the challenge.
- The first programming teacher who made the abstraction of programming seem so real.
- The enjoyable experiences in the high school computer club and at computer camp.
- The CS-highlights at a college fair.

There is a pattern of interest in math and programming from an early age and immersion in these areas. These early experiences, which were pre-tertiary, were pivotal in attracting the women to the area.

The facts of *daily life of a career in CS* are seen as intriguing, exciting and challenging. 75% of the respondents mentioned that this factor had some influence on their choices in education and career, though a few of the participants admit that they did not consider daily CS work-life until after graduation. They were comfortable with the research, mathematics and coding, so the rigours of daily life were not

Table 8. The percentages influenced/not influenced by experiences in CS

q. Significant Experiences Leading to Interest in Computer Science	**Influence**	**No Influence**
• The Women	100%	0%
• The Peers	29%	71%
r. Daily Life of a Career in Computer Science	**Influence**	**No Influence**
• The Women	75%	25%
• The Peers	50%	50%

a surprise to them. Some participants, however, note the elusive life balance, sleepless nights and the sometimes exhausting nature of the work as being a downside.

Their Peers

There was only one notable statement regarding *significant experiences leading to interest in Computer Science* since the women were not able to answer this question about their peers. For one of the women, "some friends had a good experience in physics class and that kept them in CS, while others had a bad experience in a mathematics class and that kept them out of the field". In addition, their peers who have observed *daily life of a career in Computer Science*, see it as time consuming. In the case of those who continue their studies in the area, they notice the sleepless nights. Since the hours are long and work-at-home time is needed, one respondent said that this life is not attractive to most women she knows.

P8: Societal Factors

Summary

See Table 9.

The Women

On the point of *societal difficulties*, the majority of the respondents (75% - Table 9) did not consider this factor to have influenced their choices. Only one respondent noted that she was motivated to take CS due to limited resources in her country of origin. This seems to imply that with her career success she will be able to offer needed support.

With regard to *socioeconomic background*, some responses indicated that the respondents had to work hard to achieve within the limitations of their background. The reasons for this touched on financial situations. However, most of the respondents who believed that this factor influenced them indicated that they came from middle-class families where at least one parent was educated and, in some cases, held advanced degrees. Participants noted that the education levels and lifestyles of their parents were improved by the hard work and sacrifice of their parents and other family members who were generally not educated. So they understood the privilege and products of education and did not take them lightly.

Table 9. The percentages influenced/not influenced by societal factors

s. Societal Difficulties	**Influence**	**No Influence**
• The Women	25%	75%
• The Peers	17%	83%
t. Socioeconomic Background	**Influence**	**No Influence**
• The Women	63%	38%
• The Peers	57%	43%

Their Peers

As with themselves, few of the women considered *societal difficulties* to have affected the choices of their peers. However, at least one of the women said that lack of resources influenced their peers' choices. This ties to *socioeconomic background* where, for their peers, the high economic costs limited the pursuit of education. One respondent noted that in the case of a much-chosen profession for her peers, the study of nursing was "*predictable and expected and (the) years of schooling was less*". Therefore, for immigrant women whose families are struggling in their country of origin, going into nursing was less risky, whereas CS was perceived as unpredictable. In addition, peers that may be able to handle work in CS and related fields usually gravitate to the professions that are usually promoted, such as doctor, lawyer and engineer. Those traditional professions seem to bring prestige to poor families.

ANALYSIS

The research question for this chapter is: *What are the issues that influence some Black women in Canada and the Caribbean to pursue regular academic roles in Computer Science and their peers to choose other paths?* The research presented has revealed that the women who are successful in pursuing regular academic roles and advanced education in CS in Canada and the Caribbean have some common characteristics. They were influenced in similar ways by experiences related to CS. We believe, that if they have stayed because of the factors that influenced them, it is important to analyze the differences between how they have been influenced and how their peers have been influenced.

The influences were analyzed by categorizing (and counting) the answers given by the women into dichotomous categories of "Influence" or "No Influence" on their and their peers' choice for CS and related fields. The answers from the women on their choices and the choices of their peers are two related sets of nonparametric data and are measured in the same sample. In addition the data set is relatively small. This combination of characteristics of the data means that the McNemar's Chi-square ($\chi^{2)}$ test is the suitable test for similarity (Agresti, 2013).

McNemar is a test of association that looks at the relationships between the values in the cells of a 2x2 contingency table by looking at the difference in proportions from the marginal sums of the 2x2 table: p_A=(a+b)/N and p_B=(a+c)/N (See Table 10). The test considers whether the two proportions, p_A and p_B, are significantly different.

Table 10. General structure for McNemar's test

		B		
		1	0	Totals
A	1	a	b	a+b
	0	c	d	c+d
Totals		a+c	b+d	N=a+b+c+d
p_A = (a+b)/N_T				
p_B = (a+c)/N				

The null hypothesis for each factor within a category is that the two marginal probabilities for each outcome are the same, i.e. $p_a + p_b = p_a + p_c$ and $p_c + p_d = p_b + p_d$, that is:

H_0: There is no difference in the percentage of the women that influenced by the factor and percentage of peers that are thought to be influenced by that factor.
H_A: There is a difference in the percentage of the women that influenced by the factor and percentage of peers that are thought to be influenced by that factor.

H_0 is not rejected if the p-value exceeds the significance level of 0.05 and reject it if it is below.

Findings

Based on the result of the McNemar tests shown in Table 11, H_0 was rejected and the alternative hypothesis accepted for P2, P4, P5, P6, P7 and P8. In Table 12 the focus is only on these categories and the key ways in which the women have been influenced differently to their peers within these categories is summarized.

It is interesting that the statistical analysis does not reveal a significant difference in influences in P1 for the women versus their peers, since issues like the *stereotype threat* (Redmond et al., 2013; Steele & Aronson, 1995) and the *ordeal* learning environment (Rheingans et al., 2011) which may occur in CS is known to deter women and minorities and has been identified as one of the reasons for flight from CS. However, for the women and their peers it is on average a relatively even split on being influenced by the factors in the category and not being influenced. This means that this factor though impacting both

Table 11. McNemar's χ^2 values (1 degree of freedom) and p-values for the factors

Category	McNemar's χ^2 value (1 degree of freedom)	p-value
P1	a. 0.0082 b. 0 c. 0.4812	a. 0.9279 b. 1 c. 0.4879
P2	d. 12.0143 e. 98.01	d. 0.0005279 e. 2.2e-16
P3	f. 0.3364 g. 0 h. 0.3364	f. 0.5619 g. 1 h. 0.5619
P4:	i. 84.0116 j. 7.68 k. 55.0175	i. < 2.2e-16 j. 0.005584 k. 1.195e-13
P5	l. 12.0143 m. 48.02	l. 0.0005279 m. 4.219e-12
P6	n. 14.2604 o. 5.0633 p. 1.375	n. 0.0001592 o. 0.02444 p. 0.241
P7	q. 27.0345, r. 7.68,	q. 1.999e-07 r. 0.005584
P8	s. 35.3152, t. 3.8404	s. 2.804e-09 t. 0.05003

Table 12. The ways women have been influenced differently from their peers

Category	Characteristics of the Black women in CS Education	Perceived Characteristics of the Peers of the Black women in CS Education	Key Differences (K) in the peers.
P2: Experiences in education	Supported by pre-tertiary educators through to graduate school mentors.	Reduced support in mathematics education may have hindered access to CS education.	**K2**: Lacking early support in mathematics education may reduce potential to pursue CS.
P4: Internal motivation	Early personal goals involved excelling in CS or a related area.	Not exposed to what careers in CS have to offer and underestimated the field.	**K4:** Exposure to the range of CS careers is lacking.
	Wanted to reach "great heights" and worked to achieve that.	Focused on a stable job and family early in life. Educated, but committed less time to education.	
	Non-traditional approach to life.	Traditional approach to life.	
P5: Attraction and commitment to the field	Enjoy tasks associated with Computing and committed time required for studies and CS daily tasks.	Large time commitment within this major and career was not appealing.	**K5:** Deterred by large time commitments.
P6: Support	Received strong support from parents and spouses.	Givers of support for survival of their extended families.	**K6:** Support unavailable to surpass status quo.
	Generally from families in Canada and the Caribbean who have built middle-class lives.	Canadian immigrants expected to provide support to those in the country of origin. Caribbean background has more traditional expectations for family responsibilities.	
P7: Experiences in Computer Science and Technology	Positive experiences in CS prerequisites in early CS courses.	Good experiences in CS prerequisites kept some; unpleasant experiences deterred others.	**K7**: Deterred by negative early CS/prerequisite education experiences.
P8: Societal Factors	Often middle-class socio-economic background facilitated this non-traditional field.	Sometimes challenging financial situations hindered participation in this field.	**K8:** Hindered by financial situations may have hindered pursuit of CS.

groups somewhat has not made the difference to flight or pursuit of the field. The same can be said for the P3. However, though an issue such as the lack of role models has not revealed itself as making the difference for flight from or pursuit of the field, issues such as the difference in influence of internal motivation, require the involvement of role models in a more indirect sense to help those who are floundering to make it through the challenges based on their experiences (Gurer & Camp, 2001). So one cannot discount P1 and P3 completely, but note that the interaction between some factors must be investigated.

RECOMMENDATIONS

There are notable differences in views, experiences and hence how the two groups of women have been influenced in their choices as shown in Tables 11 and 12. The differences matter to attracting and retaining Black women who do not choose CS. These main differences are summarized in Table 13 based on the categories where the differences were significant. The issues coming out of this study are shared by women in CS and minorities in CS in the US and as such relevant concerns listed in the literature are discussed and some recommendations that have been supported by previous work presented. The recommendations are intended to help move the peers' view towards those of the women who have pursued CS. The key differences are grouped where the recommendations are similar.

CONCLUSION

This chapter explores the nature and experiences of Black women and their peers in Computer Science in Canada and the Caribbean, on their paths to RA positions. The quest in conducting the study itself was quite challenging, because the population under study is quite sparse. This however further motivates the need for such studies to improve the representation that will create more role models for Black women

Table 13. Recommended solutions based on identified differences

Key Difference	Recommended Solutions
K2 and K6	Breene (Breene, 1992) states that often there is discrimination towards young women in pre-tertiary mathematics education from both teachers and guidance counselors. The issue in the Caribbean is not peculiar to women. Reduced support may occur in the case of anyone who performs poorly an early stage of study in mathematics. This can compound other issues and diminish the pursuit of careers for Black women in the Caribbean. Burge and Suarez (Burge & Suarez, 2005) recommend that in the peers' communities work should be done with the guidance counsellors, teachers and churches to improve teaching techniques for prerequisites and pre-tertiary CS to open up the pipeline to tertiary CS.
K4	Gurer and Camp (Gurer & Camp, 2001) supports the existence of lack of exposure to CS faced by the peers, and suggests that women (and by extension Black women) who have succeeded in CS should visit schools and other community locations to provide an accurate view of the nature of the variety of jobs in CS.
K5	The women pointed out that though the time commitment for computing tasks and to pursue advanced levels of education was high, they loved what they did and committed the time. It is also noted that this love was cultivated at an early age. This implies that all of the pre-tertiary and family supports are crucial early in life to cultivate interest and passion for CS. Also, families require an easily accessible home computer to support girls' interests (Leever, Dunigan, & Turner, 2002).
K7	Negative early education experiences in CS and prerequisites have pushed the peers away from the field before may have considered CS fairly. Deliberate initiatives are necessary such as the Female Education in Mathematics and Science initiative in Africa (FEMSA) which implements programs to facilitate mathematics and science education specifically for girls. This is done through teacher training for enriched mathematics instruction; improved mathematics curriculums; sensitizing teachers to the needs of girls; increased capacity in teaching staff; remedial classes for girls and facilitating community support for schools(Alumanah, 2005).
K8	Margolis states that women in CS have disproportionately wealthy upbringings (Margolis & Fisher, 2003). This may be a generalisation, however, the finding is congruent with the fact that most of the women in our study who succeeded in the pursuit of CS to advanced levels were from middle class families. Though not wealthy, their families were able to provide the financial support necessary to explore a non-traditional path. One needs to examine the extent that the peers relied on their families for the survival and the real financial limitations of the peers who are immigrants in Canada. This is especially so since in the Caribbean the peers may have their children earlier in life. *Funding sources that do not require repayment such as scholarships are therefore critical* in the engagement in CS and advanced studies(Burge & Suarez, 2005). For immigrants and mothers, this will ease financial burdens. Scragg says that women may want to raise a family, but a career in CS may be mismatched with this preference (Scragg & Smith, 1998). So mentors and role models can often offer solutions based on experience in finding solutions. Additionally, the support of parents and families is key. This support can be encouraged through *community outreach to families by role models*. So though role models may not be directly highlighted in the results of the study, they are valuable because the can share knowledge for success.

who pass through classrooms in CS. As result of the small quantity of Black women participating in our study who are RA staff or pursuing advanced degrees, this work is not presented as a comprehensive analysis of the situation. However, our findings in assimilating their reflections show that most of the race and gender related issues experienced by Black women in Canada and Caribbean in CS are similar issues experienced by women and minorities in CS in the US and Africa. Hence many of the solutions recommended in other studies can be applied to their concerns. However, there are specifics such as the reduced support for lower performers in mathematics for the Caribbean, and familial responsibilities of immigrants that must be treated differently. The issues must be remedied within educational systems and communities.

FUTURE WORK

In future work the study will be extended to non-English speaking countries in the Caribbean. The period of invitation to participate in the study will also be extended. The period for respondent solicitation and participation was six weeks, but responses will be solicited over six months to try to gain additional participation and increase the study size, that is, if there are other persons in the population that can be reached. In addition, a pilot study will be conducted with open-ended questions to poll for a more general set of factors that the respondents have experienced as influences.

REFERENCES

Agresti, A. (2013). *Categorical data analysis* (3rd ed.). Hoboken, NJ: Wiley.

Alumanah, J. N. E. (2005). Access and Use of Information and Communication Technology for the African Girl-child Under Cultural Impediments. In *Proceedings of the International Symposium on Women and ICT: Creating Global Transformation*. New York, NY: ACM. http://doi.org/ doi:10.1145/1117417.1117430

Berenson, S. B., Slaten, K. M., Williams, L., & Ho, C.-W. (2004). Voices of Women in a Software Engineering Course: Reflections on Collaboration. *Journal of Educational Resources in Computing*, *4*(1), 3, es. doi:10.1145/1060071.1060074

Breene, L. A. (1992). Women and Computer Science. *Initiatives*, *55*(2), 39–44.

Burge, J. D., & Suarez, T. L. (2005). *Preliminary analysis of factors affecting women and African Americans in the computing sciences*. ACM Press. doi:10.1109/RTCDC.2005.201646

Camp, T. (2012). Computing, We Have a Problem…. *ACM Inroads*, *3*(4), 34–40. doi:10.1145/2381083.2381097

Canes, B. J., & Rosen, H. S. (1995). Following in her footsteps? Faculty gender composition and women's choices of college majors. *Industrial & Labor Relations Review*, *48*(3), 486–504. doi:10.1177/001979399504800308

Caribbean Community (CARICOM) Secretariat. (2011). *CARICOM Member States*. Retrieved May 15, 2015, from http://www.caricom.org/jsp/community/member_states.jsp?menu=community

Crabtree, S. (2013, October 8). *Worldwide, 13% of Employees Are Engaged at Work - Low workplace engagement offers opportunities to improve business outcomes*. Retrieved May 15, 2015, from http://www.gallup.com/poll/165269/worldwide-employees-engaged-work.aspx

Frenkel, K. A. (1990). Women and Computing. *Communications of the ACM*, *33*(11), 34–46. doi:10.1145/92755.92756

Garnett, K. (1999). Girl Power Clicks Here. *NEA Today*, *18*, i2.

Gurer, D., & Camp, T. (2001). *Investigating the incredible shrinking pipeline for women in computer science*. Final Report–NSF Project, 9812016. Retrieved from http://womendev.acm.org/archives/documents/finalreport.pdf

Leever, S., Dunigan, M., & Turner, M. (2002). The Power to Change is in Our Hands. *Journal of Computing Sciences in Colleges*, *18*(2), 169–179.

Lopez, A. M., Jr., & Schulte, L. J. (2002). African American Women in the Computing Sciences: A Group to Be Studied. In *Proceedings of the 33rd SIGCSE Technical Symposium on Computer Science Education* (pp. 87–90). New York, NY: ACM. http://doi.org/ doi:10.1145/563340.563371

Margolis, J., & Fisher, A. (2003). *Unlocking the clubhouse: Women in computing*. MIT Press.

Mody, P. N., & Brainard, S. G. (2005). Successful International Initiatives Promoting Gender Equity in Engineering. In *Proceedings of the International Symposium on Women and ICT: Creating Global Transformation*. New York, NY: ACM. http://doi.org/ doi:10.1145/1117417.1117422

National Center for Science and Engineering Statistics. (2013). *Women, Minorities, and Persons with Disabilities in Science and Engineering: 2013*. National Science Foundation. Retrieved from http://www.nsf.gov/statistics/wmpd/2013/pdf/nsf13304_full.pdf

Nurse, K. (2007). *Science, Technology and innovation in the Caribbean*. Paper presented at the Meeting on Technology Policy and Development in Latin America of the Economic Commission for Latin America and the Caribbean, Santiago, Chile. Retrieved from http://vi.unctad.org/files/papers/nursepaper.pdf

Ojokoh, B. A., Adeola, O. S., Isinkaye, F. O., & Abraham, C. (2014). Career Choices in Information and Communication Technology among South Western Nigerian Women. *Journal of Global Information Management*, *22*(2), 48–77. doi:10.4018/jgim.2014040104

Peters, A.-K. (2013). Identity Development of CS and IT Students: What's the Role of Higher Education? In *Proceedings of the Ninth Annual International ACM Conference on International Computing Education Research* (pp. 187–188). New York, NY: ACM. http://doi.org/ doi:10.1145/2493394.2493427

Redmond, K., Evans, S., & Sahami, M. (2013). *A large-scale quantitative study of women in computer science at Stanford University*. ACM Press. doi:10.1145/2445196.2445326

Rheingans, P., Brodsky, A., Scheibler, J., & Spence, A. (2011). The Role of Majority Groups in Diversity Programs. *Trans. Comput. Educ.*, *11*(2), 11:1–11:15. http://doi.org/10.1145/1993069.1993075

Science, Technology and Innovation Council. (2013). *2012 State of the Nation Canada's Science, Technology and Innovation System: Aspiring to Global Leadership*. Ottawa, ON, Canada: Science, Technology and Innovation Council Secretariat.

Scragg, G., & Smith, J. (1998). *A study of barriers to women in undergraduate computer science*. ACM Press; doi:10.1145/273133.273167

Statistics Canada. (2012, May 3). *Full-time teaching staff at Canadian universities, by rank and sex*. Retrieved May 15, 2015, from http://www.statcan.gc.ca/tables-tableaux/sum-som/l01/cst01/educ68a-eng.htm

Steele, C. M., & Aronson, J. (1995). Stereotype threat and the intellectual test performance of African Americans. *Journal of Personality and Social Psychology*, *69*(5), 797–811. doi:10.1037/0022-3514.69.5.797 PMID:7473032

Trauth, E. M., Quesenberry, J. L., & Yeo, B. (2008). Environmental Influences on Gender in the IT Workforce. *SIGMIS Database*, *39*(1), 8–32. doi:10.1145/1341971.1341975

Zweben, S. (2014, November 30). *Computing Degree and Enrollment Trends from the 2010/11 Taulbee Survey*. Computing Research Association. Retrieved from http://cra.org/uploads/documents/resources/taulbee/CS_Degree_and_Enrollment_Trends_2010-11.pdf

Zweben, S., & Bizot, B. (2014, May). *2013 Taulbee Survey*. Retrieved May 15, 2015, from http://cra.org/resources/crn-archive-view-detail/2013_taulbee_survey/

Chapter 6
Race and Gender Inequalities in Medicine and Biomedical Research

Djana Harp
Morehouse School of Medicine, USA

Jamil A. Harp
Morehouse School of Medicine, USA

Ruth S. Shim
Lenox Hill Hospital, USA

William Clyde Wilcox
Georgetown University, USA

Japera Johnson
Morehouse School of Medicine, USA

Judith K. Wilcox
Boston University School of Medicine, USA

ABSTRACT

There is a critical need to develop initiatives aimed at expanding and diversifying the healthcare workforce, beginning with medical education. This chapter addresses racial and gender disparities in academic medicine and biomedical research. Racial and ethnic minorities and women remain underrepresented in medicine, biomedical research, and healthcare leadership. These disparities are national issues and have far-reaching effects which translate into significant educational and healthcare-delivery disparities. Research has shown that health care professionals from underserved backgrounds are more likely than others to work in underserved areas which would address current shortages in health professionals for the medically underserved. The authors describe their theory for the existence and persistence of these disparities. They offer evidence of these disparities while concluding with current initiatives to address these disparities, calling for innovative approaches to training underrepresented minorities and women as physicians and biomedical research scientists.

DOI: 10.4018/978-1-5225-0174-9.ch006

INTRODUCTION

Of all the forms of inequality, injustice in health care is the most shocking and inhumane. (Martin Luther King, Jr., 1966)

In 2015, African Americans and women remain underrepresented as senior medical faculty and among biomedical federal grant recipients (Carr et al., 2015; Ginther et al., 2011). There has been an attempt in recent literature to address these disparities that centers on how African American women's perception of social identity and belongingness affect their expectations and experiences in medicine and biomedical research (Purdie-Vaughans et al., 2008; Steele et al., 2002; Murphy et al., 2007; Cheryan et al., 2009; Good et al., 2012; Burgess et al., 2012). This literature provides a useful point of departure to understand individual experiences and also to identify institutional practices that shape the experiences of African American women's experiences in academic medicine and biomedical research. However, while the literature acknowledges the historical devaluation of women and African Americans in a broader U.S. context, it stops short of explicitly connecting the history of academic medicine to current patterns of gender and ethnic representation in the field. Toward that end, using a diversity science framework, the authors seek to link the history of academic medicine to the persistent underrepresentation of African American women and African Americans more globally. Further, the authors offer suggestions to address the continued barriers for women and minorities.

Diversity science framework[1] is a sociocultural analytic tool useful in understanding how an individual's behaviors are a reflection of the inscription of social, cultural and historical ideologies (e.g., ideas, beliefs or cultural assumptions), institutional practices (e.g., laws, language, or organizational policies) and daily experiences (Plaut, 2010). The diversity science framework suggests that the underrepresentation of African Americans in medicine has its roots in a report by Abraham Flexner in 1910, which would lay the foundation for the transformation of 20th century medical education. The Flexner report led to wide-sweeping reforms and shifted the paradigm of medical education toward a more professional model with well-trained professors, quality laboratories, and access to hospitals (Duffy, 2011; Hiatt & Stockton, 2003; Sullivan & Mittman, 2010; Savitt, 2006; Steinecke & Terrell, 2010; Barkin et al., 2010). However, the costly requirements were out of reach of most African American medical schools. As a result of his recommended reforms in medical education, most of the schools that provided medical education to women as well as African Americans were closed (Flexner, 1910; Duffy, 2011). The reduced opportunities for African Americans and women have contributed to the current inequities in the representation of African Americans in U.S. medical education and, ultimately, biomedical research (Hiatt and Stockton, 2003; Sullivan & Mittman, 2010; Savitt, 2006; Steinecke & Terrell, 2010; Barkin et al., 2010). Consistent with the theory of cumulative disadvantage (DiPrete TA, 2006), we posit that differences in treatment over a prolonged period have cumulated into substantial inequalities in race and gender representation in both medicine and biomedical research (Sheltzer & Smith, 2014).

BACKGROUND

In 1910, Abraham Flexner issued a report evaluating the state of medical education in the U.S. Flexner sought to increase the quality of physicians by *normalizing* medical education for the next generation of physicians. Flexner evaluated medical schools along objective criteria, including quality of professors, laboratory equipment, and affiliation with a hospital, but he was also influenced by the epidemic of racism and sexism of that era. Not surprisingly, his vision of *normal* centered primarily on a vision where Caucasian men constituted the vast majority of the medical profession. Furthermore, Flexner went on to argue that women showed a *decreasing inclination* to enter the profession, and that there was simply no strong demand for women's medical training (Hiatt & Stockton 2003). Instead of allowing the existing seven medical schools for women to build a constituency gradually as women overcame gender stereotypes and discrimination, he recommended that they be closed. By 1930 there was only one left in operation, the Women's Medical College of Pennsylvania (Barkin, Fuentes-Afflick, Brosco, & Tuchman, 2010). In 1993, the college merged with Hahnemann Medical School and, in 2003, the two colleges became part of Drexel University College of Medicine.

Similarly, Flexner urged the closing of most *medical schools for Negroes* because in his judgment, they were ineffective. Many of these schools were private schools with few resources, inadequate laboratory equipment, and had limited if any access to hospitals. Nevertheless, many of these proprietary schools filled deep needs for minority patients, but Flexner's steep financial requirements for a solvent medical school were beyond the grasp of most of these institutions. As a result, Flexner recommended that only Howard University College of Medicine and Meharry Medical College remain open to serve the needs of the minority community. He recognized that both Howard University College of Medicine and Meharry Medical College were *of course, unequal to meet the need and the opportunity*, but recommended that they be kept open because he recognized a need for a minimal number of African American physicians as a service to their *own race*. Ignoring the many African American physicians already caring for blacks and other minorities in rural and even urban areas across the south and elsewhere, Flexner posited that whites needed to teach black doctors to *feel a sharp responsibility for their race, and* he went on to predict that the *medical care of the Negro race will never be wholly left to Negro physicians* (Stockton, 2003). A second and more important reason for keeping Howard and Meharry in operation was the need to teach hygiene to African Americans. Flexner believed that Blacks needed to be educated in hygiene because of *the tremendous importance that belongs to a potential source of infection and contagion*. For this reason, he urged that these schools teach *the more promising of the race... a substantial education in hygiene rather than surgery*. In short, African American doctors were needed to serve minority populations and to prevent disease from spreading to whites.

The Flexner report was transformational in establishing the foundation for twentieth century medical education in the U.S. and it pushed medical education toward a more scientific and rigorous approach to understand human existence and disease (Duffy, 2011). It also established funding practices for medical education; for example, after the report many foundations only provided financial support to medical schools that adopted the Flexner model (Barkin, Fuentes-Afflick, Brosco, & Tuchman, 2010). Further, it had considerable influence in establishing patterns for the exclusion of women and minorities in medical education. In response to the Flexner report, most of the schools that were chiefly responsible for the education of women and minorities in medicine subsequently closed (Hiatt and Stockton, 2003; Sullivan & Suez Mittman, 2010; Savitt, 2006; Steineck & Terrell, 2010; Barkin, Fuentes-Afflick, Brosco, & Tuchman, 2010).

Flexner did not set out to close women's or African American medical schools – indeed he addressed their existence in brief appendices to his main report. But embedded in the racism and sexism of his age, he could not imagine a world where the demand for women physicians would increase and where women patients might prefer that at least some of their services be provided by a woman. Similarly, he did not think that keeping a robust supply of African American doctors was sufficiently important to try to rescue the schools that provided so many minority and minority-serving physicians. And although he did seek funding for the two remaining African American medical schools, it never occurred to him to bolster the resources of the other minority-serving schools in order to improve the health care opportunities of underserved populations. In some ways, these proprietary colleges fit more closely with Booker T. Washington's vision of education. Washington argued that historically black colleges and universities (HBCUs) should focus on providing a vocational and industrial education instead of the traditional liberal arts education advocated by northern intellectuals. Washington envisioned large numbers of Blacks acquiring skills needed to be teachers, nurses, and skilled workers, and then accumulating economic resources necessary to build separate community institutions. Teaching African Americans rudimentary medical skills and sending them out to rural areas to build institutions to serve a population that had little access to medicine fit well with Washington's vision.

Flexner in contrast recommended a smaller number of medical schools serving the best African American students, a strategy more consistent with the vision of WEB DuBois. DuBois proclaimed that *The Negro Race, like all races, is going to be saved by its exceptional men.* The problem of education then, among Negroes, must first of all deal with the *Talented Tenth* (Clyde Wilcox, 2014). *It is the problem of developing the best of this race that they may guide the mass away from the contamination and death of the worst.* Washington saw HBCUs as providing large numbers of Black southerners with the skills to earn a decent living, to attain occupational status that was beyond most African Americans, and to accumulate economic resources. DuBois instead saw the function of these institutions to provide the Black elite with the background and skills that they would need to advocate for African Americans, to contest politically with whites, and to direct further political and economic development of the community. For more than 50 years, Howard University School of Medicine and Meharry Medical College were the predominant option for medical education for African Americans (Strelnick, Lee-Rey, Nivet, & Soto-Greene, 2008). Flexner, however, did seek to bolster these two remaining schools. He advocated for the construction of a teaching hospital for Meharry, and talked with Andrew Carnegie about a sizable grant to start construction.

The Flexner report saw women and minorities as either unimportant or playing a very limited role in the medical enterprise. In 1910, the majority of medical schools did not accept African Americans or women. Indeed, the prevailing social conditions of Jim Crow segregation in the South prevented African Americans from studying with Caucasian students as a matter of law. Jim Crow segregation did not end until the Civil Rights Act was passed in 1964. By accepting Flexner's evaluation about the unimportance of women and minorities in medicine, discrimination was silently built into the structure of medical education in the United States. Flexner's normalized education for Caucasian males became the rule, and groups outside of this norm have faced and continue to face considerable challenges being recruited to, being retained in and advancing in academic medicine and biomedical research.

However, much has changed since Flexner wrote his report. African Americans now teach in top medical schools in the country. Women now constitute more than half of the entering medical school classes. But the indifference that Flexner expressed toward the role of women and minorities in medical education remains pervasive in the majority of institutions. Toward that end, we argue that the persistent underrepresentation of women and minorities in medicine and as a result biomedical research is in part a result of an uncontested discriminatory ideology dating back to the turn of the twentieth century. In this chapter, we seek to clearly identify the remaining artifacts of an ideology that systematically undervalues women and minorities in medicine and biomedical research and offer suggestions to address the continued barriers for women and minorities in these spaces.

We suggest that as a result of limited access to a medical education, fewer African American physicians were trained. This thereby fueled a significant reduction in numbers of physicians at all levels of the medical training establishment, including, not only clinicians, but academic faculty as well (Sullivan & Mittman, 2010). Smaller classes resulted in fewer academic faculty members being required and fewer physicians being generated. With lower levels of funding and less cutting-edge equipment, biomedical research lagged behind larger well-equipped and -funded majority institutions. Therefore, as a result, a platform was created that not only resulted in fewer African American physicians and faculty, but fewer African American biomedical researchers as well.

Importantly, concurrent to the closing of women's medical colleges, there was a societal shift to expand women's rights. This shift encompassed a decade long women's suffrage movement, culminating in women nationally earning the right to vote in 1920. The landscape for medical education of Caucasian women rapidly changed. It would not be until 1965 when African Americans were effectively enfranchised and civil rights legislation was enacted that desegregated schools and afforded African American men and women opportunities to not only become physicians, but to also engage in biomedical research at resource- and research rich institutions. The hard-won fight for women's suffrage, however, did little to counter the fact that women comprised only 3% of the physicians at the turn of the century. This trend would persist with minimal increase until the 1970s when women comprised 8% of physicians in the U.S. (American Medical Association, 2006). In 1972, the U.S. government passed Title IX legislation prohibiting gender discrimination in education and a large number of women entered science fields. Women now comprise 33% of all active physicians (American Association of Medical Colleges, 2014). As a result of federal legislation, there have been significant strides with women now comprising 50% of incoming medical students, and nearly 50% of all residents and fellows. Despite these demographics, few women attain the rank of professor or are appointed to higher level academic and executive leadership positions in academic medicine and biomedical research (Sheltzer & Smith, 2014). Similar to findings for African American physicians, women continue to remain underrepresented in top levels of the biosciences despite having greater parity at the undergraduate and graduate levels. There appears to be clear pipeline leak (Sheltzer & Smith, 2014; Soper, 2014).

MAIN FOCUS OF THE CHAPTER

Issues, Controversies, Problems

Underrepresentation of African Americans in Medicine

Underrepresented minorities in the United States today remain below the national average in education, employment, and health status (Wilson & Balotin, 1999; (Cook, 2015b); (Cook, 2015a). We suggest that the impact of the Flexner Report on medical educational opportunities for African Americans has had a lasting impact as a result of cumulative disadvantage on the number of African American physicians as well as biomedical researchers. As a result of the impact on the education, training and attrition of minority physicians and indirectly biomedical researchers (Schelter & Smith, 2014), there are segments of the U.S. population that face significant barriers to healthcare access and quality (i.e., there are only four historically black medical schools in the US). Meharry Medical College was founded in

1876 in Nashville, TN; Howard University College of Medicine founded in 1868 in Washington D.C.; Charles Drew University of Medicine and Science was founded in 1966 in Willowbrook, CA; and the Morehouse School of Medicine founded in 1975 in Atlanta, GA. The size of the graduating classes range from 24-105 with a median size of 61, although both Meharry and Morehouse schools of medicine plan to increase their class sizes to 70 and 100, respectively, over the next few years. Nevertheless, three of these institutions account for 85 percent of all minority physicians in the U.S. In 2014, African American enrollees in medical school rose to 1.1% (for a total of 1,412). This may, in part, explain the shortage of a diverse faculty in many U.S. medical schools. As a result of this lack of diversity in faculty, research topics addressing healthcare disparities are less likely to be pursued in health care.

The historically black medical schools draw a significant proportion of their students from historically black colleges and universities (HBCUs). In 2006, about 21% of all Blacks accepted to medical school had earned their undergraduate degrees from HBCUs (Wilcox, Welles, Haddad, & Wilcox, 2014). Similarly, during the period 2002 to 2006, 16% of all Black students who graduated from medical school received undergraduate degrees from HBCUs (Treadwell, 2009). The HBCU pipeline is also important for producing Black graduates who go on to receive terminal STEM degrees. The National Science Foundation identified that 21 of the top 50 institutions for producing black graduates with doctorates in science and engineering are HBCUs. Nevertheless, these HBCUs must continue to stabilize and grow their student bodies, attract and retain highly qualified faculty and students, adjust to expanding oversight of state and federal government, find ways to improve graduation and retention rates, provide education at a cost that is within reach to a population that largely hails from lower income communities, compete for grants and contracts against large institutions, and grow endowments during a time of decreased state allocations and rising inflation (Toldson, 2014).

Demographics of African American Physicians

Despite the fact that African Americans account for 16% of the U.S. population, only 3% of all faculty members at U.S. medical schools are African American (Association of American Medical Colleges, 2011). Although, the numbers of minority faculty and students have increased over the last 20 years, the proportions of underrepresented minority faculty and students remain far below proposed targets (Rodriguez, Campbell, & Pololi, 2015). A 2010 study showed that once hired, African Americans were less likely to be retained than any other demographic group (Sullivan & Mittman, 2010; Ginther, Schaffer, Schnell, Masimore, Liu, Haak, & Kington, 2011). African American physicians are also less likely than their Caucasian counterparts to be promoted, hold senior faculty or administrative positions, or receive research awards from the NIH (Sullivan & Mittman, 2010).

There are also pipeline challenges in medicine and biomedical research for African American students (Marchant & Omary, 2010). The percentage of black men among all medical school graduates has declined over the past 20 years. In 2012, there were just 517 black men among the more than 20,000 graduating students at U.S. medical schools (Cuyjet, 2009; Association of American Medical Colleges, 2012). The country's traditionally black medical colleges – Howard, Meharry, and Morehouse – still graduate the majority of African American physicians in the US despite efforts by various organizations (e.g., the National Medical Association, the American Medical Association, and the American Association of Medical Colleges) to increase the representation of racial/ethnic groups in medicine, only limited gains have been made (Jeffe, Yan, & Andriole, 2012).

Racism and Discrimination against African American Physicians

Despite diversity efforts at many institutions across the country, many African American physicians encounter racism, discrimination, and inequities in academic medicine (Carr et al., 2007). Racial bias is most evident during the promotion process. This is particularly true across gender lines and is clearly reflected in gender distributions in medical school faculty as well as in basic science departments that perform biomedical research (Sheltzer & Smith, 2014; Soper, 2015). African American faculty members are promoted at lower rates compared with Caucasian faculty (Fang, Moy, Colburn, & Hurley, 2000). Therefore, African American faculty members in medical schools are more likely to hold junior faculty positions rather than senior faculty or leadership positions. In fact, 31% of the 84,195 faculty members at U.S. medical schools were full professors in 2011, as compared with just 11% of 3,952 African American faculty members (Nunez-Smith et al., 2012).

In 2006, Dr. James Sherley, an African American biological engineer, with expertise in adult stem cell research, made national headlines. Dr. Sherley received a B.S. from Harvard University and an M.D. and a Ph.D. from Johns Hopkins University School of Medicine. He was denied tenure at the Massachusetts Institute of Technology (MIT). Sherley was one of 23 African American professors at MIT. At the time about 17% of MIT's faculty was reported to be ethnic minorities. Dr. Sherley expressed his view that his race was the crux of the issue. Despite Dr. Sherley's recognition as a thought-leader, MIT refused to act favorably on the evidence put forth and refused to call for an independent review. Although Dr. Sherley's treatment at MIT drew criticism from some MIT faculty and alumni from diverse racial/ethnic backgrounds, the decision was upheld (Bhattacharjee, 2007).

Underrepresentation of Women in Medicine and Biomedical Research

Although more and more women are seeking careers as physicians in the US, there is greater ground to cover if we are to create an equal workforce landscape. While numbers in the academic and biomedical pipeline are increasing, there remains a pipeline leak, especially in the biomedical sciences. Few women are in the higher echelons of academic medicine or biomedical sciences (Sheltzer & Smith, 2014). There are historical and institutional barriers to women's ascent, beginning with our nation's centers of scientific learning.

Historical Barriers

In 2011, only 13% of women in academic medicine had attained the rank of full professor while only 11% achieved the rank of professor in 1980. Despite progress over the last 30 years with women now composing nearly half of each matriculating medical school class, little progress has been made resulting in underrepresentation in leadership positions with a lower percentage of women than men. Since 1984, the percentage of tenured or tenure-eligible faculty positions dropped from approximately 60% in 1984 to 30% in 2010 (Mayer, 2014). A similar story is true for the biomedical sciences. Although women have attained gender parity among undergraduate and graduate students, they remain significantly underrepresented in the top levels of bioscience (Soper, 2015). This has important implications for the development of future female scientific leaders. A recent study clearly demonstrated that male scientists train significantly fewer women than do female scientists (Sheltzer & Smith, 2014). In labs

Table 1. Percentage of women in biology laboratories by status

Faculty	25%
Full Professors	21%
Associate	31%
Assistant	29%
With one or more trainees	27%
Postdoctoral researchers	39%
Graduate students	49%

Modified from Scheltzer and Smith, 2014

with female heads, there is no gender bias in hiring, but women are still far less likely than men to be lab leaders. As long as male lab leaders discriminate against female applicants there will be a pipeline problem in the biological sciences. Table 1 shows the percentage of women in biology labs at various levels of career development.

Institutional Barriers

Many medical schools are now offering a variety of academic tracks outside of the traditional tenure track. These options have emerged as faculty have increasing clinical and educational demands. The need for patient care revenue is playing an ever-increasingly important role in the financial survival of academic medical centers. Female faculty are more likely than their male colleagues to be on the clinician-educator track than the traditional tenure track. Despite this trend, the, promotion rates for clinician-educators lags behind those of their traditional tenure track colleagues (Mayer, 2014).

The problems in gender bias in biology labs described above are even more acute in elite labs in top universities and research institutions. Denied access to the top research environments, women are more likely to end up in less prestigious institutions. This denies women access to resources, networks, international visibility, and mentoring. It leads to fewer publications and grants. Moreover, studies have shown that all male tenure and promotion review panels are far less friendly to women than mixed gender panels, so that the concentration of men in elite institutions poses a barrier to women seeking to advance in the field.

Demographics of Female Physicians

Women's progress in academic medicine is often impeded by variables such as traditional gender roles, sexism in the medical environment, lack of effective mentors, and gender differences in approach to career and life goals. Women are viewed less favorably than men as potential occupants of leadership roles (Pololi et al., 2013). Today, 38% of full time faculty in academic settings are women, which is an increase from 25% two decades ago. However, only 21% of women are full professors in clinical departments and 23% are full professors in basic science departments. Interestingly, despite being overrepresented in certain clinical departments, the percent of senior level appointments remains disappointingly low. For example, although women account for 57% of obstetrics/gynecology faculty and 53% of pediatrics faculty, only 22% of obstetrics/gynecology department chairs are women, and only 20% of department chairs in pediatrics are women. This disparity is even greater and more glaring in surgery and surgical subspecialties. At the time of this writing, there are currently no female department chairs in orthopedic surgery, one in general surgery, three in otolaryngology, and eight in ophthalmology (Lautenberger, 2015). Furthermore, women comprise only 24% of division heads, 24% of department vice chairs, 15% of department chairs, 46% of assistant deans, 39% of associate deans, 33% of senior associate deans, and 16% of medical school deans (Lautenberger, 2015). While many women achieve the rank of assistant professor, fewer are promoted to associate professor, and even fewer to full professor. The same holds true for promotion to senior level deans.

Demographics of Female Biomedical Researchers

While women are awarded >50% of both bachelor's and doctoral degrees in biology, they represent only 36% of assistant professors and 18% of full professors in biology-related fields (Nelson, 2007). The federal government and its cadre of scientists are not immune to gender inequity. Data demonstrates a disparity in gender distribution of research leadership in the federal government. Women comprise 33% (Agency for Healthcare Research and Quality, and 27% (NIH of top research leadership. Women are also underrepresented as principal investigators. Men have an eight-fold greater chance of obtaining a Cooperative Studies Program trial award compared with women, which can not be explained by clinical trials publication history (McCarren & Goldman, 2012).

Discrimination Based on Gender

Gender bias is an unconscious barrier that contributes to low expectations for women in science, and is often perpetuated by both men and women at high ranks within medicine and biomedical research. Approximately 70% of lay citizens surveyed in 34 countries revealed stereotypes associating science with males more than with females (Nosek et al., 2009). Women who become leaders in academic medicine are often met with skepticism and prejudice. When women exhibit behaviors that fulfill the duties of a leader, this is viewed less favorably compared to men. Several consequences then ensue: First, the prevailing attitudes are less positive toward females than males as potential leaders and, second, it becomes more difficult for women to become leaders and achieve success in leadership roles (Burgess et al., 2012). While holding fewer leadership positions, women also earn less than their male colleagues in medicine and biomedical research. A recent study determined that among 800 physicians who received a highly competitive early career research grant, women earned approximately $12,194 less than men a year when all other factors were held constant. This translates into 30-year career devaluation of more than $350,000 (Jagsi et al., 2006). Gender bias also exists within groups of women. For example, a 2007 study by Correll et al. (2007) indicated that mothers in the U.S. are 79% less likely to be hired and are offered a salary $11,000 less than women with no children.

Gender bias also crosses over to editorial positions held and publication quality, quantity, and prestige. Evidence demonstrates that men dominate in prestigious first and last author positions. Moreover, women are significantly underrepresented as authors of single-authored papers. Institutions should be aware of the subtle ways that gender disparities can seep into scholarly authorship (West, 2013).

In addition to women having lower rates of publication and lower success rates for high-status research awards than men, women must obtain a higher level of achievement than identically credentialed men to be deemed hirable and to confirm competence as scientists as well as physicians (Kaatz, Gutierrez & Carnes, 2014). Wenneras and Wold (1997) determined that female applicants for postdoctoral research fellowships required more than twice as many publications to achieve the same scores as comparable males. A recent study published in the Proceedings of the National Academy of Sciences found that high-achieving male scientists train fewer women in their laboratories by an order of 10 – 40% which limits the number of female candidates who can become competitive for faculty positions (Sheltzer & Smith, 2014). This gender bias was not observed in top labs headed by women. While this may seem encouraging, there are just not enough female scientist leaders to train women. Of the 24 Nobel Laureates included in the study, only two were women.

These unconscious biases hurt women in the workplace and have a psychological as well as economic impact. Women in academic medicine as well as biomedical research report feelings of isolation and invisibility, feeling like cultural outsiders, being undervalued, carrying the weight of gender role expectations, and being in a work-centric culture that does not value their family responsibilities.

Challenges and Benefits of a Diverse Physician Workforce

African American medical students are more than twice as likely as Caucasian students to express a desire to care for underserved communities (Talamantes et al., 2014). Evidence shows that minority physicians and women, in particular, more often provide care in a manner that promotes patient satisfaction and meets the needs of an increasingly diverse U.S. population (Wong et al., 2001). Data also indicates that patients seeing physicians of their own race rate their physician's decision-making style as more participatory. Thus, when there is improved cross-cultural communication between physicians and patients there is greater patient involvement in care, higher levels of patient satisfaction, and better health outcomes (Cooper-Patrick et al., 1999). However, the disbursement of African American students and physicians is often concentrated in urban areas, leaving rural areas with an even less diverse group of physicians.

Many African American faculty members feel compelled to support diversity efforts, but these efforts are often time-consuming and not as valued as other scholarly activities that help advance career toward promotion. This likely contributes to time delays in academic promotion and to lower numbers of African American faculty who attain the level of professor. African American physicians are often more willing to work in indigent and health resource scarce areas, often expressing a desire to serve their communities (Mahoney, Wilson, Odom, Flowers, & Adler, 2008). African American faculty also express a strong desire to serve as role models and to mentor future minority physicians and biomedical researchers.

Faculty diversity and inclusion has multiple benefits for learners such as fostering greater civic engagement, enhancing the ability to recognize racism, providing greater exposure to a variety of pedagogical approaches, and increasing cognitive complexity. There are also health benefits which includes greater access to care for low income patients, racial and ethnic minorities, including non-English speaking patients, and greater access to care for persons with Medicaid (Colleges, 2015) . There are also hidden costs to this inclusion. Physicians who care for underserved and poorer patients not only tend to see sicker patients with multiple co-morbidities, but often see less clinical revenue as a result of serving the indigent. This translates into less pay than physician peers who provide services for more financially affluent patients and less cachet in the faculty practice because of diminished revenue returns.

SOLUTIONS AND RECOMMENDATIONS

Effective Mentorship and Faculty Development

Studies indicate that medical learners prefer mentors who are similar to them and that role modeling is more facile when the learner resembles the mentor. This indicates that underrepresented minority and women physicians may better mentor minority and women learners. When that is not possible, institutions should make an extra effort to embrace diversity and practice in a culturally informed and competent manner (Wright & Carrese, 2003). Although cultural sensitivity training is required at many institutions, the institutions will nevertheless need to monitor outcomes of their practices to determine whether there is gender and racial imbalance, which can foster potential discrimination (McCarren & Goldman, 2012). Simple procedural requirements such as taking an online sensitivity course will not likely yield high returns or intra-institutional cultural changes.

The presence of minority peers and faculty positively impact graduation rates for African American students adding significant value to collective participation in higher education and bolstering the earning of degrees (Stewart, 2012). To address pipeline challenges, early educational initiatives will need to be strengthened and enhanced. Exposing young people to careers in science, technology, engineering, mathematics and health care early at the onset of education will result in a more robust pipeline of future medical practitioners and researchers. There is strong evidence that racial diversity also enhances the experience of majority students at the post-secondary level. Survey data from the Cooperative Institutional Research Program showed that exposure to racial and ethnic diversity and cross-racial interaction for Caucasians (non-Hispanic whites) had such long-lasting effects as enhancing leadership skills and their level of pluralistic orientation. This experience encourages students to acknowledge and respect other races and social classes outside of their own (Jayakumar, 2008).

Medical schools as well as biomedical research departments need to attract and retain a diverse cadre of students as well as educators. They should develop a cross-cultural educational program that responds to an increasingly heterogeneous population within the U.S. In line with attracting and retaining a diverse faculty and student body, medical education should incorporate training in concepts of culture, socioeconomic status, racism, health care access, cultural competence, and enhancing doctor-patient relationships (Peña, Dolhun, Muñoz, & Grumbach, 2003).

In the medical arena, women are especially underrepresented in surgical specialties. When female surgical leaders were asked, perseverance, drive, good communication skills, passion for scholarship, a stable home life, and a positive outlook were identified as being critical attributes for success (Kass, Souba, & Throndyke, 2006). They also reported gender prejudice to be a major obstacle in their careers. Among female surgical leaders, most reported having mentors; nevertheless, greater than 50% acknowledged that they had not been effectively mentored (Kass, Souba, & Throndyke, 2006). Thus, it appears that a crucial component of success for African Americans and women is having effective and strong mentorship.

Sometimes, the traditional mentorship relationship fails. This may occur for a number of reasons, including lack of connection between mentor and mentee. Some academic centers formally assign mentors to junior faculty, while others may leave mentoring relationships to develop more organically by chance. Regardless of how the mentor-mentee relationship develops, it is important to encourage and support effective mentoring relationships for junior minority and female faculty members. In academic settings, physicians have devised creative and innovative techniques that provide needed junior faculty support. At Emory University, for example, an innovative peer mentoring model has been implemented that encompasses a core curriculum (professional development, research development, and teaching development), peer support (feedback, encouragement, and accountability), and group projects and presentations (Bussey-Jones et al., 2006). There are many models in the literature that have been evaluated for effectiveness.

Strategies that some minority women faculty employ to cope include self-silencing, creating microenvironments in which to work, focusing on work-life balance, and assuming the dual identities of succeeding in the current environment while also trying to change the status quo (Pololi & Jones, 2010). In order to address some of these coping strategies, institutions may also want to establish programs to specifically target minorities by providing opportunities for mentorship, networking, sponsorship, and career development. On-site childcare is also a helpful consideration to helping support the success of minority women in medicine and biomedical research (Bauman, Howell, & Villablanca, 2014).

Institutional culture must change in order to allow the growth and success of minorities as well as women. Enhancing the campus environment and learning community with faculty development, mentorship, and networking opportunities will cultivate faculty job satisfaction and help with recruitment, retention, and advancement (Wagner et al., 2014). Effective strategies for reducing and eliminating bias and discrimination against female and African American physicians and researchers include educating leadership about the organizational benefits of gender and racial / ethnic diversity in the workforce and promoting open and ongoing dialogue about gender and racial / ethnic discrimination and the effects on healthcare delivery and health outcomes. In addition, these strategies include encouraging double-blind peer review for publication and grants, creating formalized mentoring programs, and providing greater transparency in hiring practices and pay structures.

Institutional Solutions for Eliminating Race/Gender Bias in Biomedical Research

Ginther, Schaeffer, Schnell, Masimore, Liu, Haak, and Kington, (2011) reported that African American applicants for R01 awards from the NIH were far less likely than Caucasian applicants to receive funding, as shown in Figure 1.

Figure 1. Probability of NIH RO1 awards by race, blacks and whites, Ginther, Schaeffer, Schnell, Masimore, Liu, Haak, and Kington, Science 2011

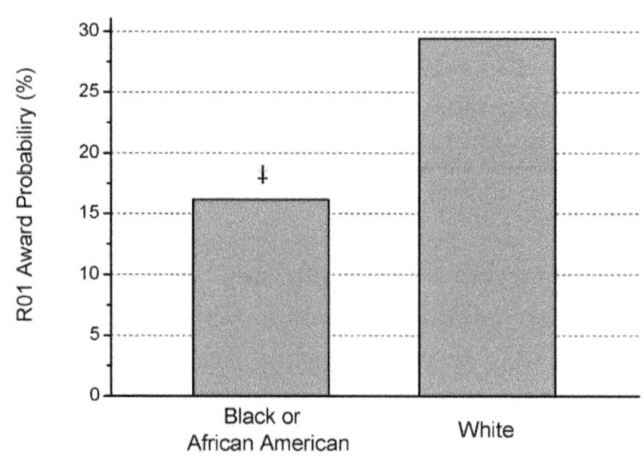

Figure 2. Effects of affiliation and previous research on R01 award probability. Ginther, Schaeffer, Schnell, Masimore, Liu, Haak, and Kington, Science 2011

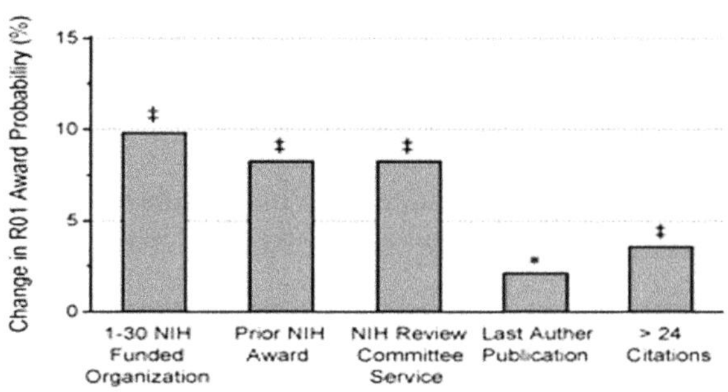

After a complex set of statistical controls, they concluded that even among applicants who had participated in NIH-funded training programs, the race gap persisted. As a result of this research, NIH has developed and tested new approaches to address biomedical workforce diversity by expanding biomedical research and mentoring programs on a nationwide level (Ginther et al., 2011). In 2012, NIH launched a 10 year, $500 million initiative to train and mentor minority scientists. One of the funding mechanisms is called the Building Infrastructure Leading to Diversity (BUILD) program. The program funds experimental training awards that aim to attract and retain students from diverse backgrounds to the biomedical research workforce (Thompson, Pattillo, Stiles, & Schatten, 2014). NIH has also launched a National Research Mentoring Network (NRMN) to train and provide mentors from diverse disciplines for minorities in medicine as well as biomedical research.

In addition to mentoring and faculty development opportunities, the National Institutes of Health (NIH) is also evaluating current processes to better detect instances of racial or gender bias that might occur during the funding process. Figure 2 shows the factors that influence R01 award success, and African Americans are unlikely to be disadvantaged on many of these factors. For example, African Americans are far less likely to serve on NIH review committees as only those who have received past R01s, and who have international reputations are invited to serve. This makes it difficult for African Americans to break through the cumulative racial barrier they face. Publications and citations are a critical predictor of R01 funding, but it is difficult to attain these publications from less resource-rich environments.

All mentoring and faculty development programs will now be evaluated by the Coordination and Evaluation Center (CEC) to clearly determine program effectiveness. Additionally, the NIH Director of the Center for Scientific Review will perform internal exercises, such as de-identifying proposals by race and institution prior to review, to observe the impact on funding scores. Furthermore, NIH will look into whether there are perceptible differences in the way R01 reviewers describe proposals submitted by female scientists and researchers of different races (Kaatz, 2015).

FUTURE RESEARCH DIRECTIONS

There is a critical need to address the culture of institutionalized gender bias and racism, which is inhospitable to the requisite development of underrepresented groups of students, faculty, physicians, and biomedical researchers. This culture of institutionalized racism is a scion of separate and unequal sociopolitical constructs regardless of whether they are unintentional (as may be the case for the Flexner Report) or deliberate and sanctioned as with Jim Crow laws enacted for over 70 years in the South.

The issue of women in medicine and biomedical sciences is more nuanced as women have made noteworthy progress in medicine over the last 40 years. Women have less of an early pipeline issue as they are now entering medical school in equal or even higher numbers than men. However, the challenge remains as to how to translate this progress into meaningful strides in the leadership hierarchy and how to achieve parity in publication opportunities, salary, and research funding.

There is also an urgent charge to diversify the medical and biomedical research training pipeline and to plug leaks in order to address the needs of a rapidly growing diverse U.S. population. While there is a clear understanding of the demographic inequities for African Americans within the medical and biomedical research field, there are also clear inequalities in the demographics of women -- whether African American or not -- in leadership positions in medicine and research arenas. These data need to be better understood so that we can create effective methods that can be enacted to eliminate these disparities. Concerted efforts will need to incorporate early education initiatives all the way through post-secondary education. Colleges and universities will also need to evaluate institutional practices and culture to train and retain underrepresented women and minorities. Obtaining parity in publication opportunities, career, and research funding are good places to start.

CONCLUSION

The challenges faced by African Americans will not be solved in a vacuum. It is in our nation's best interest to foster racial, ethnic, gender, and sexual orientation diversity in the health care workforce as well as biomedical research. The current census data indicates that by 2050, there will be no racial or ethnic majority in the US. In fact, between 2000 and 2050, new immigrants and their children will comprise 83% of the growth in the working-age population (Colby, 2014).

Supreme Court Justice Sandra Day O'Connor (2003), writing the majority opinion in the 2003 Grutter v. Bollinger ruling explained:

Diversity promotes learning outcomes and better prepares students for an increasingly diverse workforce, for society, and for the legal profession. Major American businesses have made clear that the skills needed in today's increasingly global marketplace can only be developed through exposure to widely diverse people, cultures, ideas, and viewpoints (p. 18).

In addition to the ethical goals of promotion of workforce equity and fairness, there are considerable economic benefits to advancing diversity, including increased creativity, innovation and productivity. There is much to be gained in creating an environment that supports the development of research scientists, physicians, as well as leaders from diverse backgrounds and experiences. Enhancing equality for minorities and women can only enhance the fields of medicine and biomedical research, and will foster a stronger culture of innovation and inclusion.

REFERENCES

American Association of Medical Colleges. (2014). *2014 Physician Specialty Data Book*. Author.

American Association of Medical Colleges. (2015). *Altering the course: Black Males in Medicine*. Washington, DC: Association of American Medical Colleges.

American Medical Association. (2006). *Physician Characteristics and Distribution in the U.S., 2006 edition and prior editions*. Retrieved March 8, 2015, from www.ama-assn.org/ama/pub/category/12192.html

Association of American Medical Colleges. (2011). *Table 5. Distribution of U.S. Medical School Faculty by Degree and Race/Hispanic Origin.* Retrieved March, 7, 2015, from https://www.aamc.org/download/271904/data/11table5.pdf

Barkin, S. L., Fuentes-Afflick, E., Brosco, J. P., & Tuchman, A. M. (2010). Unintended consequences of the Flexner report: Women in pediatrics. *Pediatrics*, *126*(6), 1055–1057. doi:10.1542/peds.2010-2050 PMID:21059716

Bauman, M. D., Howell, L. P., & Villablanca, A. C. (2014). The women in medicine and health science program: An innovative initiative to support female faculty at the University of California Davis School of Medicine. *Academic Medicine*, *89*(11), 1462–1466. doi:10.1097/ACM.0000000000000403 PMID:25006704

Beck, A. H. (2004). The Flexner report and the standardization of American medical education. *Journal of the American Medical Association*, *291*(17), 2139–2140. doi:10.1001/jama.291.17.2139 PMID:15126445

Bhattacharjee, Y. (2007). Academic tenure. MIT colleague quits to protest Sherley dismissal. *Science*, *316*(5830), 1405. doi:10.1126/science.316.5830.1405 PMID:17556555

Burgess, D. J., Joseph, A., van Ryn, M., & Carnes, M. (2012). Does stereotype threat affect women in academic medicine? *Academic Medicine*, *87*(4), 506–512. doi:10.1097/ACM.0b013e318248f718 PMID:22361794

Bussey-Jones, J., Bernstein, L., Higgins, S., Malebranche, D., Paranjape, A., Genao, I., & Branch, W. et al. (2006). Repaving the road to academic success: The IMeRGE approach to peer mentoring. *Academic Medicine*, *81*(7), 674–679. doi:10.1097/01.ACM.0000232425.27041.88 PMID:16799297

Carr, P.L., Gunn, C.M., Kaplan, S.A., Raj, A., & Freund, K.M. (2015). *Inadequate progress for women in academic medicine: findings from the National Faculty Study*. Academic Press.

Colby, S. L., & Ortman. (2014). *Projections of the size and composition of the U.S. population: 2014 to 2060*. Current Population Reports.

Colleges, Association of American Medical. (2015). *Altering the Course: Black Males in Medicine.* Washington, DC: Association of American Medical Colleges.

Cook, L. (Producer). (2015a, August 30). U.S. Education: Still Separate and Unequal. *Data Mine Numbers You Can Use*. Retrieved from http://www.usnews.com/news/blogs/data-mine/2015/01/28/us-education-still-separate-and-unequal

Cook, L. (Producer). (2015b, August 30). Why Black Americans Die Younger. *Data Mine Numbers You Can Use*. Retrieved from http://www.usnews.com/news/blogs/data-mine/2015/01/05/black-americans-have-fewer-years-to-live-heres-why?int=a29308&int=a29308

Cooper-Patrick, L., Gallo, J. J., Gonzales, J. J., Vu, H. T., Powe, N. R., Nelson, C., & Ford, D. E. (1999). Race, gender, and partnership in the patient-physician relationship. *Journal of the American Medical Association*, *282*(6), 583–589. doi:10.1001/jama.282.6.583 PMID:10450723

Correll, S. J., Benard, S., & Paik, I. (2007). Getting a job: Is there a motherhood penalty? 1. *American Journal of Sociology*, *112*(5), 1297–1339. doi:10.1086/511799

Cressey, D. (2014). Journals weigh up double-blind peer review. *NATNews*.

Cuyjet, M. (2009). Invisible men-almost: The diminution of African American males in higher education. Emerald Group Publishing.

DiPrete, T. A., & Eirich, G. M. (2006). Cumulative Advantage as a Mechanism for Inequality: A Review of Theoretical and Empirical Developments. *Annual Review of Sociology*, *32*(1), 271–297. doi:10.1146/annurev.soc.32.061604.123127

Duffy, T. P. (2011). The Flexner Report--100 years later. *The Yale Journal of Biology and Medicine*, *84*(3), 269–276. PMID:21966046

Fang, D., Moy, E., Colburn, L., & Hurley, J. (2000). Racial and ethnic disparities in faculty promotion in academic medicine. *Journal of the American Medical Association*, *284*(9), 1085–1092. doi:10.1001/jama.284.9.1085 PMID:10974686

Flexner, A. (1910). The Flexner report on medical education in the United States and Canada 1910. New York: Carnegie Foundation.

Ginther, D. K., Schaffer, W. T., Schnell, J., Masimore, B., Liu, F., Haak, L. L., & Kington, R. (2011). Race, ethnicity, and NIH research awards. *Science*, *333*(6045), 1015–1019. doi:10.1126/science.1196783 PMID:21852498

Hiatt, M. D., & Stockton, C. G. (2003). The impact of the Flexner Report on the fate of medical schools in North America after 1909. *Journal of American Physicians and Surgeons*, *8*(2), 37–40.

Jagsi, R., Guancial, E. A., Worobey, C. C., Henault, L. E., Chang, Y., Starr, R., & Hylek, E. M. et al. (2006). The "gender gap" in authorship of academic medical literature—a 35-year perspective. *The New England Journal of Medicine*, *355*(3), 281–287. doi:10.1056/NEJMsa053910 PMID:16855268

Jayakumar, U. M. (2008). Can higher education meet the needs of an increasingly diverse and global society? Campus Diversity and Cross-Cultural Workforce Competencies. *Harvard Educational Review*, 615-651.

Jeffe, D. B., Yan, Y., & Andriole, D. A. (2012). Do research activities during college, medical school, and residency mediate racial/ethnic disparities in full-time faculty appointments at U.S. Medical schools? *Academic Medicine*, *87*(11), 1582–1593. doi:10.1097/ACM.0b013e31826e3297 PMID:23018339

Kaatz, A., Gutierrez, B., & Carnes, M. (2014). Threats to objectivity in peer review: The case of gender. *Trends in Pharmacological Sciences*, *35*(8), 371–373. doi:10.1016/j.tips.2014.06.005 PMID:25086743

Kass, R. B., Souba, W. W., & Thorndyke, L. E. (2006). Challenges confronting female surgical leaders: Overcoming the barriers. *The Journal of Surgical Research*, *132*(2), 179–187. doi:10.1016/j.jss.2006.02.009 PMID:16564542

Lautenberger, D., Raezer, C., & Bunton, S. (2015). The Underrepresentation of Women in Leadership Positions at U.S. Medical Schools. *Analysis in Brief*, *15*(2), 2.

Mahoney, M. R., Wilson, E., Odom, K. L., Flowers, L., & Adler, S. R. (2008). Minority faculty voices on diversity in academic medicine: Perspectives from one school. *Academic Medicine*, *83*(8), 781–786. doi:10.1097/ACM.0b013e31817ec002 PMID:18667896

Mayer, A., Blair, J., Ko, M. G., Hayes, S. N., Chang, Y.-H. H., Caubet, S. L., & Files, J. A. (2014). Gender distribution of U.S. medical school faculty by academic track type. *Academic Medicine*, *89*(2), 312–317. doi:10.1097/ACM.0000000000000089 PMID:24362384

McCarren, M., & Goldman, S. (2012). Research leadership and investigators: Gender distribution in the federal government. *The American Journal of Medicine*, *125*(8), 811–816. doi:10.1016/j.amjmed.2012.03.006 PMID:22579138

Merchant, J. L., & Omary, M. B. (2010). Underrepresentation of underrepresented minorities in academic medicine: the need to enhance the pipeline and the pipe. *Gastroenterology, 138*(1), 19-26 e11-13. doi: 10.1053/j.gastro.2009.11.017

Nelson, D. (2007). *A National analysis of minorities in science and engineering faculties at research universities*. Retrieved from http://faculty-staff.ou.edu/N/Donna.J.Nelson-1/diversity/Faculty_Tables_FY07/07Report.pdf

Newman, L. A., Pollock, R. E., & Johnson-Thompson, M. C. (2003). Increasing the pool of academically oriented African-American medical and surgical oncologists. *Cancer*, *97*(1Suppl), 329–334. doi:10.1002/cncr.11027 PMID:12491497

Nosek, B. A., Smyth, F. L., Sriram, N., Lindner, N. M., Devos, T., Ayala, A., & Greenwald, A. G. et al. (2009). National differences in gender-science sterotypes predict national sex differences in science and math achievement. *Proceedings of the National Academy of Sicneces*, *106*(26), 10593–10597. doi:10.1073/pnas.0809921106 PMID:19549876

Nunez-Smith, M., Ciarleglio, M. M., Sandoval-Schaefer, T., Elumn, J., Castillo-Page, L., Peduzzi, P., & Bradley, E. H. (2012). Institutional variation in the promotion of racial/ethnic minority faculty at US medical schools. *American Journal of Public Health*, *102*(5), 852–858. doi:10.2105/AJPH.2011.300552 PMID:22420820

Peña Dolhun, E., Muñoz, C., & Grumbach, K. (2003). Cross-cultural education in U.S. medical schools: Development of an assessment tool. *Academic Medicine*, *78*(6), 615–622. doi:10.1097/00001888-200306000-00012 PMID:12805042

Pololi, L. H., Civian, J. T., Brennan, R. T., Dottolo, A. L., & Krupat, E. (2013). Experiencing the culture of academic medicine: Gender matters, a national study. *Journal of General Internal Medicine*, *28*(2), 201–207. doi:10.1007/s11606-012-2207-1 PMID:22936291

Pololi, L. H., & Jones, S. J. (2010). Women faculty: An analysis of their experiences in academic medicine and their coping strategies. *Gender Medicine*, *7*(5), 438–450. doi:10.1016/j.genm.2010.09.006 PMID:21056870

Rodriguez, J. E., Campbell, K. M., & Pololi, L. H. (2015). Addressing disparities in academic medicine: What of the minority tax? *BMC Medical Education*, *15*(1), 6. doi:10.1186/s12909-015-0290-9 PMID:25638211

Savitt, T. (2006). Abraham Flexner and the black medical schools. 1992. *Journal of the National Medical Association*, *98*(9), 1415. PMID:17019906

Sheltzer, J. M., & Smith, J. C. (2014). Elite male faculty in the life sciences employ fewer women. *Proceedings of the National Academy of Sciences of the United States of America*, *111*(28), 10107–10112. doi:10.1073/pnas.1403334111 PMID:24982167

Sheltzer, J. M., & Smith, J. C. (2014). Elite male faculty in the life sciences employ fewer women. *Proceedings of the National Academy of Sciences of the United States of America*, *111*(28), 10107–10112. doi:10.1073/pnas.1403334111 PMID:24982167

Soper, S. C. (2015, June 18). What It's Like as a "Girl" in the Lab, Op-Ed. *The New York Times*. Retrieved from http://www.nytimes.com/2015/06/18/opinion/what-its-like-as-a-girl-in-the-lab.html?_r=0

Steinecke, A., & Terrell, C. (2010). Progress for whose future? The impact of the Flexner Report on medical education for racial and ethnic minority physicians in the United States. *Academic Medicine*, *85*(2), 236–245. doi:10.1097/ACM.0b013e3181c885be PMID:20107348

Stewart, P. (2012). *After decades, revolving door remains for black, latino scholars in the academy*. Retrieved from Diverse: Issues in Higher Education website: http://diverseeducation.com/article/17227/

Stockton, M. H. C. (2003). The Impact of the Flexner Report on the fate of medical schools in North America After 1909. *Journal of American Physicians and Surgeons*, *8*(2), 4. Retrieved from http://www.jpands.org/vol8no2/hiatt.pdf

Stockton, M., & Hiatt, C. (2003). The Impact of the Flexner Report on the Fate of Medical Schools in North America After 1909. *Journal of American Physicians and Surgeons, 8*(2), 4.

Strelnick, A. H., Lee-Rey, E., Nivet, M., & Soto-Greene, M. L. (2008). Diversity in academic medicine no. 2 history of battles lost and won. *The Mount Sinai Journal of Medicine, New York*, *75*(6), 499–503. doi:10.1002/msj.20080 PMID:19021213

Sullivan, L. W., & Suez Mittman, I. (2010). The state of diversity in the health professions a century after Flexner. *Academic Medicine*, *85*(2), 246–253. doi:10.1097/ACM.0b013e3181c88145 PMID:20107349

Talamantes, E., Mangione, C. M., Gonzalez, K., Jimenez, A., Gonzalez, F., & Moreno, G. (2014). Community college pathways: Improving the U.S. physician workforce pipeline. *Academic Medicine*, *89*(12), 1649–1656. doi:10.1097/ACM.0000000000000438 PMID:25076199

Thompson, W. E., Pattillo, R. A., Stiles, J. K., & Schatten, G. (2014). Biomedical research's unpaid debt: NIH's initiative to support and implement fairer competition for minority students is a welcome step to redress the exploitation of African Americans by science. *EMBO Reports*, *15*(4), 333–337. doi:10.1002/embr.201338274 PMID:24652854

Toldson, I.A., & Cooper, G. (2014). *Historically Black Colleges and Universities Data Dashboard*. Academic Press.

Treadwell, H. M., Braithwaite, R. L., Braithwaite, K., Oliver, D., & Holliday, R. (2009). Leadership development for health researchers at historically Black colleges and universities. *American Journal of Public Health*, *99*(S1), S53–S57. doi:10.2105/AJPH.2008.136069 PMID:19246669

Wagner, J. M., Fleming, A. E., Moynahan, K. F., Keeley, M. G., Bernstein, I. H., & Shochet, R. B. (2014). Benefits to faculty involved in medical school learning communities. *Medical Teacher*, 1–6. doi:10.3109/0142159X.2014.947940 PMID:25159341

Wenneras, C., & Wold, A. (1997). Nepotism and sexism in peer-review. *Nature*, *387*(6631), 341–343. doi:10.1038/387341a0 PMID:9163412

West, J. D., Jacquet, J., King, M. M., Correll, S. J., & Bergstrom, C. T. (2013). The role of gender in scholarly authorship. *PLoS ONE*, *8*(7), e66212. doi:10.1371/journal.pone.0066212 PMID:23894278

Wilcox, C. W., Wells, J. V., Haddad, G., & Wilcox, J. K. (2014). The Changing Democtratic Functions of Historically Black Colleges and Universities. *New Political Science*, *36*(4), 556–572. doi:10.1080/07393148.2014.954793

Wilson, D. E., & Balotin, J. (1999). Has the well run dry? Priming the diversity pump in PhD programs. *Journal of the Association for Academic Minority Physicians*, *10*(2), 27–33. PMID:10826006

Wong, E. Y., Bigby, J., Kleinpeter, M., Mitchell, J., Camacho, D., Dan, A., & Sarto, G. (2001). Promoting the advancement of minority women faculty in academic medicine: The National Centers of Excellence in Women's Health. *Journal of Women's Health & Gender-Based Medicine*, *10*(6), 541–550. doi:10.1089/15246090152543120 PMID:11559451

Wright, S. M., & Carrese, J. A. (2003). Serving as a physician role model for a diverse population of medical learners. *Academic Medicine*, *78*(6), 623–628. doi:10.1097/00001888-200306000-00013 PMID:12805043

KEY TERMS AND DEFINITIONS

Disparities: Inequalities that exist when members of certain population groups do not benefit from the same status as other groups.

Diversity: The characteristics that make one individual or group different from another, including race, ethnicity, gender, national origin, religion, disability, sexual orientation, socioeconomic status, education, marital status, language, and physical appearance. It also encompasses different ideas, perspectives, and values.

Gender Bias: Unequal treatment in employment opportunity (such as promotion, pay, benefits, and privileges), and expectations due to attitudes based on the sex of an employee or group of employees.

Flexner Report: A study of medical education in the United States and Canada, written by Abraham Flexner and published in 1910 which recommended changing admission and graduation standards, and contributed to low rates of underrepresented minority and women physicians in the United States.

HBCU: Historically Black colleges and universities

Implicit Bias: A positive or negative mental attitude towards a person, thing, or group that a person holds at an unconscious level, also known as implicit social cognition.

Inequalities: Differences in status or the distribution of determinants between different population groups.

Mentor: A more experienced, trusted adviser or counselor who offers helpful guidance to less experienced colleagues.

Talented Tenth: W.E.B. DuBois approach to minority education by focusing on elite students who could then help educate the rest of the population.

Theory of Cumulative Disadvantage: Consistent small differences in treatment can cumulate to substantial disadvantages in chances of success.

Pipeline Leak: Attrition of women and minorities at higher levels of achievement despite parity at lower levels

R01 Award: Highest NIH award for independent investigators

STEM: Science, technology, engineering and mathematics

Underrepresented Minorities in Medicine: Those racial and ethnic populations that are underrepresented in the medical profession relative to their numbers in the general population.

ENDNOTE

1 Referred to as the framework throughout the remainder of the text

Chapter 7
African American Women in STEM Education:
The Cycle of Microaggressions from P-12 Classrooms to Higher Education and Back

Susan Ferguson Martin
University of South Alabama, USA

Andre Green
University of South Alabama, USA

Melissa Dean
University of South Alabama, USA

ABSTRACT

This project examines the critical issues of stereotypes and gender bias that face students, particularly women, of color in higher education and into their careers, as well as the lack of representation of women of color in higher education STEM related disciplines. Interviews with three, African American females from their start as undergraduate students in Biology, into graduate programs, and eventually with careers as science educators in public science education classrooms, as well as a review of current literature, highlights the need for women of color within STEM disciplines, as well as strides that have been made toward increasing the number of both females and males of color in STEM related areas.

INTRODUCTION

Each year positions in science classrooms go unfilled, particularly in rural and urban areas where teacher staffing is, as in many content areas, inadequate. Often these positions are filled with those who are underqualified teaching outside their area of expertise. Many are also staffed with those who have limited formal training as educators and are serving as long term substitute teachers with no background in STEM disciplines. The deficit is so great that both President Obama and former President George

DOI: 10.4018/978-1-5225-0174-9.ch007

W. Bush created budget lines and initiatives to increase the number of new science and math teachers. Because of this shortage, the past two decades have seen a decline in the global position of the United States in terms of STEM education. In order to meet the need for high quality educators, the interest must be initiated at the elementary level, kindled in the middle grades, and reinforced in secondary education. Relevant also to this need is the diversity of people pursuing science degrees. However, as Ginther and Kahn found in a 2013 longitudinal study of NSF and related data, Women of Color are less likely than white women to graduate from college and to enter tenure track positions. With this trend in the United States, an increase in minority representation in STEM professions must take place, and that means P-12 classroom promotion of STEM disciplines to minority student populations by minority role models.

Noyce Pathway to Science (PTS) is a collaborative program between a local university's College of Education, College of Arts & Sciences, and a public school system. Pathway to Science addresses the need to not only increase the number of science teachers in the school system through enabling recent science bachelor's degree graduates to complete secondary science certification in an intensive four-semester program that culminates with certification and an earned master's degree, but it also recruits scholars from minority backgrounds. To support the success of these PTS Noyce Scholars, support is provided throughout the graduate program and after graduation. Since the program's inception in 2012, 21 students have completed the program, the goals of which are to put more qualified, certified science teachers in the classroom and to provide an alternative career path for those desiring to meet the critical need for dedicated educational professionals.

Three of the initial PTS scholars are African American females, all undergraduate Biology majors, who were recruited to the alternative graduate program in General Science Education at a Southeastern public institution. all share common threads with faculty encounters, peer bias, goals for future development and advancement and roadblocks, and outlooks for career success. And while they may find themselves situated within a female heavy profession as secondary educators, their academic peers in science disciplines sometimes suggest that their knowledge of science is best relegated to the confines of the middle school and high school classroom—a gender bias that remains prevalent even in STEM disciplines. Also considered are the choices and experiences that brought the three educators to the point at which they find themselves and how creating similar experiences for students in P-12 education and in the community at large might be viable options for encouraging more African American youth to consider STEM professions. Although they all report a generally supportive peer group within their graduate programs at their particular university, as of the 2013-2014 *University Fact Book* data, only 40 of 774 rank holding faculty identified as African American. That's right around a mere 5% of total rank holding faculty. Only 11 faculty identified as African American of the almost 300 faculty in Arts and Sciences and Engineering combined. Therefore while the University is situated in a geographical area in which African Americans account for over 35% of the population, and where of the 15,065 students enrolled, 3,019 identified as African American, faculty representation is low and has remained consistently low for more than a decade.

Pittman (2012) defines this low employment of full-time African American faculty as contributing to racial microaggressions, or subtle and sometimes more overt reactions to a person due to their race or origin. Examples might include students of color being passed over excluded from class discussions or the lowered expectations from faculty for their non-White students. Sue, Capodilupo, Torino, Bucceri, Holder, Nadal, and Esquilin (2007), explained the differences in outright racial slurs and more subtle forms of microaggression, although they admit the lines can be blurred at times. Essentially the accumulation of microaggressive behaviors takes an emotional toll on recipients. As Soloranzo, Cejo, and

Yosa (2000) emphasize, "without documentation and analysis to better understand microaggressions, the threats that they pose and the assaults that they justify can be easily ignored or downplayed" (p. 72).

VOICES OF EXPERIENCE

Sharing their experiences from an early age, all the way up to the present where they are in their own classrooms teaching science courses, they have common threads that run through their paths. For the experiences that differ vastly among them, they have explanations and postulations as to why those particular experiences deviate. In particular, they discuss their early education, their college careers, and their experiences as teachers in the secondary science classrooms.

Tonia grew up in an area that remains one of the most impoverished areas of the state. A former port for the African Slave Trade in the Southeast, the area is now home to the descendants of formerly enslaved persons—many of whom purchased or worked through share-cropping systems, to attain their homes and land from white landowners. Though the area still carries historical significance within the region, issues of environmental protection and access to fair education still weigh heavily on the area as one that continues to lack its rightful voice among local stakeholders. Those from the community who can get into magnet schools due to adequate grades tend to do so, while still carrying some of the baggage that goes along with being part of a community isolated from the larger part of the county by a bridge. Tonia discusses some of this disadvantage in her analysis of her experiences. While she currently teaches in a rural school that mirrors the socioeconomics of her own neighborhood, the school is predominantly white. She attributes much of what she currently experiences as a teacher to prejudices toward her being a female in science education, and the disbelief by the community that she has credibility in what she teaches.

Janice, who attended the same magnet school as Tonia, has a different experience. Being a triplet, Janice saw the experiences of her siblings as well as those around her, learning about equity and access within a race. Although she attended an HBCU for her undergraduate degree, she recalls limited prejudices against her in P-12 school. Some of this she attributes to her lighter skin color, which she discusses at length throughout her discourse. While she and Tonia attended the same magnet middle school, attitudes toward her were similar to those toward white students. Her graduate education, she reports, was fairly equitable, though she did experience less assistance and personalization from faculty than she did at her HBCU during her undergraduate degree. She currently teaches in a magnet middle school, but she hopes to soon be in a school where students need her more academically and are less independent, or, in her words, a school with "children who need a role model or a cheerleader."

Jenny grew up in a different state within the southeast. She recalls being part of a diverse community with "Black, White, and Asian friends," and does not feel that her race had an overtly adverse affect on her education. While Jenny feels as though her college experiences were equitable, her many experiences with fellow P-12 faculty and students have shown her how microaggressions can impact students in terms of confidence and with regards to faculty perceptions of student abilities. Jenny currently teaches in an urban school that has an African American student body of over 99%. The school where she teaches was one where she witnessed microaggressions toward students during her internship experience. In fact, much of her impetus for working in the school now is to change the climate for those students so that they have a different experience going forth than they had in the past.

Growing up African American Female in the South

The three science teachers began by discussing their early experiences. While each had unique stories, they were more than familiar with the experiences of their peers, and while they didn't all suffer from overt racism, they certainly witnessed microaggressions on their peers. In fact, Tonia and Janice attended the same magnet middle school, yet they recall their middle grades years as vastly different. While Janice felt the support and encouragement of her peers and by her teachers, Tonia had quite a different experience.

Tonia, who grew up in a part of the community known for the highest level of poverty, reports overtly racist comments, as well as an expectation by administration at the middle school that she would be an unsuccessful student. Tonia recalls,

The principal at my middle school at the time, she really didn't like black people. I mean she didn't say it and put out their saying "oh I don't like black people," but her actions spoke really loudly. Any type of discipline issue—she literally took it to the severe, like suspension type severe. She changed the curriculum, and I don't think she even sent out an email or letter to the parents letting them know. So when the black kids failed, guess what—they automatically got sent out of the school since it was a magnet school. I mean she tried her best to weed us out—uniform policy, discipline areas—any type of things to kick you out; she would try to get rid of you. I don't even think she even knew any of our names. She didn't try to speak to us. She wasn't trying to be friendly or open or anything like that. And that was like my real first encounter in dealing with racism—period.

Janice tells a slightly different story. She recalls,

I feel like my experience growing up was a bit different from Tonia's just because my elementary school was majority white and the magnet middle school was—it should have been 50/50 but I think it had more whites at that time. And so I never really experienced any racism but I can remember my mom saying, "Oh they just take to you and your sister because ya'll are light skinned." And my brother though, he's darker than I am and he had a difficult time there. He really did, and my mom again would say, "That's because he's darker than you all." To this day even if a white person comes up to me and my sister and talks to us or says, "Oh you guys are pretty," she'll say, "They're just saying that because of your skin color."

One of the researchers asked her to elaborate on this sentiment and its implications, to which Janice replied,

She says that and—I—she's—I think she has a complex. I know her mother had it—a complex with the paper bag test. You had to pass a paper bag test. If you were the color of a paper bag or lighter, you were accepted in society. If you were darker than a paper bag, you were not accepted. And she says that most white people take to me and my sister because we're lighter. And—I kind of think—I know—that's true. If anything I've had blacks to mistreat me because they see me and say, "Oh she's light skinned; she's stuck up." You know I've had black females not even get to know me but dislike me. Even the students do it, when I walk in the classroom like in the first couple of days in school, they do it. My students will say, after about the first quarter, they'll come up to me and say, "We thought you were stuck up," or "We thought you weren't approachable." When I ask, "What gave you that impression?" and they respond, "Just the way you look," I know instantly what they're referring to.

While both Tonia and Janice attended different high schools after leaving the same magnet middle school, they both report limited rigor in their high school experience. Janice went on to a magnet high school, whereas Tonia went to the high school in her neighborhood. For different reasons, they both felt their experiences were subpar in preparing them for college. Janice felt that most of what she did in high school was "a joke," while Tonia reported an attitude of status quo maintenance recounting,

My middle school experience was rough, and then when I got to high school I was back in the hood, back in the urban area but I realized as I ventured out . . .that some schools had different programs than others. And I was kind of upset because I'm like, "Why are the black kids not being prepared for college like some of the majority white schools are?" Like you can go out to (other local high schools) and they have ACT prep classes starting in the 10th grade. Why didn't we have that? They didn't tell us about the ACT until our senior year. They were just too busy trying to push us to pass the graduation exam. I did that my 10th grade year. I was done. And it was just us getting prepared for college. We didn't have college prep classes. The only one we had was chemistry. The black kids are not being prepared for school like the majority of the white kids are. So I'm like you're telling me that in order for my child to get a good decent education we have to go way out of our school district in order for them to be prepared? So it was kind of like, to me it's even racism even within the school zones or areas or whatever and I just didn't, like, understand that and I think that's what kind of inspired me to be a teacher. Because I'm like, I want it to be some type of equal playing field for everyone because everyone deserves to go to college, even if you feel like because this kid is in the majority group in this school, there's someone out there who wants to go to a four year university and make it.

Janice's story differs since she went to a magnet high school, yet still she recalls,

It was labeled "magnet school," but the teachers didn't care, white and black. You could pretty much do whatever. It was a cakewalk. No one was trying to prepare anyone for anything. I'm not going to say I wasn't prepared, because my parents made sure that we took those enrichment classes throughout the year and throughout the summer so I was prepared for college, and I got a scholarship to go to college but it wasn't because of my high school. Those teachers, they just did not care; they did not care. The first period—I jokingly kid; the first period was spades. That was my first period class, spades—everybody played spades first period, and we were supposed to be in social studies or math. There was one really good teacher, my sophomore year—a white lady—and she was good; she was awesome, so I can appreciate her, but all of the other teachers, they just didn't care.

Jenny tells a different story. She felt that pressure from educators, while sometimes stiff, inspired and encouraged her to achieve, as did her parents' expectations.

Well, growing up I had both my parents in the home. My mom was actually the breadwinner. I went to a culturally diverse school, so when I got to the school where I currently teach it was culture shock for me because I wasn't very familiar with being in the majority African American environment. So, I'm like "Where are the Asian people?" "Does anybody have any Caucasian friends, or anything like that? I was just used to having friends from a variety of backgrounds—so that's why for me, I chose South you know because I really didn't think that I would fit in to an HBCU environment because I was never in a majority African American environment. I love HBCUs and what they stand for; I just didn't that was the best choice for me.

In middle school, I struggled in science. I always struggled in science. I just could not really wrap my mind around it and I think that's the reason why once I got to college, it just all clicked. I'm like, "Yes I know this stuff!" And so that is what really pushed me to become a teacher. I remember in middle school my teacher did this thing with lasers—with sound waves. And she hooked a laser up to speakers. And she turned all the lights off, so it was pitch black dark in our classroom. She would play one of the latest songs on the radio and so we had this cool laser light show, and I thought that was just so cool. So at that moment I'm thinking, science is just too cool. Coming into college I just thought it was so neat how the professors knew so much and how they were so knowledgeable and they did all this research. And I had to work for it. It was the one thing that I really wanted to work for. I wanted to push—now my English class, I'm like this is---write a paper, I mean write a paper, yeah you have to write a paper and that's necessary, even in science. But reading those articles, even those boring science journals, I read them. I read them and I went through and I said you know what, this is interesting to me.

However, Jenny recalls a time when she was a preservice teacher in the school where she is currently chair of the science department. She believes that the attitude some students experience from teachers is as much a result of beliefs about socioeconomic factors in learning. She says of an attitude expressed to her from a veteran teacher,

With him, I don't think it's necessarily race. I think it's more so the community—the poverty stricken community. I guess he felt like as far as their learning level's concerned they can't learn that much. It was crazy because he said it to them right to their face. He said, "They're not going to college—none of them—none of them are." I could not believe he had said that to me, and that was another reason why I felt like I had to be there. These kids—they don't know what they want to do yet. I mean I didn't know what I wanted to do. You have to mold them and guide them and push them and give them that drive, you know. And if they don't have that—if they have somebody, an adult saying, "You're not going to college, you're not smart enough"—you're killing the little hope they had. That was the past. Now, I think with our new set of administrators, they're really pushing to bring that school up so I love it—I absolutely love it there.

Once again the microaggressions, perhaps not intended by teachers to demean students, cannot serve to encourage progress and growth, but only to validate an opinion they may already hold of themselves due to the prevalent attitude at their school. Unfortunately such attitudes are not atypical. As Sue, et. al. explains, a facet of microaggression is microinvalidation, where someone dismisses another's feelings as hypersensitive rather than acknowledging them (p. 274).

While they have educational experiences that differ slightly, the three echo sentiments about their interactions with students they now teach and the attitudes of those students toward them. They seem to face some discrimination in the form of doubt—doubt by students that they are knowledgeable about their field. In some cases students go so far as to challenge them on their content knowledge—even doubting their clarification and evidence. Tonia expresses,

"You know, for the most part they are sweet kids, they really are, I just have a few, mainly white male students that challenge everything I say." Janice agrees, "Oh, that's because Daddy—daddy does it at home." Tonia Continues,

I think it's because, yep, daddy does it at home and a black female, she's not supposed to know anything about science, you know. They undermine anything I say about science. And you know I encourage the questions because I like students to ask questions. But I can feel the difference when you're trying to undermine me versus curiosity, and I have one or for sure, who undermine me, and I can pick up that it the behavior at home from dad.

Janice agrees but also expresses,

But you know what, I actually get a lot of that undermining from my black kids, it's sad to say. Like some them try to punk me—they try to ask questions like "why is that or where did you get that from" or "why you know so much or why you use those type of words"—or whatever it is. And so, I had to ask them one day because they felt like they wanted to talk while I was talking. So I asked them nicely, who in here has graduated—I said who in here has left elementary school. of course everybody raised their hands. I said who has completed 6th, 7th, and 8th grade science. I said who here has completed 6th, 7th, and 8th grade math. I said who in here has graduated from high school with an advanced diploma—I said all advanced classes, all advanced classes. I said who in here has completed a degree in biology and did it in four years. I said who in here has completed a masters in a year. None of them could say anything. I said, so while you are trying to talk, thinking that you are better than me, I'm the only one here with the credentials to talk. And they couldn't say anything to me after that and so, you know, they kind of just hushed away. But I mean, I got it a lot from my black kids, with an attitude like "you young, you black, why you trying to teach us something? You can't teach me nothing."

Pittman, 2012, explains this microaggression through the lens of higher education, describing the challenges faced by African American faculty who feel either invisible or as though their credentials are continually challenged. Pittman sites one of the major concerns being excluded from "networks that were crucial for receiving resources such as mentoring and research funds" (84). Several studies also point to documented patterns of racial microaggressions wherein White students question the credibility of African American educators, so the issue is far from isolated (Harlow, 2003; McGowan, 2000; Pittman, 2010). From the perspective of any educator, whether in P-12 or higher education, support from senior faculty, as well as access to resources to remain productive and competitive, are essential to mental wellbeing and career success. Limiting these factors may result in burnout and a limited tenure as an educator.

There is also precedent for harassment by White students (Pope & Joseph 1997). Harassment certainly seems to be at play, as Tonia reports a similar attitude from students, " 'So you think you're white,' that's what I get. It's the attitude, 'You talk like a white person, so you think you're white'."

Janice continues in agreement,

And you think you're white. You think you're better than me so I'm not fixing to listen to you. That's what I get a lot from a lot of—a lot of my black kids—my black girls, especially. It's just so much attitude, but I'm like, you know, like why—why you have so much attitude with me? Like I have nothing against you.

VOICES OF CHANGE

Such experiences are all too typical, and while the three had quite different stories to tell, they were all too familiar with situations akin to those of their peers. Janice recalls witnessing microaggressions toward her classmates who, in her words, had "darker skin and different hair." The three agree that even when they did not personally experience microaggressive behavior in their P-12 and higher education, they have seen students overlooked for everything from assistance that seemed available to other students to accolades for high achievement or work done well.

While there is a growing body of work on the topic of microaggressions, there is a need to personalize incidents of prejudice—to put faces and lives with the situations that oppress individuals that have the same—or higher—qualifications as their White peers so that the need for change becomes more apparent to the mass population. Starting from Pre-K through higher education experiences, the need for educators to recognize their own prejudices is tantamount to effecting such change. Facing ones own prejudices is the first step since many do not realize the harm they are doing because much of it is deep and subconscious, such as beliefs about student abilities due to racial or socioeconomic situations and misconceptions about the household situations and aspirations of students. In *White Teachers, Diverse Classrooms*, Landsman and Lewis address these subtleties through challenging the reader to consider what they believe about student goals and aspirations and whether those differ among cultures, races, and socioeconomic sectors. In their vignettes they challenge teachers of all backgrounds to consider situations that differ from their own, leading the reader through experiences from multiple perspectives—those of both teachers and students. Through heightened awareness of the lasting impact of prejudice on individuals, educators may be able to help stop the cycle of microaggressions.

REFERENCES

Espinosa, L. L. (2011). Pipelines and pathways: Women of color in undergraduate STEM majors and the college experiences that contribute to persistence. *Harvard Educational Review*, *81*(2), 209–240. doi:10.17763/haer.81.2.92315ww157656k3u

Ginther, D. K., & Kahn, S. (2013). Education and academic career outcomes for women of color in science and engineering. In K. Rapporteur (Ed.), *Seeking solutions: Maximizing American talent by advancing women of color in academia: Summary of a conference* (pp. 71–92). National Academy Press, National Academy of Sciences.

Harlow, R. (2003). 'Race doesn't matter, but . . .:' The effect of race on professors' experiences and emotion management in the undergraduate college classroom. *Social Psychology Quarterly*, *66*(4), 348–363. doi:10.2307/1519834

Landsman, J., & Lewis, C. (2011). *White teachers/diverse classrooms: Creating inclusive schools, building on students' diversity, and providing true educational equity*. Sterling, VA: Stylus Publishing.

McGowan, J. M. (2000). Multicultural teaching: African American faculty classroom teaching experiences in predominantly White colleges and universities. *Multicultural Education*, *8*, 19–22.

Pittman, C. T. (2010). Race and gender oppression in the classroom: The experiences of women faculty of color with White male students. *Teaching Sociology*, *38*(3), 183–196. doi:10.1177/0092055X10370120

Pittman, C. T. (2012). Racial microaggressions: The narratives of African American faculty at a predominantly white university. *The Journal of Negro Education*, *81*(1), 82–92. doi:10.7709/jnegroeducation.81.1.0082

Pope, J., & Joseph, J. (1997). Student harassment of female faculty of African American descent in the academy. In L. Benjamin (Ed.), *Black women in the academy: Promises and perils* (pp. 252–260). Gainesville: University of Florida Press.

Soloranzo, D., Ceja, D., & Yosso, T. (2000). Critical race theory, racial microaggressions, and campus racial climate: The experiences of African American college students. *The Journal of Negro Education*, *69*, 60–73.

Sue, D., Capodilupo, C. M., Tonino, G. C., Bucceri, J. M., Holder, A. M. B., Nadal, K. L., & Esquilin, M. (2007). Racial microaggressions in everyday life. *The American Psychologist*, *62*(4), 271–286. doi:10.1037/0003-066X.62.4.271 PMID:17516773

Chapter 8
Are Female STEM Majors Academic Risk Takers?

Tara L. R. Beziat
Auburn University – Montgomery, USA

Brooke A. Burks
Auburn University – Montgomery, USA

Kristin M. McCombs
Wheeling Jesuit University, USA

Jennifer Byrom
Auburn University – Montgomery, USA

ABSTRACT

The existing literature does not focus on risk-taking differences within females and how these differences may influence academic choices in Science, Technology, Engineering, and Mathematics (STEM) fields. The current project examined differences in academic risk-taking between STEM and non-STEM female students. A total of 272 undergraduates from 3 universities in the United States participated in a total of 2 studies. Results from the first and second studies indicated differences between STEM and non-STEM females in academic risk-taking. Future studies should explore these academic risk-taking differences between STEM and non-STEM female students.

INTRODUCTION

As each semester begins, students discuss which courses they are taking and which ones they perceive to be difficult. Students may feel the course will be difficult because of the amount of work expected, the structure of the course, or poor instruction (Carter & Brickhouse, 1989; Ornek, Robinson & Huagan, 2008). For some students, these perceived difficulties might lead them to enroll in different sections or different courses all together. Others are willing to take on the possible challenges of the course and the potential risk of doing poorly in the course or even failing the course. Science, Technology, Engineering and Math (STEM) courses, which are often viewed as difficult or challenging (Carter & Brickhouse, 1989; Ornek, Robinson & Huagan, 2008; Stodolsky, Salk & Glaessner, 1991), have historically been dominated by males. Are females avoiding the potential risks associated with these courses? There is a considerable amount of research on gender and risk-taking (see Byrnes, Miller & Schafer,1999 for a review); however, most studies do not consider how gender differences in risk-taking may influence academic choices in the STEM fields. It is important to explore why females may see these courses as too much of a risk.

DOI: 10.4018/978-1-5225-0174-9.ch008

BACKGROUND

Before examining the relationship between academic risk-taking and female STEM majors, we need to define both risk-taking and academic risk-taking. Slovic (1964) defines risk-taking as a chance of loss. Building on Slovic's definition, Furby and Beyth-Marom (1990) define risk taking as the "action (or inaction) that entails a chance of loss" (p.3). Additionally, Slovic notes that risk is multidimensional in that a variety of factors influence how individuals assess risk. Risk must be measured looking at different dimensions including the probability of the risk, the magnitude (variance) of the risk, and the expected value of the risk. Of particular importance in this definition is the chance of loss weighed against the expected value. When students choose to enroll in certain courses, they may weigh their chances of not doing well and therefore not earning credit against the expected value of completing the course. Also, they may forfeit money and time by enrolling in a course they do not feel they will succeed in or pass. Despite these chances of loss, students may choose to take the course because it is a requirement for their major, which will eventually lead to a job in a lucrative field (e.g., engineering or medicine).

In the academic setting, when someone makes a choice to do a particular activity or take a particular course, he or she may be calculating the chances of earning a good grade, the benefits associated with the task or course, and how it will affect his or her self-esteem. If the student believes there is only a small chance of success despite the value associated with the task or course, he or she may opt not to take the chance. This calculation of risk, which involves educational choices, is academic risk-taking. It is defined as the selection of school-related achievement items or tasks varying in difficulty (Clifford & Chou, 1991). Put differently, a student may select a social sciences course over a physical science course because of the difficulty associated with the latter.

Clifford (1988) developed a measure, the School Failure Tolerance (SFT) scale, to predict students' risk-taking in academic settings. The SFT is comprised of 36 Likert-type questions with a total of three subscales: feelings about failure, actions following failure, and preferred task difficulty. Individuals who score below 3.5 on this scale are categorized as less tolerant of school failure. Clifford (1991) found SFT scores (0 = no tolerance for failure; 6 = tolerance for failure) were lower for students in upper grade levels as compared to elementary age students. According to these studies, elementary students are more likely to take academic risks. As students moved through the school system, it is hypothesized they are less tolerant of failure and, therefore, less willing to take academic risks.

Many U.S. government organizations including the National Academy of Sciences and the U.S. Department of Education are concentrating their efforts and resources on STEM education. These organizations and others are investigating how to improve enrollment and retention in STEM fields. Between 1995-1996, males outnumbered females enrolled in STEM majors two to one. As of 2003-2004, 14% of undergraduates in the United States were enrolled in a STEM major (Chen, 2009). By 2007-2008, females were awarded 57% of bachelor's degrees in (the U.S.), but males still dominated the STEM majors (Aud, Hussar, Planty, Snyder, Bianco, Fox… & Drake, 2010).

Though women do enter STEM fields, they are more likely to switch majors than their male counterparts (Chen, 2013). Chen (2013) notes there are numerous "contextual and climate factors" on college campuses and in future workplaces that might lead to the attrition of females in STEM fields. Some factors associated with such attrition is the lack of female peers, support and role models, and a distain for the competitive environment associated with STEM fields (Chen, 2013; Espinosa, 2011; Blickenstaff, 2005). The lack of females in these areas could lead other females to perceive they are taking a risk by entering STEM areas. The possible risk would be fulfilling the stereotype that women are not as competent as men in math, science, and technology. Individuals are concerned that if they act in a certain way or partake in a certain activity, then they may confirm stereotypes about their subgroup (Aronson, Quinn & Spencer, 1998).

It is possible that females perceive these factors as risks and choose not to pursue STEM degrees. In two studies, Schmader, Forbes, Zhang and Mendes (2009) found that tasks that are viewed as stereotype threat tasks expend a considerable amount of the working memory capacity of the individuals. When women are confronted with stereotype threat tasks, their mental resources are often exhausted (Schmader, Forbes, Zhang, & Mendes, 2009). Women in these studies were using their mental resources to complete the task, monitor their performance on the task, and monitor the possible perceptions of individuals. Metacognition, or knowledge monitoring, becomes an added challenge for those dwelling on negative stereotypes to perform at the levels at which they may actually be capable.

Other plausible theories have been examined to explain the lack of females in STEM fields. Some have hypothesized connections between pursuing such courses or fields and the individual's expectations of potential success or failure (Kost, Pollack, & Finkelstein, 2009). Another view examines how much the individual appreciates the course or field (Young, Fisher, & Vrongistinos, 2009; O'Shea, Heilbronner, & Reis, 2010). The sex differences in STEM and non-STEM majors have also been attributed to different motivational or attitudinal differences (Eccles, Adler, & Meece, 1984). Researchers, though, have not fully examined the impact of academic risk-taking to explain the considerable discrepancy between the number of male and female STEM majors.

Numerous studies have examined the relationship between gender and risk-taking behavior (Charness &Gneezy, 2012; Croson & Gneezy, 2009; Eckel & Grossman, 2008; Byrnes, Miller, & Schafer, 1999). An overview of the research on gender and risk-taking finds that women are more risk averse than men; however, gender differences could be mitigated by the context of the experiment and individual differences such as knowledge, wealth (Eckel & Grossman, 2008; Byrnes, Miller & Schafer, 1999) and profession (Croson & Gneezy, 2009). Though there is extensive research on the differences between genders on risk-taking, the current literature does not examine differences in risk-taking within a gender.

To gain a better understanding of females in STEM, research by O'Shea et al. (2010) examined the characteristics of females who succeeded in STEM areas. Participants were identified as gifted because they scored higher than the 95th percentile on the quantitative section of the SAT. In addition to being identified as gifted, other broad categories emerged from the data. These female students had strong quantitative skills and leadership abilities. Also, their family life supported their math education endeavors. Lastly, this group of female students had social and emotional traits that helped them both in the classroom and beyond. Most of these students (57%) attributed their success to ability. These students also valued math and were persistent when challenged by it.

Research by Piotrowski & Hemasinha (2012) explored career aspirations of undergraduate math majors. The authors noted that while many other studies report an underrepresentation of females in STEM fields, their own sample had a slight majority of female math majors; however, few of these majors planned to enter into math fields full-time upon graduation, with most seeking teaching positions at the high school level.

While females are outnumbering males in college enrollment and degree completion, and while females outperform males in math in elementary and high schools in the United States, females continue to be underrepresented in STEM careers, especially in physical science, engineering, mathematics, and computer science. It seems that females' perceptions about their abilities, confidence, and prior academic preparation all influence this disparity (Perez-Felkner, McDonald, Schneider, & Grogan, 2012).

Moakler & Kim (2014) cover previous research on the underrepresentation of women in STEM fields. Women account for only 24% of STEM jobs in the United States, even though women comprise 48% of the workforce. Prior research described by the authors suggests the lack of women in STEM may be a result of issues with self-image, group identity financial barriers, and influence of mentors. However, when female students are confident and successful with their mathematical attempts, it sets the stage for more involvement in STEM fields in the future.

In the academic setting, the chance of loss could result in a lower grade on the assignment or exposing oneself to others intellectually. Some may choose easier tasks because they think they can obtain a better grade or they want to protect their self-esteem or academic identity and not fulfill a negative stereotype. Though women are taking more math and science courses in high school and entering into STEM degree programs, their attrition rates remain high. As a result of the perceived stereotype threat, women may choose not to engage in certain activities in order to protect their self-esteem. If women perceive the potential risks of not doing well in a course or appearing intellectually inferior, they may choose to avoid the situation. Perception of a stereotype threat, in addition to perceived risks of failure, may influence women's avoidance of certain courses or career paths.

MAIN FOCUS OF THE CHAPTER

Issues, Controversies, Problems

The purpose of this research study is to determine if there is a difference in academic risk-taking between STEM and non-STEM female students. At this time no research could be found that examined academic risk-taking within a particular gender. However, there are numerous studies that examined the relationship between gender and risk-taking behavior. If there is a difference between the academic risk-taking behaviors of those involved in STEM programs and non-STEM programs, this may lead to a better understanding as to why women choose to go into these programs and remain in them. The goal of this study is to see if there is a difference between women who are in STEM programs and those who are not, in their reported academic risk-taking. It is the researchers' belief that STEM women will be more likely to take academic risks.

Methods, Analysis and Results

For the studies described in this chapter, the same procedure was used to collect data. In each of the studies, students enrolled in STEM and non-STEM courses at three universities in the United States were sent an email that contained an invitation to participate in the study including information on how to access it via the Internet. Approval was obtained from each of the participating universities' Institutional Review Boards. Participation in the study was voluntary.

For the following study, a quantitative survey-based design was used, and an online survey was used to collect data for this study. Participants were asked to answer demographic questions, such as, "What is the concentration of your major?" In the first study they were asked to complete the Clifford School Failure Tolerance (SFT) scale (Clifford, 1988). Currently, the SFT is the only known measure of academic risk-taking. The measure contains three subscales: feeling about failure, action following failure, and preferred task difficulty. The reliability coefficients for the overall measure and for the subscale are

SFT, 0.88; feeling about failure, 0.72; action following failure, 0.86; and preferred task difficulty, 0.82. In the second study, students completed the SFT and the 12-item GRIT scale (Duckworth, Peterson, Matthews, & Kelly, 2007). Duckworth and colleagues (2007) created a 12-point question survey to specifically measure an individual's perseverance. In a series of studies, they found gritty – those who had high levels of perseverance – individuals were more likely to pursue and attain higher levels of education and made fewer changes career wise. They noted that grittier individuals pursue their long-term goals despite obstacles. Descriptions of the individual studies are provided below.

In the first study, the sample included 117 students who were enrolled at a midwestern university and enrolled in a variety of majors. Sixty of the participants were female and 57 were male, with 66 students classified as STEM and 51 identified as non-STEM majors. Further analysis showed that of the 60 females, only 20 were enrolled in a STEM program, whereas 46 of the 57 males were enrolled in a STEM program. Additionally, 90.6% of the participants were non-Hispanic white, and the majority of the participants (49.6%) were sophomores.

Alpha reliability coefficients were again conducted for this study. The overall SFT had alpha score of 0.77; for feelings about failure, 0.85; preferred task difficulty, 0.86; and for action following failure, 0.54. Again, there were unequal group sizes and a Levene's test was conducted and the assumption of equal variance was met for each group. Analyses of variance were conducted to determine if there were any differences between the groups. When comparing males to females on the SFT, only the subscale of Preferred Task Difficulty had a significant result ($F\,(1,115) = 4.67, p = .033$). Next, males were eliminated from the sample and STEM females were compared to non-STEM females. Two of the three subscales, preferred task difficulty, and action following failure were significant (See Table 1.)

This study found differences between males and females as well as female STEM majors and female non-STEM majors. As in previous studies (Beziat, Issacson, & Was, 2010), the subscale of preferred task difficulty was significantly different for the female STEM majors and female non-STEM majors as well as between males and females. It appears that males and those enrolled in STEM programs prefer more difficult tasks.

In the second study, the researchers administered the GRIT survey in addition to the SFT. In the first study, females differed on the Preferred Task Difficulty scale, which meant that females enrolled in STEM majors preferred more difficult tasks. This could be viewed as a measure of perseverance or GRIT. For the current experiment, Chronbach's alpha for the 12 point GRIT was 0.765, n=117.

The sample for the second study included 155 students who were enrolled at either a midwestern university or a southeastern university and enrolled in a variety of majors. A total of 117 females completed the study, and 68 were enrolled in STEM majors.

Analyses of variance were conducted to determine if there were differences between the female STEM majors and non-STEM majors on the SFT and GRIT scales. There were no significant differ-

Table 1. Differences between female non-STEM majors and STEM majors in academic risk-taking

	F	*p*
Clifford SFT	6.10	.016
Action Following Failure	6.46	.014
Preferred Task Difficulty	.042	.838
Feeling about Failure	6.83	.011

ences between the groups on the GRIT measure. There were significant differences between the groups on the SFT Total (F (1,115) =10.31, p=.002), the SFT affect subscale (F (1,115) =9.26, p=.003), and the SFT difficulty subscale (F (1,115) =8.41, p=.004).

This experiment confirmed that female STEM majors and non-STEM majors differ in the area of academic risk taking; specifically, STEM majors are willing to take on difficult assignments. Interestingly, there were no differences between the groups on the GRIT measure, and both groups showed average "grittiness" (M=3.38, SD=.522, on a 5 point scale). Though STEM majors were more likely to take on difficult tasks, this did not translate into a long-term perseverance, as measured by the GRIT scale.

SOLUTIONS AND RECOMMENDATIONS

The results of these studies indicate there are differences between STEM majors and non-STEM majors in academic risk-taking. Individuals who identify as being in a STEM major are more likely to take on difficult tasks, despite the risks of failure. Both males and females in STEM majors show signs of taking on academic risks. Additionally, the results of this research showed differences within females on academic risk-taking. Females who identified as STEM majors were more likely to take academic risks than those who identified as non-STEM majors. A body of research shows females are less likely to take risks than males. A fear of failure is a plausible reason for the lack of females in STEM majors.

As Ceci and Williams (2010) describe, despite increased attention addressing issues of gender inequality in STEM fields and remarkable gains in such gaps, there is still limited representation of females in certain fields, such as computer science, chemistry, and engineering. Women are making gains in some of the STEM fields, such as biological sciences, but in other areas their numbers remain low. Of concern is that women are taking courses in high school and college in STEM fields but are not selecting career paths in those fields. Evidence suggests that attrition for women in STEM fields is greater as compared to attrition for males. As women move along in their STEM courses, they may see fewer and fewer females in the courses. Also, the professors are less likely to be female.

Similarly, in considering the aspects of stereotype threat, females may simply be in the mindset that they cannot perform as effectively in the STEM fields as their male counterparts. Research by Stout, Dasgupta, Hunsinger, and McManus (2011) describes how stereotype inoculation was effective in bolstering females' perceptions and performances in STEM environments in domain identification. That is, if females were exposed to a same-sex expert in their STEM field, they had greater self-efficacy and a better self-concept. Depending on how familiar an individual is with a certain situation or topic, he or she may perceive something as more or less risky (Figner, 2011). If females work with other females in STEM fields, they may perceive the field to be less risky for females. The ongoing challenge, however, is increasing the number of female role models in the STEM fields and providing these inoculation experiences in the classroom and career environments.

FUTURE RESEARCH DIRECTIONS

One finding from this study and previous studies shows that task difficulty appears to be an indicator of risk-taking. Those who are inclined to take academic risks are willing to challenge themselves. The STEM majors were more likely to choose more difficult tasks. However, in the second study we found that that STEM majors are not "grittier" than non-STEM majors. Though the immediate risks of academic challenges drive some away, those who accept the challenge are likely to benefit in the long run. Females may reach a point where many of them no longer feel comfortable taking on the immediate risks associated with many of the STEM fields. If educators can work with females at this critical point, it may be possible to encourage them to pursue STEM careers.

Additional research is needed to further explore these gender differences in STEM-related academics and careers. Future research could examine risk-taking in academics in greater detail. In particular, this research could focus on reasons why students avoid new academic risks or challenges. Furthermore, students' behaviors with regard to challenge and acceptance of risk or potential failure could be explored. It is important to understand these topics to help females bridge the gap and to attempt to equalize the playing field in such male-dominated fields.

CONCLUSION

The purpose of this research was to examine differences in academic risk-taking. The goal was to explore if there were differences within the female population that provided some explanation of the lack of females in the STEM fields. Previous research generally examined differences between males and females in relationship to enrollment in STEM fields. Additionally, research in risk-taking focused on differences between males and females. This research has shed light on the value of exploring academic risk-taking within females as it relates to their career choices.

REFERENCES

Aronson, J., Quinn, D. M., & Spencer, S. J. (1998). Stereotype threat and academic underperformance of minorities and women. In J. K. Swim & C. Stangor (Eds.), Prejudice: Target's perspective (pp. 83-103). San Diego, CA: Academic Press. doi:10.1016/B978-012679130-3/50039-9

Aud, S., Hussar, W., Planty, M., Snyder, T., Bianco, K., Fox, M., & Drake, L. (2010). *The Condition of Education 2010 (NCES 2010-028)*. Washington, DC: National Center for Education Statistics, Institute of Education Sciences, U.S. Department of Education.

Beziat, T. L. R., Issacson, R., & Was, C. (2010, October). *Differences within Female Education Majors on Risk Taking*. Paper presented at the meeting of Mid-Western Educational Research Association, Columbus, OH.

Blickenstaff, J. C. (2005). Women and science careers: Leaky pipeline or gender filter? *Gender and Education*, *17*(4), 369–386. doi:10.1080/09540250500145072

Carter, S. C., & Brickhouse, N. W. (1989). What makes chemistry difficult? Alternate perceptions. *Journal of Chemical Education*, *66*(3), 223–225. doi:10.1021/ed066p223

Ceci, S., & Williams, W. (2010). Sex differences in math-intensive fields. *Current Directions in Psychological Science*, *19*(5), 275–279. doi:10.1177/0963721410383241 PMID:21152367

Charness, G., & Gneezy, U. (2012). Strong evidence for gender differences in risk taking. *Journal of Economic Behavior & Organization*, *83*(1), 50–58. doi:10.1016/j.jebo.2011.06.007

Chen, X. (2009). *Students Who Study Science, Technology, Engineering, and Mathematics (STEM) in Postsecondary Education (NCES 2009161)*. Washington, DC: National Center for Education Statistics, Institute of Education Sciences, U.S. Department of Education.

Chen, X. (2013). *STEM Attrition: College Students' Paths Into and Out of STEM Fields (NCES 2014-001)*. Washington, DC: National Center for Education Statistics, Institute of Education Sciences, U.S. Department of Education.

Clifford, M. M. (1988). Failure tolerance and academic risk-taking in ten- to twelve-year-old students. *The British Journal of Educational Psychology*, *58*(1), 15–27. doi:10.1111/j.2044-8279.1988.tb00875.x

Clifford, M. M., & Chou, F. (1991). Effects of payoff and task context on academic risk taking. *Journal of Educational Psychology*, *83*(4), 499–507. doi:10.1037/0022-0663.83.4.499

Duckworth, A. L., Peterson, C., Matthews, M. D., & Kelly, D. R. (2007). Grit: Perseverance and passion for long-term goals. *Journal of Personality and Social Psychology*, *9*(6), 1087–1101. doi:10.1037/0022-3514.92.6.1087 PMID:17547490

Eckel, C. C., & Grossman, P. J. (2008). Differences in the economic decisions of men and women: Experimental evidence. In C. Plott & V. Smith (Eds.), *Handbook of experimental economics results* (Vol. 1, pp. 509–519). New York: Elsevier. doi:10.1016/S1574-0722(07)00057-1

Espinosa, L. L. (2011). Pipelines and pathways: Women of color in undergraduate STEM majors and the college experiences that contribute to persistence. *Harvard Educational Review*, *81*(2), 209–240. doi:10.17763/haer.81.2.92315ww157656k3u

Figner, B., & Weber, E. (2011). Who takes risk when and why? Determinants of risk-taking. *Current Directions in Psychological Science*, *20*(4), 211–216. doi:10.1177/0963721411415790

Furby, L., & Beyth-Marom, R. (1992). Risk taking in adolescence: A decision-making perspective. *Developmental Review*, *12*(1), 1–44. doi:10.1016/0273-2297(92)90002-J

Gysler, M., Kruse, J. B., & Schubert, R. (2002). *Ambiguity and gender differences in financial decision making: An experimental examination of competence and confidence effects*. Working Paper. Center for Economic Research, Swiss Federal Institute of Technology. Retrieved from http://www.cer.ethz.ch/research/wp_02_23_paper.pdf

Kost, L., Pollack, S., & Finkelstein, N. (2009). Unpacking gender differences in students' perceived experiences in introductory physics. *AIP Conference Proceedings*, *1179*(1), 177–180. doi:10.1063/1.3266708

Moakler, M. Jr, & Kim, M. (2014). College major choice in STEM: Revisiting confidence and demographic factors. *The Career Development Quarterly*, *62*(2), 128–142. doi:10.1002/j.2161-0045.2014.00075.x

National Center for Educational Statistics. (2011). *Table 301 Bachelor's degrees conferred by degree-granting institutions, by sex, race/ethnicity, and field of study: 2009-2010* [Data file]. Retrieved from http://nces.ed.gov/programs/digest/d11/tables/dt11_301.asp

O'Shea, M., Heilbronner, N., & Reis, S. (2010). Characteristics of academically talented women who achieve at high levels on a scholastic achievement test-mathematics. *Journal of Advanced Academics*, *21*(2), 234–271. doi:10.1177/1932202X1002100204

Ornek, F., Robinson, W. R., & Haugan, M. P. (2008). What makes physics difficult? *International Journal of Environmental and Science Education*, *3*(1), 30–34.

Perez-Felkner, L., McDonald, S., Schneider, B., & Grogan, E. (2012). Female and male adolescents' subjective orientations to mathematics and the influence of those orientations on postsecondary majors. *Developmental Psychology*, *48*(6), 1658–1673. doi:10.1037/a0027020 PMID:22390658

Piotrowski, C., & Hemasinha, R. (2012). Career aspirations of undergraduate mathematics majors: An exploratory study. *College Student Journal*, *46*(1), 46–50.

Planty, M., Hussar, W., Snyder, T., Kena, G., KewalRamani, A., Kemp, J., . . . Dinkes, R. (2009). The Condition of Education 2009 (NCES 2009-081). National Center for Education Statistics, Institute of Education Sciences, U.S. Department of Education.

Schmader, T., Forbes, C., Zhang, S., & Berry, W. (2009). A metacognitive perspective on the cognitive deficits experienced in intellectually threatening environments. *Personality and Social Psychology Bulletin*, *35*(5), 584–596. doi:10.1177/0146167208330450 PMID:19228598

Schmader, T., Forbes, C. E., Zhang, S., & Mendes, W. B. (2009). A metacognitive perspective on the cognitivedeficits experienced in intellectually threatening environments. *Personality and Social Psychology Bulletin*, *35*(5), 584–596. doi:10.1177/0146167208330450 PMID:19228598

Stodolsky, S. S., Salk, S., & Glaessner, B. (1991). Student views about learning math and social studies. *American Educational Research Journal*, *28*(1), 89–116. doi:10.3102/00028312028001089

Stodolsky, S. S., Salk, S., & Glaessner, B. (1991). Student views about learning math and social studies. *American Educational Research Journal*, *28*(1), 89–116. doi:10.3102/00028312028001089

Stout, J., Dasgupta, N., Hunsinger, M., & McManus, M. (2011). STEMing the tide: Using ingroup experts to inoculate women's self-concept in science, technology, engineering, and mathematics (STEM). *Journal of Personality and Social Psychology*, *100*(2), 255–270. doi:10.1037/a0021385 PMID:21142376

Young, S., Fisher, W., & Vrongistinos, K. (2009). Calibrating a measure of gender differences in motivation for learning technology. *Journal of Instructional Psychology*, *36*(3), 259–272.

Chapter 9
Fortitude:
A Study of African Americans in Surgery in New York City

Robert S. Kurtz
SUNY Downstate Medical Center, USA

ABSTRACT

In New York City, from the 1990s to the present, covert racism is alive and well in the field of medicine and medical education. This racism largely manifested itself as inequitable treatment of illness. The most heavily impacted are African American and Caribbean American females and males. These inequities engendered results such as unwarranted criticism in residency education, forced changes of medical occupations and jobs, and false attributions of behavioral health issues. Combating these challenges requires fortified character armor, seeking percipient well positioned minorities, white and off-whites allies, and a willingness to maintain continued vigilance. With persistence and tenacity, success is possible in terms of protecting minorities both in the educational process, and subsequent medical career.

INTRODUCTION

In this chapter, we will examine, by means of case studies, how women of color in medicine navigate the pathways on the journey from training in residency to fully trained attending physicians, and beyond. The variety of these pathways is the result of differences between requirements in different fields of medicine, individual personality issues, physicians' family, ethnic and cultural backgrounds, and the availability of allies. These allies must support, remove barriers, advocate, encourage, and insofar as possible, protect them from harm. We will examine literature on gender and racial equity issues in medical training. Developing white allies is one method by which racial minorities withstand the trials of medical school and residency education, so we will look at how one might be trained or developed to be an ally and real life experiences of those who are already allies. Five vignettes will be presented which illustrate the vicissitudes and tribulations of African American females in premedical and medical education, as well as in residency and later years as attendings and faculty. Plans for organized institutional support of minority medical students and residents will be discussed, and directions for future research will be examined.

DOI: 10.4018/978-1-5225-0174-9.ch009

BACKGROUND

Medicine like any other social institution is always in large part a manifestation of the culture in which it exists and traditionally racism and sexism have regrettably been endemic to the society in the United States (Loewy, 2008). When examining the complex issues of women of color in medicine, the depth and breadth of the research cut across a number of intersections and weave in and out of continuums from medical education to residency and beyond. The examination begins with a look at the perception of female medical students of color and their perceived gains in gender and racial equality (Heever & Frantz, 2011).

Gender bias has been ingrained in healthcare education, research and clinical training. In 2002, the World Health Organization implemented a gender policy committing itself to advancing gender impartiality and equity in health and to set right health inequities that are a consequence of gender roles and unequal gender-relations in society. Building on earlier international research, the goal of the study was to play a part in the understanding of women's encounters with gender discrimination and inequality while partaking as learners in health education. The research question was whether the medical educational structures could be enabling the inequalities or contributing to the re-establishment of the discrepancies known to be real.

The WHO study included all fourth-year female medical students at a university located in South Africa. Of the overall number of questionnaires disseminated, 48/72 fourth-year female medical students responded, generating a response rate of 68%. Of the respondents, 32% reported having the experience that they were not taken seriously by patients because they were women. In addition, 24% state that they were not taken seriously by their male peers. The findings point toward the inevitable need for supplementary support for women in medicine as well as addressing the gender role supposition apparent in the educational experience through curriculum development.

What does this mean as the field considers the impact on the changing dynamic and composition of medical schools in the western world (Grbic & Brewer, 2012)?

Analysis In Brief (AIB) looks at: (1) the latest patterns in number and proportion of female applicants to medical school by first-time, repeat, and total applicants; (2) the patterns in medical school matriculation; and (3) whether the undergraduate conduit is in some countries, though not all, the reason for stasis in the percentage of female applicants. In certain 3rd world countries the gender obstacles in undergraduate education, especially lack of mentors, role models and support subverts the equalization of gender distribution in medical school.

It is essential to examine specific issues and in some cases women in the field as they are questioning themselves (Hinze, 2004). This study examines the everyday lives of women and men resident physicians to understand the context within which harassment unfolds. The narratives explored in that study reveal how attention is deflected from the problem of sexual harassment through a focus on women's 'sensitivity'. Women resist by refusing to name sexual harassment as problematic, and by defining sexual harassment as 'small stuff' in the context of a rigorous training program. Ultimately, both tactics of resistance fail. Closer examination of the relations shaping everyday actions is key, as is viewing the rigid hierarchy of authority and power in medical training through a gender lens.

As more women enter the field of medicine despite gross underrepresentation, women have a perception that they can serve in an inspirational role to enhance access and visibility (Shankar and Singh, 2012). In Nepal, a developing nation in South Asia, the amount of women medical students has increased considerably. The study was conducted to explore perceived causes for this increase, and to study the impact on teaching-learning activities, medical school framework and probable supposed changes in doctor-patient relationships. Participants felt more female students were taking up medicine in the country because of more colleges opening in the cities and communities making it easier for female students to enroll in the course. The study shows that the participants perceived an increasing number of female medical students may be susceptible to changes in Nepalese society like increasing education and socioeconomic advancement. The students and the respondents were predominantly from the urban, educated sector of Nepalese society. They felt there may be opportunities for teaching-learning activities and for medical schools and the doctor-patient relationship to become more patient-centered. Participants were of the attitude women may be less likely to work in rural areas and more likely to work part-time and in an adaptable manner.

This need to serve and enlighten is dampened and hampered by inequities of the macro level and micro level sort (Beagan, 2001). Based on the survey of third-year student interviews and faculty administrator interviews at one Canadian medical school this research article goes further than the identification of unconcealed forms of bias to examine micro level interactional practices of inclusiveness and omission that cumulatively convey messages about who does and who does not belong in medical school. These micro inequities felt in everyday life, construct an institutional climate that may marginalize and distance some students reproducing an inappropriate pecking order notwithstanding the institutions expressed obligation to prescribed parity.

To further undergird this pattern, physicians of color, as women, are disproportionately subjected to these inequities (Liebschultz, et al, 2006). A study was conducted to describe black residents' perception of the impact of race on medical training. There were nineteen participants, ten of whom were male, distributed evenly among medical and surgical fields. Four key themes surface from the narratives: discrimination; different expectations; social isolation; and consequences. Participants' awareness of being a highly discernible minority permeated each theme. Participants perceive blacks to be punished more harshly for the same offense and expected to perform at lower levels than white residents. Participant suspicion of racism as an incentive for individual and institutional behaviors was tempered by self-doubt. Black physicians face complex social and emotional trials during postgraduate training, and deep frustrations and anxieties, not present in their white peers, from constantly having to perceive their real selves and yet act to accommodate or counter racist stereotypes about themselves.

These inequities exacerbate the confidence level and perception thereof in the environment and minds of medical students. Blanch, Hall, Roter and Frankel (2008) sought to review gender differences and issues of self-confidence of medical students; as well as present original research on observation of medical students' confidence. Female medical students were viewed as significantly less confident than male students especially at the beginning of the interaction. Test data indicates that despite performing equal to the male peers female medical students consistently reported decreased self-confidence and increased anxiety particularly over issues related to their competence. In a standardized patient interaction examination situation female students also appear to have significantly limited confidence versus the male students. To the independent observer this implies that medical educators should concentrate on issues of female confidence, increasing faculty sensitivity and publicly recognizing and discussions perceptions of self-assurance.

This can further reflect distress in medical students (Dyrbye, 2005). On a professional level studies suggest that students stress contributes to skepticism and consequently may affect students care of patients, relationships with faculty and ultimately the culture of the medical profession. In this research the manifestations and causes of students distress are potentially adverse and include personal and professional consequences. Proposed institutional approaches such as provision of mentors, role models, and increased faculty sensitivity, must be instituted to decrease student distress.

To further identify the complexities within a mesosystem of a specialty, in this case concerning women surgeons and their experiences, Longo and Straehley (2015) looked at gender discrimination during residency time and surgical programs. These phenomenon is widespread and sometimes blatant much more so than during medical school or later in practice. A questionnaire was disseminated among two hundred female surgeons and surgical residents asking them to converse about the experience as women in the field of surgery to include medical school, residency and their careers as surgeons. The study was primarily based on participants' observation interviews and extensive surveys of women surgeons. The surgeons, whose average age is below 45 years and their responses to the questions, describe their personal path in surgery and how they overcame the obstacles they encountered. Seventy-five percent of the reported women experienced gender discrimination.

Within that same mesosystem one can further unpack the perceptions of racism by black students experienced at predominantly white institutions of medical education (Bullock, et al, 1987). Thirty one black medical students attending five white medical schools were seen in individual interviews of 1 to 2 hours to evaluate their discernment of racism in their medical school education. The interviews were focused on racism experienced in high school, college and medical school. Over one half of the population came into contact with racism during high school and college education while 30 of the 31 subjects reported racist experience in medical school education. The students report a multiplicity of methods when coping with racist experiences and emphasize the importance of fellow minority students, faculty, and the minority office in managing the stress of racist experiences. It is recommended that those offering counseling services to minority students should recognize the actuality of racist experiences in medical education.

Butler's (2010) study aims to provide a written analysis of particular womens' medical education experiences. The participants included 25 practicing women physicians who graduated from US medical schools. The author identifies the following themes, which extensively impacted womens' physician experiences, as medical students, in practice, and as primary caregivers. Marginality in women's health education exists in all echelons of the medical training core curriculum specific to reproductive and psychiatric women's health. It impacts physicians' attentiveness to treating female patients. Physicians are accountable to the existing chain of medical command during training and practice. Fundamental to this issue is the pervasiveness of sexual harassment, which eight physicians of the 25 reported having experienced or witnessed during their medical training. Finally 21 of the 25 physicians identify mentorship programs in medical school and residency programs as worthwhile. However, haphazard assignment of mentors and students took away from the meaningfulness of such programs.

EXPERIENCES OF WHITE ALLIES

From the studies cited above, particularly those of Liebschultz et. al., Blanche, Hall, Roeter and Frankel, and Dyrbye, one may conclude that residents of color would benefit greatly by having beneficent mentors, if not African American, then strong Caucasian allies. Such allies are few in number, but essential, and may have followed uncommon life trails of their own, to come to this role.

For those seeking equity and fair treatment of black and other underrepresented minorities in medicine, there is no clearly delineated path or protocol to follow. One such ally on a medical school faculty relates his experience. His concern with interracial fairness and equity began early in life. He was raised by parents who were militantly anti racist, and actively involved in the civil rights struggle in their own neighborhood, when, at a time of de facto segregation, they allied with others of like mind, and helped an African American family move in to their housing development. The move was actively resisted by the landlord company, a large corporation, which ultimately succeeded, amid some civil conflict, in evicting the black family. This struggle, and his parents insistence that their conduct was the norm, and that it was the unremarkable family practice always actively to support the oppressed, was formative. He was sent to interracial summer camps, and had friends, male and female, both white and black in his teen years. He himself participated in campus movements supporting the sit ins in the South while an undergraduate, and, at the end of medical school, gave medical assistance in the Poor People's Campaign, serving at the encampment near the Lincoln Memorial in 1968, that followed the murder of Rev. Martin Luther King Jr. In residency training, he witnessed and intervened against police brutality in the Emergency Room in Baltimore where he was assigned. This thread was one he followed during his entire life. His upbringing and life experiences forced him to reflect on what it meant to be a white ally of African Americans in the struggle against racism, and resultant professional inequity.

To do the work of an ally, the prime requisite in addition to the foregoing meant to examine the meaning of white privilege, and to acknowledge that however much he despised a system where such a thing existed, he had it. To be a white male is to have white privilege, and there is no use in denial or defensiveness about it. Confronting it, and owning it made it more meaningful to consider those areas of professional life in which African American physicians, not having it, were at a disadvantage. He found this was best done by strong, lively and ongoing association with black friends, family and colleagues. He also chose to work in an institution 90% of whose patients were black and where African Americans constituted the leadership of the institution. These all represented conscious choices. By such voluntary associations, a willing subject can begin to appreciate what life looks like through the eyes of black peers, and to see how many lifestyle and political choices of the white male and female population seem quaint, and, with reason, inapplicable, or senseless, to African Americans. He can walk down a street in a black neighborhood without feeling threatened, but at the same time realize that Caucasians without his experiences might be so inclined. In a sense, he has to try to acquire, albeit without having to face the life threatening realities of black life in America, an appreciation of the double consciousness, of two conflicting identities that cannot entirely merge, within African Americans, explicated by WEB DuBois, in "The Souls of Black Folk". Understanding WEB DuBois' concept of the veil, as developed in the same book, that to Caucasians, black life is lived behind an impenetrable veil, whereas African Americans are all too familiar with the requirements, practices, and even intimate personal lives of Caucasians.

With that superstructure of ideas, he was able to take note of the different requirements, and interpretations of performance imposed on and expected of African American residents as distinguished from Caucasians. He also tried to use his white privilege to put his thumb on the scale to balance out unfair negatives when evaluation of residents of color came up.

As a senior member of his department, he participated routinely in these periodic evaluations of residents in his field. The ratings of each involved faculty member for his or her assigned residents were reviewed collectively by the professors in the department over the years. A pattern emerged in the form of especially critical commentary from the same two or three faculty leveled at all residents of color. The commentary was often couched in sarcastic and humorous terms, but ones familiar to any African American in the US. Residents were described as slow, having inadequate funds of knowledge, being too flamboyant, or uncertain, or paranoid, or unable to function without extra crutches. None of these charges bore any resemblance to reality, and as a white ally it was each time incumbent to intervene. Intervention required speaking up, in unpleasant circumstances, and overcoming a sort of white backlash phenomenon, wherein less attentive or more dubious department members would be skeptical and unsupportive of the intervention, even though they themselves would not initiate the negative remarks. Ally faculty would consistently point out that the negative remarks did not describe the facts, and were unsupported by evidence. The differences of evaluative opinion would then have to be resolved at the level of the chair, sometimes one way, sometimes another. The negative faculty would never couch their remarks in racial or racist terms, but they only cut in one direction.

ISSUES, CONTROVERSIES, AND CONCERNS

The following five case studies illustrate the practical working out of the racial inequities experienced by African American medical students, residents and attendings, both Caribbean and American, delineated in more general terms in the literature search covered in the foregoing section.

Case #1

This doctor, an African American woman, whom we will refer to with the pseudonym Kate, and a few of whose details will be altered to protect anonymity, was born in the urban upper Midwest. Her parents were educated. Her father was the first African American junior high school principal in their city, and her mother was also well educated.

Beginning in elementary school, Kate has no recollection of her white teachers, though they were the majority, but still, many decades later, remembers her only two black teachers, who noting her intelligence and precocity, told her that she was "indeed her parents' daughter, and as such she was expected to perform well". This advice, while daunting, was not given in an intimidatory way, and was to her inspiriting, leading her to feel it was her role to achieve highly. Most other children in her school were white, but with the inspiration given by her two black teachers, Kate felt empowered to work hard and to excel.

In Junior High School, Kate did very well academically, generally getting "E" for excellent in most of her grades. However, her high scores did not prevent her white guidance counselors from directing her to the district high school, where she could learn how to be a very good housekeeper. Fortunately for her, she had parents with the education and standing in the black community who were not misdirected by the school, and told her in no uncertain terms that being a housekeeper for white families was definitely not in her future. Yet even they could not take steps to compel the guidance counselors involved to refrain from giving such destructive advice to black students.

However, they were able to encourage Kate successfully to apply and be admitted to a well known city high school which was the local equivalent to New York's Bronx High School of Science and the High School of Music and Art combined. There, Kate flourished in multiple subjects, and having skipped 2 years in grade school, managed to get through high school by the age of 15. At that time it was the custom to announce the leading students grades in a public assembly. The principal, having named her as first in her class, and having remarked that it must been more than coincidence that her last name was the same as that of an American president, was shocked when she was asked to stand for recognition, and he saw that she was African American.

She wound up going to an in-state private liberal arts college, where she was the first African American female student. She was not permitted to room in a college dormitory, and was directed to an off campus cooperative dorm. Despite the unwelcome intrusion of living quarter racism, she did well scholastically. Nevertheless, she was given a friendly warning from an older white house mother figure not to pledge a sorority, since all the white sororities were de facto segregated, and she took the advice. Eventually, Delta Sigma Theta, the second oldest black female sorority behind Alpha Kappa Alpha, and an offshoot of it, set up a special state chapter for her and other black girls on different campuses, who would otherwise have been excluded from sorority life entirely. This membership was especially important to her because of family legacy, her mother and grandmother both having been members. Despite sorority life, it was nearly impossible to date on campus, her white male peers having no interest in dating African American girls. Nevertheless, Kate survived college with excellent grades, to enter medical school at age 18, having compressed four years into three.

At medical school, in a state institution, which she entered at age 18, as one of the very early black women students, she did not encounter major racial strife and tribulations. She had finished college in three years, with her actual college degree granted after the first year of medical school. Her accomplishments were in fact prodigious, and she was considered emotionally mature enough to do medical studies. In her class at that time, out of approximately 75 students, there was one other female student, who was Caucasian, and two black males. It was a help to her that her father, who by then had become Deputy Superintendent of Schools in the Midwestern city where she was raised, was personal friends with the Dean of the medical school. She passed through medical school relatively uneventfully, considering the place and time, but, with some obeisance to gender trends, was eventually shunted into one of the stereotypical female specialties, pediatrics.

Kate's residency education began at a large urban hospital, then newly affiliated with the nearby medical school. She was among the last to do a rotating internship, without salary, for room and board only. Board was a small room with a bed and sink, in what further back in history had been the nursing school dormitory. This was the only place on or near campus considered suitable for the few female house staff then present. It was not, however, suitable for Kate, who made the move off campus, to a basement apartment, the best she could afford, shared with a white female intern. The remainder of internship was tolerable, and she followed through getting a pediatric residency at one of the major nationally prestigious teaching hospitals elsewhere in the city.

There, she experienced both sexual and racial harassment. A well-known husband of a high profile white mother of one of her patients hit on her physically. Kate, a tall, imposing and beautiful woman, had to fend him off physically. She succeeded readily, but the incident left her quite upset. At that time there were no counseling or support groups. It was left to the violated woman to work it out herself. The experience left her with lasting memories, but she had developed strong enough emotional armor that she continued residency with no overt hindrance.

Racial disrespect also reared its head. It was the custom at that time for pediatric residents to scrub in when their patients went to surgery, for continuity of care, to keep the surgeon apprised of up to date and necessary pediatric information, but mostly to give the resident herself a realistic idea of what surgery looked like and meant to their patients. The surgical relief and physiologic stresses the pediatric population faced in surgery, how to relate the experience to the family and a realization of what an eventual referral to surgery meant for the pediatrician were just some of the forms of knowledge transmitted by this experience. What did not need transmitting was the first experience Kate had in the neurosurgical operating room. As she entered the OR, the pediatric neurosurgeon, a man with a national reputation, called out on seeing her "Hey-there's a nigger in the room"! This offensive racist remark shocked Kate beyond measure, but in the spirit of the times, and in view of her junior and powerless status, she felt constrained not to say anything, let alone confront a senior, renowned and patriarchal figure, who held her medical life in his hands. She was equally shocked when he then invited her over to the operating table, to participate in the case, and thereafter displayed tact and courtesy in addressing her for the rest of the time in the OR. From the medical point of view the whole case was very instructive. However, racist thinking was so normative to the professor he had no understanding or qualms about the effect of his language on an actual African American woman. This experience hung over Kate during her residency, reminding her if a reminder were needed, where the power lay, and of the inner feelings and beliefs of those set over her, and whom she otherwise would have looked up to. This type of experience was not a first for her, but coming as it did from a revered teacher, while in a way confirmative, was also continuously disturbing to her peace of mind, and tremendously disappointing as showing lack of faith in her abilities.

Her time as a pediatric resident finally over, she rethought her interests, and realized that though she was not at all fazed by the physical aspects of medicine, even including surgery, she was much more strongly attracted to the inner lives of patients, and to understanding their emotional makeup and how their felt experiences influenced their behavior and future. Even more important was the opportunity to help her patients to cope with the exigencies of life arising from and having their principal effects in their minds and thoughts. Thinking along these lines, the only choice for her was psychiatry, and so she obtained not one but two psychiatry fellowships. The first was in adult psychiatry, and following that, she did another in child psychiatry. These were at a municipal teaching hospital with a close medical school affiliation.

Even there, in the heart of a sophisticated urban area, Kate could not rent an apartment, or even a room in a house, due to unacknowledged racial bias, in the mixed Italian and Jewish neighborhood surrounding the hospital. She was compelled for a time to resort to a much more remote black neighborhood in a different part of the city before she could find living space. Because of the difficult commute, she even considered moving back to the dormitory, but then, and surprisingly, her chairman made a genuine offer of a room in his well-appointed house, and that offer solved her housing problems. This offer was touching, and jibed with the pride the chairman and the personnel felt in a very diverse department, with faculty and residents from white Anglo Saxon Protestant, Jewish, African American and Hispanic backgrounds. However, even this ethnically diverse department was not free of gender related issues, with many episodes of various supervisors hitting on her, despite or because of the power differential separating her and her older male leadership.

Kate survived these vicissitudes, and went on to a long and storied career in her chosen field. She became the first African American Chief of Service in the group of hospitals where she worked, but even this distinction was achieved only after conflict with a department chair who coveted her future job for himself, and came up with a peculiar rationale that since her patients would be 95% minority, a Caucasian chief was required! Kate, along with community and political allies she had acquired, was eventually able to move into the chief position on her own. Without these allies, she would have been lost, since it took a threat to the chairman from the Board Chair of the medical school before he saw the light. She was compelled to develop acquaintances and colleagues of considerable influence in the political realm. Through these connections she also entered public service, outside her academic job, in a regulatory field where she spent decades in a position of benefit to thousands of underrepresented minorities who passed under her purview. It took a lifetime to build into a position of respect, comfort and leadership in her chosen field, but not without intense strain and the need continually to meet and overcome frustration, and covert opposition on racial and gender bases. By the time Kate "retired", a relative standing, since her life in retirement has been busier than that of most people working full time, she felt she was very tired, that she had had enough of being the first black female in each of her important positions. She has taken great satisfaction in being a major contributor in her field, but at a high and enervating price to her peace of mind.

Kate had an unusual background, starting as a child reared in what Margo Jefferson has called "Negroland", the province of African Americans reared in the very small very educated world of those with parents with a higher education, with advanced degrees and positions to match. These advantages of birth, however, especially in the era in which she grew up, did not insulate her or protect her against gross racial and gender discrimination which could easily have resulted in her fall from grace, through no fault of her own. She avoided such a fate by inner moral and physical strength, character armor, tenacity, and a nurturing faith. She had tremendous support early on from her parents, and had absorbed from them the knowledge that she had the ability to excel. Those lacking such a background may well never have succeeded in Kate's world, where most of her professional efforts were pioneering.

Case #2

This study looks at a female African American surgery resident, Sara (a pseudonym), an American graduate of an American medical school, of Haitian parentage originally. Her childhood was challenging in that she had little contact with her birth parents, and was raised in a foster home. Her foster grandmother constitutes her primary family allegiance. Despite this challenge, she completed college, graduated from a Midwestern medical school, and completed her surgical residency in an academically oriented medical school program.

Here she encountered a problem befalling many African American physician trainees. She was a very conscientious resident, displaying a balanced compulsivity characteristic of the best resident and attending physicians. She kept careful lists of each of her patients with all the appropriate data, in addition to information kept on the computer. She regularly made use of this information in meticulous patient care. She was deft and quick in the operating room, where she was quite talented. Her eventual interest was to do plastic surgery, which requires meticulous surgical technique and very skillful handiwork.

Her very meticulousness however became a target in regular resident evaluations, wrongly taken by several white senior faculty to mean she was slow and uncertain of herself, neither of which was true. Two of the senior faculty in particular, often were hypercritical of black residents, never in overtly racial tones, but giving vent to personal criticisms which took some of the resident's best qualities and unjustly turned them from the virtues they were into faults. Their conduct was consistent from one black resident to the next, frequently finding fault when none was present, and presenting their remarks in pointed satirical terms which drew laughter for their wit, but discomfort from the content of what they said.

During the penultimate fourth clinical residency year, the faculty criticism reached such a crescendo, with doubts raised by the critical duo, based on what they considered wrongly was poor information management and control over patients, and their remarks clad in scathing terms, that this resident panicked, thinking her life as a physician was over. She lacked any family or other strong personal backup, her grandmother not having the training or educational experience to be able to offer concrete help.

Sara had acquired a strong white faculty ally, who at that time was unavailable, with the result, that, not knowing where to turn, she did not come to work for a period of three days. Her program director reached out to locate her, but she wasn't at home, and it required an actual search attempt to discover her where she was staying outdoors in her own neighborhood. On discovery, she was extremely upset and emotional and had no one to turn to for practical or spiritual support.

One of her supervisors, a Caucasian woman, not a psychiatrist, jumped to the conclusion that she had a psychotic condition. This was unwarranted by the facts, or by the evaluator's professional standing, but others in the department, less involved than she, went along with it. The result was that Sara was put in the hands of the State Medical Society Committee on Physician Fitness.

This group usually deals with doctors who have become drug addicts, acquired a severe mental illness, or are involved in actual criminal conduct. The personnel assigned to take a stewardship interest in those referred with psychiatric issues actually are senior experienced psychiatrists, who examine the referee, initially and repeatedly, diagnose the problem, give treatment where indicated, and monitor progress. However, the information they start with comes from the referring program and the referee, and therefore may be limited, flawed and inaccurate. The result of this combination of circumstances faced by her faculty ally on his return was that the resident faced dismissal from her program. Dismissal would have destroyed her life in surgery, and probably in medicine, quite apart from the psychological damage it would have left in its wake.

She explained the situation to her ally three weeks after the start of the affair. It was immediately apparent that her detractors had willfully, and with a complete lack of understanding, misinterpreted her diligence and compulsiveness in close follow up of her patients, keeping lists, referring to them regularly, as paranoid behavior. Her faculty ally took the initiative to explain this to her psychiatric overseer, who had meanwhile been receiving biased information as to the meaning of the resident's interpretation of her work. He was shocked to hear the extent of the mockery of the truth he had been receiving, stating that the accurate facts given him put the resident's behavior in a whole new light.

The end result of the process was a rapid termination of the psychiatric supervision, and the return of Sara to her surgical program, which she successfully completed as Chief Resident a year and a half later. Subsequently she secured a fellowship in her preferred field of plastic and reconstructive surgery, at a Research One university, is now in the successful practice of plastic surgery, and has married and become a new mother. She does not have and never had any true psychiatric illness.

In this case, Caucasian supervisors, ill informed as to the stresses incumbent on a black resident mistook appropriate and admirable behavior, which would not be questioned or denigrated in a white resident as paranoid in a resident of color. There was no empathic attempt to explore why the resident was stressed enough to stay out of work in the first place. Their assumption was unspoken that the resident had enough of a support network as has been the case for the Caucasian residents in the program, to have avoided this problem. Failure to understand the slender nature of support available to many residents of color, or of the need for a certain level of defensiveness, and of a demonstrated need to excel, to do better than the Caucasian residents at the same level lest they be overlooked, led to an indefensible ascription of the resident's behavior to psychopathology. Their lack of understanding initially misled her assigned psychiatric monitor. Fortunately, he had the sensitivity, security, and objectivity to be able to reverse his understanding of the case quickly in light of facts contradicting what he had originally been told. The consequences of his failure to understand would have been devastating to the resident.

Remedies to this sort of situation are easy to suggest but hard to achieve. They include increasing the number of minority residents admitted to residency programs in surgery and all other fields, which has been a desired but unachieved goal for a generation. African Americans constitute 14% of the US population, and only recently have arrived at the level of 7% of the medical profession. Further progress is critical and essential to best medical training and to quality medical care in the US, but we are still far from the goal. Eventually, equalization would produce a faculty body better able to understand the background of residents of color, to acknowledge their powerful desire to succeed, to see their training life through their eyes, and to avoid mischaracterization of motives that stems from viewing black American as "the other".

Residents of color also should be taught, if they have not already learned them, a multitude of survival strategies, including acquiring peer as well as faculty allies, showing strength combined with humility, and learning to navigate racist obstacles by weaving around them, when direct confrontation is impossible due to the power differential. The point of all this is to make available to our whole population first rate medical care by a diverse cadre of physicians who are a better match for the public than now.

Case #3

This 28 year old African American female surgery resident, whom we will call Paula, had a short career in her surgical program. She was quiet, methodical, kept careful account of all events and lab data in her patients, and maintained a thoughtful involvement with each of them. Her demeanor was self effacing, but her control of information and planning for her patients was excellent. Her faculty detractors in reviews made much of her quietness, and careful record keeping, and construed these wrongly as evidence of timidity and weakness of character. Some of these points were made in faculty reviews with acerbity, wit and sarcasm. No overt expressions of racism were ever heard in these discussions, but the same two Caucasian faculty members who criticized her could be counted on to raise objections about the ability of every African American resident in the program when the time came for their evaluation . White faculty allies contradicted these claims, but in the long run they carried the day, and Paula was encouraged to look into other fields than surgery. To her grief, she had to leave her chosen field, and entered a residency in anesthesiology. She succeeded in this realm, and in time became a successful attending in anesthesiology, and no longer mourns her original attachment to surgery. Her attachment to surgery was cut short by rebarbative and acerbic criticism seldom leveled at Caucasian residents. The weight of her allies was insufficient to overcome that of her critics.

This is another instance where alienation of Caucasian faculty from residents of color is so extreme that it prejudices objective performance evaluation. Remedies such as diversity training are unlikely to succeed with older Caucasian males, although they may be worth a trial. Better screening of faculty candidates regarding their attitudes toward diversity is essential, and presently plays no part in faculty selection. Surgical skills and experience, research accomplishments and potential, recommendations from respected mentors, academic achievements in the younger faculty are important, but diversity screening is nonexistent. The time has come to surface this notion as job interview requirement, and to begin selecting faculty with a welcoming attitude to diversity as a necessity. Absent this orientation, we will remain at dead center in terms of ability to evaluate objectively residents of color, and will waste talented people even when they come forward, and surmount obstacles to get to the point of becoming residents in surgery in the first place.

Case #4

This case is that of an African American resident, whom we will call Vivian, a graduate of an American medical school, with very high ambitions in surgery, to enter one of the least populated and rarefied specialties in surgery. Her surgical training proceeded through its first two years, during which she had bimodal evaluations, at one pole emphasizing her devotion to patients, her high degree of skill, her good results, and quickness to catch on to difficulties and changes in her patients, and remedy them.

At the other pole, she was considered by her faculty detractors as difficult to get along with. Vivian's explanation of the circumstances leading to the negative commentary was that she was unwilling to take degrading criticism, and talked back at the time. She was definitely not supine, but she also faced a level of criticism not experienced by her Caucasian peers in like circumstances. Her analytical abilities and her quick intelligence in her job were excellent, and she continued on the usual path for a surgical resident until she was involved in life threatening trauma that resulted in her having to take a year off from the program before she was able to resume residency. Already prior to her return, the white detractors formerly ready to criticize had, without a shred of supporting evidence, and no knowledge of her medical condition, begun talking about her having had a traumatic brain injury, which had never occurred. In advance of her return they were already discussing what to do about it and whether she should continue in the program.

Vivian faced critical scrutiny on resuming in her slot. She extracted herself from her difficulty by keeping her eyes on the goal, immensely hard work, and success at what she was doing. She also spent time in the research laboratory, due to her interest in the research and also because it enhanced her chances for a fellowship in her chosen field. Despite her success as a resident and in the lab, questions continued to be raised about her mental capacities and temperament being altered by her health issues. These questions were not stilled until she succeeded in gaining one of the most competitive fellowships to which surgical residents can aspire. Subsequently, she has become an attending surgeon in her chosen field.

Case #5

This case is that of an African American surgeon, whom we will call Jo, a pseudonym, the Director of Surgery at a teaching hospital. Jo was born and raised in the American South, cared for principally by her African American mother alone, who herself had undergraduate education and an advanced degree in neurobiology. Her mother's education level at the time put her in the upper 3% of black women to begin with. Jo was able to attend a private high school, where she excelled, but did not escape racial pressures. A classmate, acting anonymously, but whose identity she eventually deciphered, sent her a letter requesting a contribution to the Ku Klux Klan. She had the strength, fortified by her mother's upbringing, to dismiss the solicitor by telling him any such contribution was contingent on his donation to the Black Panthers. That was the last she heard from this classmate, but the experience reminded her of the continuing racial divide. She did well enough in school to get a paid up education at her state university. Jo majored in biology, and had some surcease from racial pressures at a university with a surprisingly diverse student body, for a southern state institution, not historically black. She was able to identify more openly with African American culture and styles, especially by finding a small but supportive group of friends. It was at University that she formulated the idea of going to medical school, and had a competitive enough GPA that she was admitted to the University medical school.

Her attachment to surgery arose from her desire to do a field of medicine in which she could intervene, and get quick, measurable and satisfying results. The act of intervening appealed to her much more than the conversation and contemplation that characterized the fields of internal medicine, a fairly common story in those attracted to surgery. She was able to find supportive mentors at medical school, not immediately available to black students of medicine, depending on geography and era. Jo was accepted into a surgery residency at a major national university medical school, and due to a gregarious personality, and having learned how to guard and preserve her identity, could discover and eventually consort with older Caucasian attendings who had power, and national reputations, and who could advance and protect her.

Their mentoring while helpful could be top heavy, in their appreciation of her racial background. One professor, explaining to her the protocol in going to a department function at a club, drew her a detailed plan showing how all the other residents could enter by the front door, while he diagrammed the way she would be obliged to enter through the back way. She realized that this was his heavy handed attempt to show her his understanding of the black experience, but his mode of expression could easily have been misinterpreted and not seen as a joke. It was based on the assumption that she could be expected to share his perspective, and not so much that he might share hers.

As her surgical residency proceeded, and the time came to consider fellowships or jobs, her chairman, a nationally famous surgeon, suggested she had nothing to worry about in her choice, since, as a black female, she would automatically be granted whatever she wanted. This misperception on the part of Caucasian leaders in the field, certainly not confined to medicine, and stemming from a certain degree of repressed discomfort with African American advancement, is not news. That blacks in medicine, or any profession have an easier and protected route, compared with their white peers, certainly is news to anyone who has lived or lived with the black experience.

Eventually, Jo acquired direct and fellowship experience in two different areas of surgery, and having put in her time, done research, and connected with a number of influential academic surgeons, was brought to a different area of the country by a former colleague, to take up a directorship of surgery, with responsibility for her own and multiple other surgical subspecialties It took her 17 years to get to this point, on a journey that began in the early post-segregation south, and to reach a pinnacle of surgical success. That she was under a little less racial pressure, but far from none, was the result of being twenty five years younger than Kate, whose experience was much more raw than hers. Even so, overt and barely concealed racist attitudes accompanied her from high school through residency in surgery.

In the cases of the specific residents, Sara was misrepresented as having a paranoid break by white faculty who had no other way of conceiving her actions in very distressing circumstances. Her white ally was unavailable for three weeks around this period when she was referred for psychiatric assistance, and it was only the return of this ally that reversed the whole unnecessary torture.

In Paula's case, the same faculty ally could not prevail.

In Vivian's case, prolonged intervention by her ally with numerous attendings over the years eventually helped her graduate and achieve a very competitive fellowship. The same faculty ally carried each of these cases. The reason appears to be that being outspoken, impervious to criticism, and aware of the modern forms of race oppression, was in his DNA, and those qualities were invaluable in intervening in evaluations of residents of color. While his personal background is unusual, it is not rare, but it is difficult to conjure up people with that sort of background on demand. It is uncomfortable to do what white allies do in intervention since sharp differences with colleagues especially in racial matters, create tensions and anxieties which mostly go unresolved. Nevertheless, there are rich satisfactions inherent in sticking up for the underdog, especially in racial matters, and succeeding in achieving equity.

CONCLUSION

Lessons to be Learned and Conclusions to be Drawn from These Cases

The challenges to be faced by minority surgical residents do not abate readily across a lifetime of professional commitment. They begin in high school, with advice that is uneven, limited and which does not open the way to high aspirations for the student, This advice may not be ill intentioned, but is circumscribed by the advisor's own limited opportunities and experiences. Some students may receive inspiring advice, while their peers are sent in more quotidian directions.

The same situation obtains in college, where many minority students are given little hope of achieving the highest goals in placement for professional schools. Those surmounting lack of support or misdirection either have exceptional inner strength or allies. Neither of these is easy to replicate, but a substitute may be sought in setting up mentoring structures, at high school, college and graduate or medical school levels.

Such programs first require achievement of major funding initiatives, because they require manpower, implying salaries, and money to pay them. While these programs must of necessity grow in size and complexity from high school to college to graduate to medical school, they should all possess certain features.

Such programs require inspired and committed leadership. Leadership does not have to be minority, but nevertheless, students are likely to be more readily trusting and forthcoming if they can see that the leadership may have surmounted some of the same challenges in life they are being asked to deal with. Otherwise, and this is certainly possible, trust will have to be earned but at the expense of great effort on both sides.

Next there will have to be enough staff to aid in sponsorship of new initiatives at all levels. Getting other faculty involved, with knowledge of particular subject matter, whether the teachers involved are minority or not, takes time and effort to cultivate recruitment targets. Creating the initiatives themselves involves brainstorming and solitary thought and cannot be done or effectuated by a sole dedicated individual.

Activities of such organizations have to cover outreach to incoming students, recruiting multiple mentors, matching them up with recruits, and nurturing and monitoring relationships. They also have to create ongoing programs to enrich scholarly experience, to give a kind of tutoring when necessary that is done with genuine sensitivity and respect for the candidate, so as not to diminish the candidates self confidence. It's also necessary to select out outstanding candidates as recipients of recognition, to develop leaders, and to inspire all with programs that draw in and elevate the group experience intellectually and socially. All these tasks also require a functioning office, with staff to answer phones, provide information, keep records, and facilitate access of students to faculty and diversity managers.

These responsibilities do not begin and end at the borders of the institution itself. Success in the long run requires recruitment of suitable candidates in the first place, of such a caliber as to make the institution proud and the candidates successful, integrating them into the community life and giving them a heartening start.

And such responsibility does not end without attending to the advancement of the student beyond institutional walls. Placement for jobs or schooling at the next highest level is a critical need impacting on the success of the institution, and the future professions and lives of the students. Minority students in medicine, while making recent advances in entering OBGYN, and to a lesser extent a few surgical specialties, such as neurosurgery and urology, by and large have gotten where they are totally without the types of program discussed above.

For high school students who are unsure of their ambition and are searching for a role in the world, such organized help at the high school level is essential.

For minority college students who have an interest in medicine, testing their goals against reality, getting encouragement in their dreams, and getting mentors in their assumed fields who can explain what life is like in their chosen area are all essential. Knowing accomplished people in their field who will merely talk to them is tremendously inspiring, and will raise the level of self-confidence.

Once in medical school, having allies and defenders, true also in residency, is critically important. There are many instances, most crippling in the early years of training, when the resident is most defenseless, and that is when allies and especially minority mentors, are most important. They can defend the candidate against unjust criticism and the tendency of Caucasian seniors if not consciously then unconsciously to ascribe untrue stereotypical and derogatory attributes to minority residents, and then to use these misattributions to damage their career development and future chances in medical life.

FUTURE RESEARCH DIRECTIONS

What is needed for a more fruitful future for African Americans and other minorities in surgical and medical training is fundamental, and requires a structured environment in which talented students, who abound in the community, can learn and thrive. Useful research will be important to measure, monitor, and improve their performance, happiness, and habilitation in their (often new) environment. Doing the research implies that there is a new experimental condition to be inspected, viz., the setting up of such a structure to begin with, so that its effects can be measured in the first place. Without such a structured environment there is nothing new to measure.

So the primary task here, to promote the doing of future research into the growth of students of color in medicine and surgery is to develop a framework within which survival becomes more ordinary, and the crises faced by the students of color are of a similar order to those faced ordinarily by Caucasian students. The desirable framework is the one outlined above. Granted the establishment of such an organization, or department at a given educational institution, it will become possible to measure the effects of interventions into student lives and studies.

For example, it should at least be possible to do retrospective studies examining the effects of talent enhancement, or tutoring programs, on admitted students, comparing outcomes with those before program establishment, i.e. using historical controls. Another such example would be to look at the improvement in performance, if any, flowing from a program encouraging the recognition and development of the abilities of the most outstanding students in the group, as opposed to outcomes absent such a program.

Yet another possibility is a type of program used at the University of Texas (NYT April 15, 2014 "Who Gets to Graduate?"), in which minority Latino and black students, formerly streamed into remedial classes, with resulting mediocre graduation rates, were given intensive and rigorous tutoring by senior professors. They were invited into a program (TIP- the Texas Interdisciplinary Plan) characterized by four offerings:

Rigorous classroom instruction, identical with that offered the best students; two hours a week of extra instruction; close contact with individual advisors, ready to intervene in case of trouble; and infusing the students with the critical notion that they were a high achieving part of a close learning community. Although the precise particulars of such a plan have to be modified to fit individual schools, and such a model cannot be universal, such a plan is flexible, adaptable in different university and medical school learning environments, and, very important, subject to either retrospective research, or randomized controlled trials of matched groups of students who have either had such a program or have had the usual professional school education. Graduation rates, USMLE scores, in surgery, ABSITE scores, selection rates into desirable fellowships, accession into academic positions, achievement of tenure track jobs, obtainment of funding, are all measurable variables that can be used to gauge the success of such a program. The retrospective results at UT, were exceptional, in that students in TIP achieved higher not lower than average test results, and graduation rates, and improvement of their performance in other subjects than just the one (chemistry) in which the program at first operated.

Such programs are premised on the correct notion that minority students have the full mental endowment of Caucasian dominant groups, do not waste time debating the question of capacity, and just need to have past educational program deficiencies remedied, and their self respect and esteem given credence, to achieve at the highest levels. These suggestions for future research are relatively easy to state. Accomplishing them would be the capstone of an entire career.

REFERENCES

Beagan, B. (2001). Micro Inequities and Everyday Inequalities: "Race," Gender, Sexuality and Class in Medical School. *The Canadian Journal of Sociology / Cahiers Canadiens De Sociologie, 26*(4), 583-610.

Blanch, D. C., Hall, J. A., Roter, D. L., & Frankel, R. M. (2008, September01). Medical student gender and issues of confidence. *Patient Education and Counseling*, *72*(3), 374–381. doi:10.1016/j.pec.2008.05.021 PMID:18656322

Bullock, S. C., & Houston, E. (n.d.). *Perceptions of Racism by Black Medical Students Attending White Medical Schools*.

Cooke, C. L., & Dawson, P. L. (2002). Book Reviews - Forged by the Knife: The Experience of Surgical Residency from the Perspective of a Woman of Color. *Signs, 27*(2), 584.

DeCastro, R., Griffith, K. A., Ubel, P. A., Stewart, A., & Jagsi, R. (2014, January01). Mentoring and the career satisfaction of male and female academic medical faculty. *Academic Medicine: Journal of the Association of American Medical Colleges*, *89*(2), 301–311. doi:10.1097/ACM.0000000000000109 PMID:24362376

Dyrbye, L. N., Thomas, M. R., & Shanafelt, T. D. (2005, January01). Medical student distress: Causes, consequences, and proposed solutions. *Mayo Clinic Proceedings*, *80*(12), 1613–1622. doi:10.4065/80.12.1613 PMID:16342655

Grbic, D., & Brewer, R. L. (2012, April01). Which Factors Predict the Likelihood of Reapplying to Medical School? An Analysis by Gender. *Academic Medicine*, *87*(4), 449–457. doi:10.1097/ACM.0b013e3182494e54 PMID:22361796

Hinze, S. W. (2004). 'Am I being over-sensitive?' Women's experience of sexual harassment during medical training. *Health (London, England: 1997), 8*(1), 101-27.

Jefferson, M. (2015). *Negroland. A Memoir*. Pantheon Books New York.

Liebschutz, J. M., Darko, G. O., Finley, E. P., Cawse, J. M., Bharel, M., & Orlander, J. D. (2006, January01). In the minority: Black physicians in residency and their experiences. *Journal of the National Medical Association*, *98*(9), 1441–1448. PMID:17019911

Meharry Medical College. (2004). *Journal of health care for the poor and underserved*. Nashville, TN: Institute on Health Care for the Poor and Underserved, Meharry Medical College.

Office of Minority Health U.S. Department of Health and Human Services. (2002). *Teaching Cultural Competence in Health Care. A Review of Current Concepts, Policies and Procedures.* Contract Number: 282–98–0029 Task Order #41 Task 2: Synthesis Report. Retrieved from: http://minorityhealth.hhs.gov/assets/pdf/checked/1/em01garcia1.pdf

Ramanan, R. A., Taylor, W. C., Davis, R. B., & Phillips, R. S. (2006, April01). Mentoring matters: Mentoring and career preparation in internal medicine residency training. *Journal of General Internal Medicine*, *21*(4), 340–345. doi:10.1111/j.1525-1497.2006.00346.x PMID:16686809

Williams, M. A., McGuire, D., & Rizzo, M. (2012). *Practical ethics in clinical neurology: A case-based learning approach*. Philadelphia: Wolters Kluwer Health/Lippincott Williams & Wilkins.

Chapter 10
Black Women and Science in Higher Education:
Not Much Has Changed – Still the Struggle

Bonita Flournoy
Year UP Greater Atlanta, USA

ABSTRACT

This chapter gives an account on the experiences of a female STEM faculty member's journey from high school through graduate school, and then entering the workforce as she navigates the high and low points of being a black woman in a science field, as a science educator that will mentor other black women scientists, while also trying to sustain a viable family life. Her introduction to science, the profiles of mentors that provided her a support network, and the barriers that continue to plague black women in their preparation for a career in science, remains as blockades to access, to what is currently considered the most needed competency area to fill the workforce of the 21st century.

INTRODUCTION

She was inspired by her female high school physics teacher, of whom she never told. She was the first female science teacher she encountered. It was probably because she attended an all-girls high school in Philadelphia, where their mission was to develop leadership in women and provide opportunities for women that may not be chosen if men were present. Although she was Caucasian, she made an indelible mark in her mind that "I can do science". This brought to mind a statement made by Dr. Jewel Plummer Cobb, Ph.D, "when I was a sophomore in high school, I had a microscope given to me in a laboratory to look through, and that was it! I said, that's for me, biology". Her tenth grade teacher at Englewood High School, one of the finest high schools in Chicago, motivated her to take classes in botany, chemistry and zoology.

Later in high school, the health professions were sought out, being a physical therapist was a desire, but soon in her first year of college in Texas, decided to pursue a career, which she thought was even greater; to be a physician. She then decided to major in biology. After the first semester, she found that

DOI: 10.4018/978-1-5225-0174-9.ch010

biology was not challenging for her, and with a hard nudge from the Chemistry advisor, soon thereafter changed her major to chemistry. He would always say, "Biology, that's too easy, now Chemistry is the real science". Yes, it was **the** challenge. Her chemistry professor, Dr. "J" made a competition out of the chemistry courses and always pitted the women against the men, just by saying male student(1) made an 85 and female student(1) made an 80; she made an 88. It was always in a competitive and collegial spirit. She never knew how much that would impact her life in almost everything she did. Dr. "J" was a true mentor and role model, and he supported young women as scientists. This was also seen in his own family, where there are four daughters, all of whom pursued science as a career; three of which have their doctoral degrees. This family also had three sons. While matriculating through a historically black college as a chemistry major, there were opportunities provided that many were not fortunate to have; in particular, to engage in research, receive a stipend, travel, and make professional presentations. There were no female science teachers at the undergraduate institution. This was a time when science and female role models were scarce, especially African American. It was also interesting that there were just as many female science majors as male, at this time at the college. These college days were enjoyable, and the degree was awarded in 3.5 years, by attending school during the summers.

In reflection, a white female and a black male shaped the early journey through her science career. However, it was always a question in mind, why there were no female science or mathematics teachers at the college; more specifically black women?

It would soon be answered.

MENTORS PAST AND PRESENT

Historically and currently science has been, and is a career chosen by white males. Too often the contributions of all women in science and invention regardless of color are still unsung.

Dr. Reatha Clark King entered Clark College in 1954. Reatha planned to major in home economics and return home to teach in her local high school. After her first chemistry course, her plans changed. Her chemistry teacher, Dr.Alfred Spriggs, recognized her special abilities and became her mentor. She soon decided to pursue a degree in chemistry. With the help of Dr. Spriggs, Reatha saw that she could excel in science. She graduated with honors and eventually pursued her Ph.D. in physical chemistry from the University of Chicago. She faced discrimination and was not offered a suitable position. In 1963 she began her prestigious career at the Bureau of Standards, eventually became a faculty member at York College, and served as the associate dean of natural science and mathematics and dean of academic affairs. She later became President of Metropolitan State University in St. Paul Minnesota, becoming one of the few African American presidents of a major college or university (Sullivan, 2002).

Irony or a serendipitous occasion is noted, because the author also landed on the campus of Clark College, and due to the mentoring and trailblazing legacy of the college, her first full-time science career appointment began. A picture of Reatha Clark King hanging on one of the walls of the Chemistry Department, and speaking to Dr. Spriggs, who was then Chair, about Dr. King, is well remembered. He was very proud and talked about how she stood out among other students and how he supported her through the achievement of a chemistry degree while encouraging her to continue to seek advanced degrees. He also talked to her about continuing her education, and obtaining a doctoral degree. Dr. Spriggs was an energetic and 'no non-sense' chair. He would let you know that this is serious business and that we were here to teach a group of students that 'can do science', and that we teach who is in the classroom, not who we want to have in the classroom. The connectivity among the staff and the students was also a critical relationship that helped students navigate through what they probably didn't think they could do. During this undergraduate education, she became a member of Beta Kappa Chi, a national science honor society. As a member of the faculty at Clark College she was heavily involved in the same organization at the college, because of the type of networking system it provided for novice and veteran black scientists to support success in STEM education and industry careers. To see someone like you was important to black students in STEM and more importantly to black women in STEM, barring that you may never encounter one during your education in this era of STEM education. It was during the organization's national meetings and conferences that you met people like you, doing science, and were successful in it. The opportunities for young black women in STEM to encounter, share, and uplift each other is what makes the difference in persistence. It was a pleasure to at last meet Dr. Reatha Clark King, during one of the alumni STEM showcase events that was sponsored by the campus Beta Kappa Chi chapter. She was gracious and shared impactful real experiences in her quest for equity in the STEM industry. Hopefully, she left knowing that she not only inspired the young scientists she encountered, but the author, as well. As she continues to reflect, it's like walking in Dr. Clark's footprints, which then seemed huge. We must remember, she pursued this path in the 1950's. This event took place in the 1980's. Thirty years had passed and the same story applies.

In 1978, 50% of all undergraduate enrollments in science were women. Women comprised nearly half of those enrolled in the biological sciences, but lesser proportions in other sciences. To explain these increases, several researchers have examined factors affecting the choice of science by women. For example, Peng and Jaffe (1979) examined 16 variables in a study designed to identify multiple factors influencing the entry of women into male-dominated fields as they entered college. The data showed that high school course work, academic ability, success orientation, and educational plans were important predictors of women's entry into male-dominated fields. The results of this study indicated that women pursuing college degrees in male-dominated fields had higher academic ability and had more high school courses in science and mathematics. In addition, women who selected science majors were more career oriented than those selecting disciplines that were traditionally feminine such as social studies, the arts, and the humanities. These findings substantiated earlier ones by Erlick and LeBold (1977).

It is not that women perform lower than men; early exposure to science and math preparation and mentoring by those already successful in the field, will impact the presence black women in undergraduate science programs.

This journey, as all do, has crests and troughs, of which occurred with some frequency. Graduate school was filled with both.

GRADUATE SCHOOL IN SCIENCE AND MAINTAINING A FAMILY

Soon after her undergraduate education, a return to the northeast occurred where she tried to secure a job with a brand new bachelor of science in chemistry, of which she applied to many companies and businesses, none of which were successful. In the chemical capital of the nation, companies like Smith Kline, DuPont, and many others, employment was not secured. What does your advisor always tell you? Go back to school! That was done and admission to the chemistry master's program at Atlanta University was successful. It was a time when coming to the south again was not on the radar, but being married required making sure that her husband had a productive life. Both pursued their graduate education. The navigation through a crest and trough began. The first semester, resulted in a first pregnancy, however, completing the semester satisfactorily was managed. There was one very supportive faculty member who was a physical chemist, a male, and for some reason understood the challenges of a pregnant woman while working on a graduate degree. The rest of the professors were all male; not one female. There were two African- American women faculty members in Biology. The opportunity to take their classes never occurred. A support system consisting of peers and friends, helped sustain through school and the pregnancy. Women have the blessing to be able to withstand the pain of childbirth, but it is the insensitivity of others during this creation time that gives her the most unbearable pain. It is refreshing to be in a current era were an African American President and his partner in life understand the family and its impact on education and the workforce. President Barack Obama's administration's efforts to provide program flexibility to allow the propagation of family-friendly practices that allow women to remain in the workforce while balancing the demands of caring for their family is demonstrated by the National Science Foundation's Career Life Balance Initiatives.(White House government,2013).It provides programmatic policies, aimed at creating more flexible environments for grant recipients – including no-cost grant extensions, year-long deferrals for child birth or adoption, and increased opportunities for virtual panel reviews – to a Foundation- wide level.

This type of initiative addresses barriers to women that force them to make a choice between family, career and research. Although not yet in the workforce full-time, being a full –time graduate student was similar and stopping out and returning was not an option, or at least from the perspective of the writer, would have been perceived as less committed to the program. Women who return to the workforce from taking time off to care for families could have challenges, but support is possible through this type of special program.

GRADUATE SCHOOL AND NO IMMEDIATE BLACK FEMALES: THE CAREER TWIST

In the early 1980s, Margaret Rossiter presented two concepts for understanding the statistics behind women in science as well as the disadvantages women continued to suffer. She coined the terms "hierarchical segregation" and "territorial segregation." The former term describes the phenomenon in which the further one goes up the chain of command in the field, the smaller the presence of women. The latter describes the phenomenon in which women "cluster in scientific disciplines.

Graduate school in chemistry was an interesting time. While in graduate school at a historically black university there were still limitations in having African American women scientists present. At the graduate level, there were two Black women in the biology department. Again, there were none in chemistry on the faculty. As a teaching assistant at an undergraduate college, which was part of a University Center, she was the only female teaching chemistry in the entire department. In biology, there was one full-time white female and none in physics. No other in chemistry. As a result of the teaching assistantship and love for teaching undergraduate science courses, she changed her major to science education in order to prepare for teaching science at the college level. She then encountered in the School of Education, one dynamic, enigmatic advisor. The science education advisor, Dr. A. She was the third role model. Dr. A was strong, fierce, and competent. She mentored her, and fused the gap that bonded her career. 'You can do what anyone else can do if you are determined', she always reminded. Being the only science education student that selected to write a master's thesis was decided, to be competitive and ready for doctoral level work.

Research on women's participation in the "hard" sciences such as physics and computer science speaks of the "leaky pipeline" model, in which the proportion of women "on track" to potentially becoming top scientists fall off at every step of the way, from getting interested in science and math in elementary school, through doctorate, postdoc, and career steps. The leaky pipeline is also applicable in other fields. In biology, for instance, women in the United States have been getting Master's degrees in the same numbers as men for two decades, yet fewer women get PhDs; and the numbers of women Principal Investigators have not risen. (Schiebinger, Londa (2001). Writing a thesis provided the critical, creative, analytical and technical writing skills that are needed to be successful in scientific research. Although, choosing science education as the final major, it was known that these skills would eventually be used, in research, as a faculty scholar and teach others these skills as an educator. Thesis writing was tedious and seemed to be a never ending process. The writes and re-writes were like a bog. The question of whether the major advisor knew what she was doing arose, and a loss of confidence in her and self was looming. At the time, the department head in Curriculum and Instruction was a female and the thesis advisor's supervisor. A complaint was made and a meeting occurred where encouragement to continue was made and sharing that she as well as the advisor went through the same process and it would be a lifelong rewarding experience. Perseverance must be practiced. Those were words of wisdom. They took root, especially coming from two females that also struggled, but were successful in their careers. Although they were faculty and administrators at the university they were mentors; they looked like her and shared her experiences. This was labeled the 'career twist 'because as one prepares herself to further her career it seems that going back to school and working at the same time is a twist, the act of moving in opposite circular motions as to create a spiraling action, moving through hierarchical and territorial segregation.

THE DOCTORAL LEVEL STRETCH: ARE YOU IN THE RIGHT CLASS, OR PLACE FOR THAT MATTER?

Working for several years in a four year, historically black college, Department of Chemistry teaching chemistry lectures and labs as well as serving as a research assistant was rewarding. Mentoring undergraduate students in the research lab in which were several young black women and black males was equally rewarding. Another mentoring relationship grew with a black male, Dr. T, who continued the chemistry lineage of my previous undergraduate mentor. It was Dr. T. who warned to obtain the Doctoral calling card of respect, if there was a desire to continue to be promoted in higher education. Enrollment in the largest research institution in Georgia was accomplished, with a major in science education to help continue the research work on 'problem-solving strategies of African Americans in chemistry'. Fortunately enough the previous major advisor, Dr. A from the University Center was now a professor at this university and she served as the major advisor on the dissertation committee. Experiences at this institution of higher education were unlike any that were ever encountered. For one; being the only African American in most of the classes. Another; being the only black student that this school enrolled in this program, for over twenty five years. A share of racial challenges in several classes, where students asked, were you in the right science course, to faculty members saying I don't know what to do with you in my class. The other extreme of never being called on to answer any questions, was also a part of the experience in one class in particular. You hear about these stories, but you never think they would actually happen to you. They do, and they did.

In an article entitled "Conspicuously invisible: Women of color in university science", Marion Renault desk editor at Minnesota Daily, reported that a junior biology major at the University of Minnesota, as a black female student, said she faced the double challenges of race and gender in a field historically dominated by white men. She reported that "it was isolating to be one of a handful of women in a lecture hall of hundreds, and discouraging to have never had a black science teacher. Like many other women whose ethnic groups are underrepresented in science programs at the university, she faced not only rigorous STEM coursework, but also struggled with daily feelings of isolation and a lack of faculty role models who share her background."

These challenges were also experienced this in the graduate education level, but perseverance was enabled due to being a commuter student who was able to return to family every day and work in a culturally sensitive environment.

Numerous reports and research studies have shown that the paucity of women in STEM coupled with the culture of science can create a climate that surpasses chilly to be frigid for women in those disciplines. The climate in science departments continues to be based upon masculine ideals, such as competitiveness and a belief in the objectivity of the scientist. In a study of biology and chemistry departments, Ferreira, 2003 found that men and women differed in their perception of certain aspects of climate, including whether gender was a barrier to a successful career in science, and whether women have to prove themselves more than men.

Although the men and women in Ferreira's study also had similar perceptions of the culture of science, most of the data in general pointed to science as a masculinized and inflexible career. Department climate in terms of STEM graduate education is characterized by departmental differences in the orientation and support provided to students, faculty expectations of and relationships with graduate students, and the quality of student peer relationships. One of the seminal works on doctoral degree progress concluded that department climate influences degree progress and that student commitment to earning the degree shifts constantly in response to experiences in the department .

Robert Leslie Fisher (December, 2013) draws on sample data from 1300 male and female respondents from White, Black, Hispanic, and Asian students. In his book, "Invisible Student Scientists", he shows how the student culture of graduate schools in science and engineering sees women, Black, and Hispanic students as outsiders and deprives these budding scientists and research engineers of the collaborators they need to succeed in their careers. Fisher argues that we must inspire female, Black, and Hispanic graduate students to believe they can succeed in their careers by (1) changing the student culture in graduate schools' science and engineering programs to be more inclusive, (2) removing burdensome undergraduate educational duties from graduate students so that they can concentrate on mastering the difficult subject matter of their disciplines, and (3) hiring more women and under-represented minorities as faculty to serve as role models.

It was the fierce mentor, Dr. A, that came to the rescue and fought and struggled alongside, as she too was the only black full- time female in her department in science education.

The two shared the struggle. When graduating with the doctorate from this institution, Dr. A was the first to inform that this was 1 out of the 3 blacks that graduated with a doctorate degree that year across the nation.

Having professors in the department who serve as mentors and who care about student success, is associated with decreases in the likelihood of being in the low commitment category. Specifically, an increase in the perception that one's department has professors who act as mentors and care about graduate student success is associated with a 64% decrease in the likelihood of indicating low career commitment. Additionally, the better a student's relationship is with their advisor, the lower the likelihood of being in the low career commitment category relative to high career commitment. Two factors differentiate the medium career commitment category from the high career commitment category. Students are less likely to have medium levels of career commitment relative to high commitment if they have mentoring type relationships with professors who care about their graduate students, and feel that they have good relationships with their advisor.

Thus, overall, it seems that having good relationships with faculty in one's department is highly beneficial to a graduate student's sense of career commitment.

After graduating and returning to teaching in the chemistry department, soon thereafter a transfer over to the School of Education, Department of Curriculum, ironically where the Master's degree was received took place. This was the full circle. Being a black female in the department and given the chance to inspire, educate and support others that may encounter the same or different biases, barriers, or burdens that may impede success. Tenure and promotion with no problem was secured, and several leadership roles, such as the Chairperson of the department, during tenure at this graduate Research I university was obtained. After eighteen years, experiences with a different population and community of learning, was desired.

NAVIGATING THROUGH THE CREST

A position at a state university in the College of Education as an Associate Professor was secured, which was located in even a more southern demographic than Atlanta. There were three other black females in the department, one in reading, technology education, and mathematics education. In the science departments of this university there was only one black biology full- time assistant professor, no other black females in chemistry, or physics. Both bonded and shared their experiences in higher education. She was much younger, therefore, it was believed there was a responsibility for mentoring her. She had challenges with tenure in that she outshined and performed all of her peers, and even those who already had tenure. This was a precarious position to be in, but having already securing tenure at this institution, the experience was shared experiences. She was successful in gaining promotion and tenure. Women today currently earn 41% of PhD's in STEM fields, but make up only 28% of tenure-track faculty in those fields. Reducing the dropout rate of women in STEM careers is especially important in the quest for gender equality because women in STEM jobs earn 33 percent more than those in non-STEM occupations and the wage gap between men and women in STEM jobs is smaller than in other fields. Small things like office space allotment, who gets the next priority office who serves on high profile committees were items for question. The question is, are you not selected because you are a black female in a predominantly white college or something else. Yes, you are a part of the club now, but are you a voting member? During tenure at this institution, it was clear that leadership roles were not awarded to the few black females that were in this college. It again reflects the hierarchical segregation paradigm. Yes you can follow, but not lead. Two of the black females that were there for at least eight years were looked over numerous times. Most times they had to demand equity, it was never intentionally given. Black women demanding equity are perceived as angry black females. Through years of being paid less then white males, white women, and black males, it is not anger, but disappointment. The crest of a wave is the pinnacle or peak with maximum value of upward displacement. Similarly, the movement to leadership positions, allows for career upward mobility.

Addressing the challenges of women in the STEM workforce will require the creation and support of institutional environments that are attractive to women in all stages of their careers.

After four years, and limited leadership opportunities, pursuit of these opportunities were explored and an appointment to serve in a position at an urban two year college as the Dean in the Division of Science, Mathematics, and Health Professions occurred.

THE GOOD OLE BOYS: CAN YOU HEAR ME?

A dean has the responsibility of overseeing the daily operations of the unit as well as secure funding, faculty, and first class education for the institution's students. It can be considered the blue-collar administrative position. More of what some term the managerial level that bridges the administration and the faculty.

This too was a unique situation where the white-collar executive cabinet at the institution was male dominated. A female Vice President was hired soon thereafter, but she always shared with that it was a man's world and women's expertise and opinions were not respected. This was difficult to believe, because she was a V.P. and this was position was dean in science. Another female dean was hired for the business division. Yes, this was what some may say a picture of women's arrival in society, but that is all it was; a picture. As the administration planned and executed the mission of the college, major decisions were made to the exclusion of the input from females in decision making positions. The male voice was the decisive voice. The "good ole boy" syndrome was quite apparent, even at an urban access institution. How could this be? Have you ever heard the contention that if a man says something and the women said the exact same thing, the man's ideas are accepted or heard? An ear attuned to a man's voice. This may be considered tunnel hearing.

While shadowing a white female Vice President and attending a forum on professional women and careers in higher education, the V.P. was the speaker and the floor opened for questions after the presentation. It was asked what my thoughts were about being a female in a male -dominated higher education setting. The response was, try to be yourself as only you can be; through a women's eyes. You cannot think like a man, and you should not have to. Think and filter everything through your own lens; a woman's mind, which may be different, but not wrong. The V,P, that was shadowed said she would have a problem if she did that, and that she had to think like they do. What a terrible situation in which to have to work. Do we have to "think like a man", command and control to be good leaders? This may be the usual management style of men, but women have a different make-up. The socialization of relationships seems to be inherent. Yes command and control are also attributes used, if needed, but there are more dominant characteristics of women that authentically are used to "get the job done". These two do not have to be at odds with each other, but can work synergistically within a hetero-gender unit. One, however, must be accepting and respectful of the other. Due to the tendency of men to be commanding, this is inevitably used by them, even with women who are their peers.

It was fortunate to complete an Executive Leadership training where you met very dynamic men and women eager to climb the executive ladder. Although the director for the program was female, it was noticed that all of the presenters were male. Another oddity. There was, however, something valuable to learn during this institute; there are many ways to lead; one size does not fit all. Which means; I can lead as a woman leads. But is this what has been experienced by women who lead?

Women's lower status and salaries in the scientific community are also reflected in statistics. According to the data provided in 1993, the median salaries of female scientists and engineers with doctoral degrees were 20% less than men. This data can be explained as there was less participation of women in high rank scientific fields/positions and a female majority in low-paid fields/positions. However, even with men and women in the same scientific community field, women are typically paid 15–17% less than men· In addition to the gender gap, there is also salary differences between ethnicity: African-American women with more years of experiences earn 3.4% less than European-American women with similar skills.

Women are also poorly represented in the sciences as compared to their numbers in the overall working population. Within 11% of African-American women in the workforce, only 3% are employed as scientists and engineers. Hispanics made up 8% of the total workers in the USA, and yet only 3% of that number are scientists and engineers.

Women tend to earn less than men in all industries, including government and academia. Women are less likely to be hired in highest-paid positions· The data showing the differences in salaries, ranks, and overall success between the genders is often claimed to be a result of women's lack of professional experience. But, according to the National Science Foundation research, after examining other factors such as age, experience, and education as the causes of why there is a gap in success between men and women, they concluded that discrimination is the only explanation for the poor positions and salaries of women and minorities. The rate of women's professional achievement is increasing. In 1996, the salaries for women in professional fields increased from 85% to 95% relative to men with similar skills and jobs. Young women between the age of 27 and 33 earned 98%, nearly as much as their male peers. In the total workforce of the United States, women earn 74% as much as their male counterparts (in the 1970s they only made 59% as much as their male counterparts.From 1999–2000 to 2009–10, the percentage of degrees earned by females remained between approximately 60 and 62 percent for associate's degrees and between 57 and 58 percent for bachelor's degrees. In contrast, the percentages of both master's and doctor's degrees earned by females increased from 1999–2000 to 2009–10. Within each racial/ethnic group, women earned the majority of degrees at all levels in 2009–10. For example, among U.S. residents, Black females earned 68 percent of associate's degrees, 66 percent of bachelor's degrees, 71 percent of master's degrees, and 65 percent of all doctor's degrees awarded to Black students. Hispanic females earned 62 percent of associate's degrees, 61 percent of bachelor's degrees, 64 percent of master's degrees, and 55 percent of all doctor's degrees awarded to Hispanic students. (http://nces.ed.gov/FastFacts/display.asp?id=72)

As this career continued, an appointment as an interim vice president occurred. You really begin to see the higher education game stadium. Bullying females, disrespectful language, and sexist remarks; statements such as "Women can never devote the time to administration; they have to be home with the family", to" I need to add more bench strength to the administration", meaning there is not enough room for women at the top and they don't have enough professional weight for the position. This is a form of sexism stereotyping. This sounds like the argument we heard from white males when women were fighting for equal rights to work; this was the language of the white male to the white female. The assimilation of the black male has truly been successful. Historically, black men were the head of the household, but black women also worked. The struggle for equality in the workplace impacted each gender. It is debilitating when in the workplace black women are subjected to sexism from the ones of which they live with on a daily bases, or walked side by side in the struggle for equality. A shared history should lead to a shared present and future. The struggle for equality is for all. It is difficult to comprehend some of the behaviors of black male administrators toward black women administrators. Is it a fear or threat of power takeover? The black male has had to fight this fight too long to just give the power up now to a black women. Hopefully this is not the case. This would be history repeating itself with white male dominance taking the back seat to black male dominance. This is the normalized social hierarchical dominance, exclusive of race. Surely the educated black male has not come this far to be the same as his once oppressor. The playing field should at last or least be level in the annals of higher education. Have we not gone through the same throws and snares that you have. Do black women not deserve an equal place as you do when they are just as educated, had the same training, skills, and experiences that their peers have. At institutions of higher education it is expected that we are the educated. But what does that mean?

The position was interim, and lasted a year, and was filled with a black male. Was he selected because he was male or better? We may never know.

CONCLUSION

Gender and racial stereotyping also contribute to the lack of Black women in the STEM fields. The Higher Education Research Institute found that 40% of minority women in scientific disciplines reported subtle forms of discrimination—which is far more than that of their White or male counterparts. This is the reason behind the decisions of some Black women to leave the STEM fields. The limited numbers are staggering. Black women comprise only 2% of the science and engineering workforce in the U.S. compared to 51% of White men and 18% of White women. In 2004, of the 7,488 science and 1,941 engineering doctoral degrees awarded to Americans, only 1.7% were awarded to Black women compared to 30% of science doctorates and 15% of engineering doctorates awarded to White women. By 2006, the number of Black women with doctoral degrees had declined to .34% in computer science and .58% in engineering.

According to research studies, Black women are at the bottom in the education hierarchy and are subject to racial and sexual discrimination at all levels of academic life. The majority of African American women administrators are in urban institutions and those in community colleges feel they are in dead end positions. African American women are underrepresented in community and senior level colleges and tend to serve as assistants to assistants. (Howard-Vital, M.1989).

As we turn toward the future, black women are still making advances in science and science education as well as serving in leadership positions despite the barriers. It however does seem that their paths are the same, the players have changed, but the struggle remains the same.

REFERENCES

Brainard, S. G., & Carlin, L. (2001). *A six-year longitudinal study of undergraduate women in engineering and science*. Academic Press.

Erlick & LeBold. (1977). *Website*. Retrieved from http://www.engr.washington.edu/cwd/Publications/2005-ASEE_Paper.pdf

Ferreira, M. M. (2003). Gender issues related to graduate student attrition in two science departments. *International Journal of Science Education*, *25*(8), 969–989. doi:10.1080/09500690305026

Howard-Vital, M. (1989). African American Women in Higher Education. *Journal of Black Studies*. Retrieved from http://www.jstor.org/discover/2784699?

Luckenbill-Edds, L. (2000). The 'Leaky Pipline:' Has It Been Fixed? *The American Society for Cell Biology*. Career Strategy Columns.

Peng & Jaffe. (1979). *Women Who Enter Male-dominated Fields of Study in Higher Education. American Educational Research Journal*, *16*, 285–293.

Renault, M. (2015). *Conspicuously invisible: Women of color in university sciences*. Education · Minneapolis. Retrieved from http://www.mprnews.org/story/2015/02/21/women-of-color-in-science

Rossiter, M. (n.d.). *Has Feminism Changed Science?* Retrieved from muse.jhu.edu/journals/bulletin_of_the_history_of.../77.1rossiter.html

Schiebinger, L. (2001). *Has Feminism Changed Science?* Cambridge, MA: Harvard University Press.

Sullivan, O. R. (2002). *African American women scientists and inventors. Black stars*. New York: Wiley.

Watts, R. (2007). *Women in Science: A Social and Cultural History*. London: Academic Press.

Chapter 11
STEM Fields and Ethnic Women in New Zealand:
Issues of Sexism and Racism

Charles Mpofu
Auckland University of Technology, New Zealand

ABSTRACT

A critical race theory was used to analyse policies and strategies in place to enable the participation of New Zealand ethnic women of Latin-American, Middle Eastern, and African (MELAA) origin in Science, Technology, Engineering, and Mathematics fields (STEM) in education and industry. The aim was to find out what policy – and other – levers are available for better participation in the STEM fields by the ethnic women's population. The process involved an analysis of publicly available official documents on STEM strategies at national and regional levels. The main findings were that gender issues are expressed in a generic way, either across all ethnic groups, or across the four ethnic groups where the MELAA stands not clearly identifiable in the classifications. Recommendations include the need to develop policies and strategies that account for race and gender equity as part of an agenda to eliminate marginalization of this group.

INTRODUCTION

In 2014 the New Zealand government, in its five-year strategy-Tertiary Education Strategy 2014-2019 (Ministry of Education, 2014), made a landmark announcement of the need for an increase in funding for students taking studies in the science, technology, engineering and mathematics (STEM) fields in universities. This was in line with the Government strategy of boosting the economy through supplying the industry with the required skills and by research and scientific innovation. Graduates in these subjects are in short supply in industry and therefore this was also a response to changing workforce needs (Ministry of Business Innovation and Employment & Ministry of Education, 2014). Following this decision, the state-funded universities responded by incorporating this change into their five-year investment plans. Before this policy was announced, some government and non-government organizations were already

DOI: 10.4018/978-1-5225-0174-9.ch011

involved in initiatives aimed at encouraging women and vulnerable minority groups, such as the Maori and Pacific Island communities, to enroll in STEM subjects. Little is known about initiatives targeting a group which in New Zealand is classified as 'other' or 'ethnic minorities' (consisting of Middle-Eastern, Latin American and African people (MELAA), 1.2% of the total population) to participate in STEM careers. It was against this background that a document analysis was undertaken to find out the presence or absence of statements in policy documents that target women from this ethnic minority group.

This chapter therefore utilises critical race theory to examine the current state of the STEM field for MELEAA women and to analyse organizational and policy instruments that are available to address issues of participation of ethnic women in STEM fields in New Zealand. Evidence that ethnic minority women tend to be on the fringe of policy action in documents targeting ethnic minority women will be presented. This chapter can be of significance internationally in other English-speaking Western countries, such as Canada, Australia, the United States of America and United Kingdom, as the ethnic minority women of New Zealand may be originally from similar source country backgrounds as ones in these countries.

BACKGROUND

In New Zealand and most western High Income Countries (HIC) STEM fields have been seen as critical to the advancement of national economies and competitiveness in the global economy (George, Neale, Van Horne, & Malcolm, 2001; Marginson, Tytler, Freeman, & Roberts, 2013). The New Zealand government d earmarked STEM fields as vehicles for advancing the economy. Thus, the link between universities with industry has been encouraged in tertiary education strategies: the rationale being that STEM subjects are seen as catalysts in achieving economic goals (Ministry of Education, 2014; Mpofu, 2014). For example in most reports it has been argued that mathematical sciences provide direct contributions to profits and savings and underpins a significant proportion (often over 50%) of total business, industry and government activity within New Zealand (Ministry of Research Science and Technology, 1998). This means that the skill gaps identified in STEM at tertiary education institutions and industry are an issue not only for those aspiring for jobs but for the Government's goals of boosting the economy too.

The globalised nature of the workforce leads to competition in attracting this workforce especially among HIC countries. Most western countries have a fluid labor force that flows between these countries and this adds to the competition for workers in these fields (Langen & Dekkers, 2005). Thus, governments in HIC countries are now compelled to come up with novel STEM policies in order to be competitive in this market. Furthermore, although this issue is of concern in most western countries, it has received different levels of attention in different parts of the world. For example, the European Union (EU) formed an official strategy of increasing enrolments in STEM subjects at the beginning of the 2000s (Langen & Dekkers, 2005), while New Zealand has only adopted a related major policy shift recently in 2014 (Ministry of Education, 2014). New Zealand therefore needs more STEM workers, not only to meet industry demands but to keep pace with other HICs which have already put strategies in place to increase their STEM pool. In addition to encouraging the general uptake of STEM fields in tertiary education, different countries have specific policies of addressing equity issues targeting groups such as women and vulnerable communities that have traditionally had low participation in the labor market (Marginson et al., 2013).

PHILOSOPHICAL UNDERPINNINGS: CRITICAL RACE THEORY AND MARGINALITY

The gender and race-related nature of the STEM fields participation calls for attention to be given to both issues of race (see definition in this section) in general and to those of females in particular. Critical race theory is therefore being used as a tool or a theoretical framework to analyse evidence inclusive policies for New Zealand MELAA women to participate in STEM fields. Understanding the definition of racism may help in understanding why the issue being discussed need to be framed in the lens of critical race theory. In the US the concept of race has been used to refer to physical or genetic traits shared by the group, such as skin color, facial features, and stature. However, in the New Zealand context race tends to be defined as a group of people united or classified together on the basis of not just physical or biological traits but common history, culture and nationality (Smedley, 2007). In New Zealand race is more of a social category than a biological one and hence the term ethnicity is usually preferred. In terms of racism, the definition of adopted for the purpose of this work is one used in the US context by Marable (1992). Marable defines racism as 'a system of ignorance, exploitation, and power used to oppress African-Americans, Latinos, Asians, Pacific Americans, American Indians and other people on the basis of ethnicity, culture, mannerisms, and color' (Marable, 1992, p.5). Although exploitation and oppression are too strong words to use in the current era in New Zealand where there laws barring racism, issues of power and subordination of those who are non-white still exist mainly at societal and institutional levels.. The non-white populations who tend to be subordinated include a broad group of Maori indigenous people, Pacific Islands people, Asians and other ethnic minorities who fall in the group of Latin Americans, Africans and Middle-Easterns. In the context of New Zealand racism can therefore be described as a chain of practices, beliefs and other evidence of situations that lead to the reproduction of a racial hierarchy or a social structure with evidence of superiority and privilege for some.

Examples of practices or situations of racial hierarchy are systems that may disadvantage some non-white ethnic groups from participating in social, political and economic structures. Such forms of practices can be result of either systemic racism or institutional racism. They are systemic because New Zealand as a country was founded on policies and practices that disadvantaged the indigenous Maori population and other non-whites from participating in the society (M. Durie, 2003). The systemic racism legacy currently manifests differential employment outcomes which tend to be positive for the New Zealand white population (Statistics New Zealand, 2004). These situations of racial hierarchy are also called institutional racism practices because they also manifests in differential access to educational opportunities and other social opportunities (M Durie, 1998). Institutional racism is assessed on the norms and practices within social and economic institutions, an example in the context of this chapter is the norm that STEM students in New Zealand universities are white male and not ethnic female.

Having described racism and its forms, it is important to also define critical race theory. Critical race theory – now being used in many fields such as education, feminist scholarship, social policy and indigenous studies-, has its origins in the legal field (Day, 1992). Matsuda (1991) views critical race theory as the work of progressive legal scholars of color who are attempting to develop a jurisprudence that accounts for the role of racism in American law and who work towards the elimination of racism as part of a larger goal of getting rid of 'all forms of subordination' (Dey & Hill, 2007, p. 1331). The term 'progressive legal scholars of color' is of critical importance in this definition as it implies that the people who are oppressed have a challenge to self-emancipate. Thus, any attempt which is made to eliminate forms of racism is regarded as challenging the status quo of subordination.

Other scholars who have applied this concept to the field of education have commented that critical race theory is a chain of activities that target evidence of racism in education and attempt to empower those subordinated through race, gender, class, language, and national origin with the aim of achieving self-emancipation (Day, 1992). It can therefore be seen that, broadly speaking, critical race theory is committed to social justice as it eliminates racism, and sexism and empowers the subordinated minority groups (Day, 1992; Dey & Hill, 2007). Furthermore, critical race theory is a framework or set of methods that seek to identify, analyse and transform structural and cultural aspects of tertiary institutions and employment settings that maintain the marginal position of ethnic MELAA women (Hassan Al Marzouqi & Forster, 2011). In the context of this work evidence of subordination of New Zealand ethnic MELAA women through lack of policies that encourage participation in STEM fields will be presented.

Marginality is another concept that needs to be defined in relation to any work that concerns itself with oppressed people. It can be a useful construct to understand the problem of under-representation and is used in circumstances where 'people of color' are subordinated because of their race, gender and class as in this work. The concept of marginality is important for understanding issues concerning women, as Hurtado claimed that "for women of color, race, … and gender subordination are experienced simultaneously and their oppression is not only by members of their own group but by whites of both genders" (1989, p 839). The justification for using critical race theory and marginality is therefore that use of both allows the author to analyse issues of race and gender simultaneously. This is because of what Collins called the simultaneity of oppression phenomenon or simultaneity of marginality phenomenon (Elit et al., 2011; Hassan Al Marzouqi & Forster, 2011). According to Zavella (1991) the importance of the simultaneity of oppression phenomenon is that it guides individuals to be careful about analysing issues of gender from feminist perspectives that ignore issues of color or do not put issues of color in context. Such a theories are reductionist and replicate the already existing oppression of women of color (Hewlett, 2008). The aim in using critical race theory and the construct of marginality in a discussion of issues concerning ethnic MELAA women is that these call for a movement that accounts for race and gender equity simultaneously.

After discussion of disadvantaged ethnic communities in New Zealand, the succeding sections will therefore provide the context of STEM partcipation issues for MELAA women in New Zealand by exploring the partcipation of the general women population in STEM fields. A partcicular focus on the partcipation of the ethnic MELAA women population in STEM fields will then follow. Critical race theory will be used as lens to examine each of these areas. The notion of marginalization will apply in so far as disadvantage of the ethnic MELAA group in general and ethnic MELAA women in particular are concerned.

THE DISADVANTAGED ETHNIC COMMUNITIES IN NEW ZEALAND

The categorization of ethnic communities that will be used in this work is from Statistics New Zealand – a Government department and New Zealand's national statistical office, which is a major source of official statistics and leads the Official Statistics System. According to this department, as of 2013 the ethnic categories of New Zealand are European (74%), Maori (indigenous New Zealanders) (15%), Asian (11.8%), Pacific (from neighboring Pacific countries) (7.4%), Middle-Eastern, Latin American and African (MELAA) (1.2%) (Statistics New Zealand, 2013).

In terms of disadvantage, the stance will be taken in this work that Europeans and Asians are well represented in higher education in terms of staff and students (The Royal Society of New Zealand, 2014). Europeans tend to fare well above all of these groups in socio-economic and health outcomes, followed by the Asian category, and hence in New Zealand there are few or no equity policies targeting these two groups. Similarly, in the USA, White Americans and American Asians have since the 1980s been said to generally exceed their representation in the population in terms of socio-economic outcomes (Chipman & Thomas, 1987; Cole & Espinoza, 2008).

New Zealand is an example of a country where the concept of disadvantaged ethnic communities is complex, as there are a number of categories of disadvantaged communities. There are Maori indigenous people, who are disadvantaged in the sense of suffering past colonial injustices. These issues for Maori tend to be linked to historical imbalances in the European–Maori relationship in the distribution of resources. Now these formerly disadvantaged communities have a mandated right to participate in educational, economic and social institutions through the provisions of biculturalism entrenched in the Treaty of Waitangi. Signed in 1840, the Treaty of Waitangi (the Treaty) was an agreement made between the Crown of the United Kingdom and the native chiefs and tribes of New Zealand to protect Maori against increasing colonial dominance (M. Durie, 2003). Thus, in recognition of Maori as indigenous people, as well, as the socio-economic and political injustices they suffered after the signing of the Treaty, the New Zealand government has since the 1970s introduced social reforms. These reforms were aimed at redressing the past injustices through addressing socio-political inequalities between Maori and non-Maori by adopting biculturalism in official, social and organizational practices in New Zealand.

There are also Pacific peoples or Pasifika (i.e. the broad group of ethnic groups of peoples whose ancestry is from Pacific Island nations). These are disadvantaged in the sense of socio-economic and health outcomes. Pacific peoples have worse economic circumstances than the overall population, with the majority of Pacific peoples living in areas with the fewest economic resources (Ministry of Pacific Island Affairs, 2010). The 2006 Census of Population and Dwellings showed that 35 percent of Pacific peoples had no qualifications, compared with 25 percent of all New Zealanders. The unemployment rate is higher for Pacific peoples than for any other ethnic group (Ministry of Pacific Island Affairs, 2010).

Lastly, there generic terms for vulnerable ethnic communities that include immigrants and refugees. They are known by categorizations such as 'other' or 'MELAA'. The ethnic MELAA communities arrive in New Zealand through skilled migration, family reunions and refugee status. Refugees in the MELAA group are mostly from Iran, Iraq, Afghanistan, Palestine, Syria, and Kuwait in the Middle-East. Regarding Africa they are traditionally from Tunisia, Algeria, Somalia, Djibouti, Eritrea, the Sudan, Ethiopia, Burundi, Rwanda, the Democratic Republic of Congo (DRC), Republic of Congo (Congo-Brazzaville), Sierra Leone and Uganda, while those from Latin Americans have traditionally been from Columbia and Chile (Statistics New Zealand, 2013).

WOMEN IN STEM FIELDS IN NEW ZEALAND

This section will comment of the participation level of women in general, and ethnic MELAA women in particular, in STEM fields. Data about participation in industry in general will be presented first followed by data from tertiary enrolments. Finally comments will be made about participation in other sectors in general. The school-level data (primary and secondary schools) will also be analysed. Before proceeding, it is important to describe the method used to gather and appraise the data utilised in this section.

METHODS: DOCUMENT ANALYSIS

The method used for data collection was a qualitative document analysis which is described by Bowen (2009) as a systematic procedure for reviewing or evaluating printed or electronic documents relevant to a particular topic. In terms of process followed document analysis involves finding, selecting, appraising and synthesising data contained in documents (Bowen, 2009). In a qualitative document analysis the aim of appraising and synthesising data is to gain "real," "rich," and "deep" data with more consideration being given to gaining a total context or complete picture. This differs from quantitative research in which selected, pre-defined variables are studied (Ritchie & Lewis, 2003). Validity in qualitative document analysis is achieved by the inclusion of raw data in the final report either through direct quotations or illustrative extracts. This allows the reader to see exactly the basis upon which the researcher's conclusions were made. This process of document analysis was conducted in two phases as will be described below.

Aims of Phase One of Document Analysis

The aim of conducting phase one of document analysis was to examine support for participation of women in the labor force and particularly in STEM fields in New Zealand. Such an exercise was meant to provide a context for critically analysing available support systems for MELAA women in STEM fields in New Zealand.

FINDING AND SELECTING DOCUMENTS

The main data sources on participation levels were Statistics New Zealand and from the Ministry of Education. The Statistics New Zealand tends to have more detailed information about participation levels in different sectors of the economy. The ministry of education has information about enrolments at various levels in primary, secondary and tertiary education in New Zealand. Further in other sections, sections this information was triangulated with publicly available official websites and documents, and reports on STEM subjects and careers.

APPRAISING AND SYNTHESISING DATA

Participation of New Zealand Women in Industry

The documents reviewed indicated that women in New Zealand have over the last two decades been under-represented in STEM fields and other trade-related high-wage-earning occupations (Roberts, Gardiner, Gilbert, & Vaughan, 2008). It has even been argued that women's qualifications in New Zealand have always been concentrated in a narrow range of subject areas at university; namely the arts, the social sciences and education. For example, in 2001 the most common post-school qualification for women aged 15 years and over was in the field of health (22 percent), while for men it was in engineering and related technologies (33 percent). In the same year the ratio of males to females in the field of engineering was more than 14 to 1, while in architecture it was more than 10 to 1 (Statistics New Zealand, 2005a). Further worrying statistics show that in 2006, 86.8% of architects, engineers and related professionals and 83.9% of physical science and engineering technicians were male (IPENZ, 2010).

Table 1. Participation in STEM careers among urban dwellers by sex and ethnicity in the 2013 census data

	European		Maori		Pacific Peoples		Asian		MELAA	
	Males	Females	Males	Females	Males	Females	Males	Females	Males	Females
Civil Engineers %	92.1	7.9	93.8	6.3	88.5	11.5	83.0	17.0	93.8	6.3
Quantity Surveyors %	86.7	11.7	86.4	13.6	81.8	18.2	73.4	26.6	71.4	28.6
Industrial Engineers	92.8	6.7	88.2	11.8	100.0	0.0	76.7	23.3	57.1	42.9
Mechanical Engineers	98.3	1.7	98.6	1.4	94.3	5.7	93.7	6.3	92.9	7.1
Software Engineers	90.4	8.7	87.6	12.4	87.9	12.1	87.5	12.5	90.6	9.4
Computer Network and Systems Engineers	94.8	4.9	93.8	6.3	87.5	12.5	91.6	8.4	100.0	0.0
Architectural, Building & Survey Technicians	90.9	8.3	100.0	0.0	100.0	0.0	81.8	18.2	0.0	100.0
Electrical Engineering Technicians	97.5	2.4	94.4	5.6	93.3	6.7	95.2	4.8	100.0	0.0
Fitters (General)	97.6	2.3	90.6	9.4	86.2	13.8	89.3	10.7	100.0	0.0

Suggestions have been made about examining women's participation levels by looking at employment levels in senior academic roles, such as heads of departments and deans. In this regard it has been found that participation of woman is worrying, for example, in four of the country's seven universities there are no woman participating at deans level (Ministry for Women, 2014). I has also been found that there are very few female Heads of Departments (HODs) within science faculties in New Zealand universities. Another example given was that Waikato University formerly had 42 percent female representation at HOD level in 2012, while other universities namely, AUT, Massey and Lincoln universities had no female staff at HOD level in science (The Human Rights Commission, 2008; Ministry for Women, 2014)

Another way of assessing the possible disadvantage of woman in STEM fields is by assessing the number of woman who participate in prestigious research awards such as the Marsden Fund and the Royal New Zealand Fellowship Fund. In this regard it has been found that less than one third of woman managed to secure the Marsden Fund in 2012. This fund is one of the largest funds in New Zealand for basic, non-applied research and success is seen as prestigious within the science community. Moreover in 2010, among 371 successful applicants to Royal Society Fellowship only 33 were women (Ministry for Women, 2014). Within this group of women it is not known how many were ethnic women. However, given the generally low participation levels of immigrants in the labor market, it would not be surprising to see ethnic women being far less represented in these fields.

Even current data from Statistics New Zealand shows concerning levels of under-representation of woman in STEM fields. For example, Table 1 shows that in the 2013 data on employed urban dwellers aged 15 and above, in almost all STEM fields and across all ethnicities men tended to be over-represented.

In terms of ethnic breakdown in the field of Architectural, Building and Surveying Technicians among Maori, Pacific and the MELAA ethnic group males were represented at 100% with no woman participating in those fields. Other field with 100% male participation were Electrical Engineering Technician and fitter general in the MELAA group where males were represented at 100%. It was among the MELAA group that at least three fields were represented by males at 100%. According to the household data (refer to Table 1) the mean percentage of representation of males in STEM fields was 82%, while the modal percentage representation of males in STEM fields was 100%.

Tertiary Institutions

Another way of assessing the participation levels of women in STEM fields is by looking at the enrolment data in tertiary institutions. In 2006 a total of 212,300 students enrolled in engineering and related subjects in various levels, from level one to level ten – 196,000 were male while only 16,800 were women (Statistics New Zealand, 2006a). Further gender statistics show that in all levels, from certificates to doctorates, female information technology students totalled 2,550 while males constituted 7,110 across all age groups. In the category of engineering and related technologies across all levels there were 2,730 females enrolled and 15,190 males enrolled. In mathematical sciences there were 1,730 females and 2,640 males (Statistics New Zealand, 2006c). All the statistics indeed show that women are underrepresented in STEM areas.

School-Level Data

Although school-level data is just as important as tertiary-level data in discussing issues about participation of ethnic women in STEM fields, little is known about data disaggregated to this level in New Zealand. Data available about other ethnic groups show that trends are not always predictable and hence ethnic-specific data is essential. For example data, from the 1994 Third International Mathematics and Science Study (TIMSS) indicate that Asian males outperformed their female counterparts in both mathematics and science in forms 2 and 3. The mathematics scores for Maori girls were slightly higher than for Maori boys in both forms. Pacific Island girls outperformed boys of the same ethnic group in form 2, and by form 3 they had equal scores. For European children, both boys and girls had equal mathematics scores in form 2 but by form 3 boys had a higher achievement score. By form 3 the science achievement scores were higher for boys than girls in all ethnic groups (Statistics New Zealand, 2005b). These variations indicate, firstly, that specific data about the MELAA group may be required. Secondly, the differential achievement may suggest that "shutting out" of women in STEM appears to begin even before college and workforce.

General Participation of Ethnic MELAA Women in STEM Fields

Although disaggregated data on individual ethnic groups included in MELAA was not available, data for the generic MELAA ethnic group was utilised. Surveys of Statistics New Zealand Census 2006 once indicated that, of all ethnic groups with bachelor's qualifications, the MELAA ethnic group lagged behind in full-time employment – MELAA 41%, Asian 42%, Pasifika 46%, Maori 48% and European 50.4% (Statistics New Zealand, 2006b).

Another source of data that provides worrying statistics is the New Zealand Ministry of Education (NZ Ministry of Education). The statistics from this source reveal that in tertiary enrolments in engineering and related technologies (including manufacturing, processing, automotive, mechanical, civil, geomantic, electrical, aerospace, maritime engineering and others) the MELAA had fewer MELAA students enrolled in 2013. For example in 2013 it was seen that among full-time students enrolled in these fields, from certificate to doctoral level, only 910 were MELAA, while 2,970 were Asian, 1,420 Pacific, 3,050 Maori and 11,280 European. In the category of information technology a total of 17,920 students were enrolled from certificate to doctorate level. European students constituted 5,660 while Maori constituted 1,230, Pasifika 880, Asian 2,280 and MELAA 590 with other unclassified ethnicities included in them. In the category of mathematical sciences it was seen that Europeans constituted 2,800, Maori 390, Pasifika 330, Asian 1,080 and MELAA 240 (Statistics New Zealand, 2006c).

It is important to note that the MELLA group in New Zealand consists of Latin-Americans. This ethnic group is also found in the USA and under-participation in STEM fields has been an issue for this group. Although their experiences in the US can be different from New Zealand it is important to put this issue in an international perspective. To put the issue in an international perspective, although Latinas enrol in greater numbers in undergraduate education in the US, they are under-represented in STEM fields in the US. For example, in 2005 Latinas were awarded 60% of the undergraduate degrees but only managed 37% of these degrees in STEM fields (Cole & Espinoza, 2008; National Center for Education Statistics, 2005). In the USA past enrolment trends indicate that increased involvement of under-represented groups is essential in meeting demands in STEM fields. Traditionally, STEM workers in the US have been white, non-Hispanic men (Society of Women Engineers, 2006; Hewlett et al., 2008; Frehill et al., 2009).

Having compared the USA Latin American students STEM enrolment trends with the New Zealand ones it can be seen that it would have been useful to discuss enrolment trends in other source countries of the MELAA group such as African and Middle Eastern ones. However, data about trends in these countries is scarce. Even in New Zealand discussion on specific data about ethnic minority women has been limited by lack of consistent follow-up on this group by the various government and other responsible agencies. The lack of data among specific minority groups has been found to be an issue even in the US. In this regard it has been found that some studies only include information on Hispanic American and African Americans, leaving out information on Native Americans, the reason for this being that the available statistic will be too small to make any statistical generalizations. Scholars have argued that disaggregated data is very important in order to differentiate factors that affect racial/ethnic groups and those that are generic to all students (George et al., 2001). Nevertheless, conclusions have been drawn on available data.

In summary it can be said that although the statistics that segregate participation according to gender were not available, in each ethnic group it is logical that if women in general are disadvantaged it will be more so with ethnic minority women because of the reasons already discussed. It may also be taken that if the general MELAA group is disadvantaged then the women in this group will be doubly disadvantaged, as in the general population women are under-represented in STEM fields Similarly, given that the MELAA ethnic group lags behind in employment outcomes and that women are outweighed by men in STEM fields, it can be argued that ethnic minority women are at a disadvantage in enrolling in STEM fields. It is therefore worrying that there are no overt policy platforms that specifically track this group of women in tertiary studies.

STEM INITIATIVES IN NEW ZEALAND: PHASE 2 OF DOCUMENT ANALYSIS

The primary aim of conducting a second phase of document analysis of STEM initiatives in New Zealand was to find out what policy initiatives and other levers are available for better participation by ethnic

Table 2. Key government ministries and other agencies' mention of STEM education and careers

Organization	Statement on STEM	Statement on MELAA in general	Statement on Women in general	Statement on MELAA women
Ministry of Business, Innovation and Employment	Yes	None	Yes	None
Ministry of Women's Affairs	Yes	None	Yes	None
Office for Ethnic Communities	No	None		None
Human Rights Commission	Yes	Yes	Yes	Yes
Association of Women in Sciences	Yes	None	Yes	None
Refugee and Migrant Services (RMS)	None	Yes	Yes	None
National Advisory Council on Employment of Women (Work Programme)	Yes	None		None
The State Services Commission	No	Yes	Yes	Yes

Table 3. Key documents reviewed

Document Name	Separate inclusion of Ethnic MLAA group or Ethnic MELAA Women
Literacy and Mathematics in Years 9 and 10: Using Achievement Information to Promote Success (Education Review Office, 2012	No
Mathematics in New Zealand: Past, Present and Future (Ministry of Research, Science and Technology, 1998)	No
Science in the New Zealand Curriculum: Years 5 to 8 (Education Review Office, 2012)	No
Looking ahead: Science education for the twenty-first century. A report from the Prime Minister's Chief Science Advisor. Auckland, NZ: Office of the Prime Minister's Science Advisory Committee (Gluckman, P., 2011).	No
Curriculum support in science. Patterns in teachers' use of resources. Report prepared for Ministry of Education (Hipkins, R., & Hodgen, E., 2012).	No
A nation of curious minds: A national strategic plan for science in society (Ministry of Business Innovation and Employment & Ministry of Education, 2014)	No

MELAA women in the STEM fields. The secondary aim was to find out whether any statements in policy documents target women from ethnic minority groups. This process involved an analysis of publicly available official websites and documents, and reports on STEM subjects at primary and secondary schools, tertiary institutions and employment settings.

Finding and Selecting Documents

Specific government ministries were searched for documents or statements about diversity or enablement of the participation of women or ethnic communities. Any statement that targeted empowerment or enabling participation of women or ethnic communities without necessarily mentioning STEM subjects/careers was deemed relevant. The government ministries and other agencies studied are listed in Table 2.

Publicly available official websites and documents, and reports on STEM subjects and careers were also analysed. The official and academic documents that were scrutinised and their results in relation their focus on ethnic MELAA women are shown in Table 3.

As can be seen in Table 2 and 3 the results yielded were related to both availability and lack of policies and initiatives targeting either women or ethnic minorities at school, tertiary institutions and in workplaces. The next section will therefore critique the initiatives targeting schools, tertiary institutions and workplaces.

APPRAISING AND SYNTHESISING DATA INITIATIVES TARGETING PRIMARY AND SECONDARY SCHOOLS

It was necessary to review the school-level initiatives because internationally it has been found that preparatory efforts in subjects at high school that are related to STEM careers tend to have an impact on STEM careers. In the USA it has been found that enrolment and persistence in STEM subjects has been linked to preparatory activities in high school regardless of the ethnicity of students (Cole & Espinoza, 2008). Researches that have led to such conclusions have tended to focus on the relationship between high grades in STEM subjects at school level with entry into and persistence in STEM studies (Whitley, 2005). Such researches have been used to inform other studies focusing on ethnic minorities such as Latinos in the USA, and the results have been seen to be similar (Cole & Espinoza, 2008).

Most of the New Zealand initiatives have been triggered by either evaluations or reports on the general poor performance of students in STEM areas in New Zealand. On this note, a study that gives insight into the performance of students in STEM subjects at primary school level the 1994 Trends in International Mathematics and Science Study (TIMSS), conducted by the International Association for the Evaluation of Educational Achievement (IEA) (Statistics New Zealand, 2005b). This study, which involved just under 5,000 students, focused on achievements in science and mathematics. New Zealand ranked fourteenth out of the 17 countries in mathematics achievement for grade 4 students and tenth for students in grade 8. The ranking for science was consistent at thirteenth place for both grade 4 and grade 8. Overall, the student achievement for both mathematics and science for New Zealand was lower than the international average, ranging from 38 points lower for grade 4 mathematics to 11 points lower for grade 8 science. In this study it was seen that Europeans and Asians achieved mean scores that were higher than Maori and Pacific Islands ethnicities (Statistics New Zealand, 2005b). These results reflected international trends in the disadvantages of minority students. For example, in 2005 the average Grade Point Average (GPA) score of Latino high school graduates in the US was 2.82 compared to 3.05 for white and this was used to indicate that even if they were recruited to STEM courses, ethnic minority students were less likely to persist in their studies (Cole & Espinoza, 2008; Shettle et al., 2007). Knowledge of such factors may be useful as they point educators to students with higher needs, who may benefit from retention initiatives.

Although this study gives insights about the dominance or better performance of European and Asian students over Maori and Pacific Island students in New Zealand, it is noted that the students enrolled in this group were not disaggregated into other ethnic groups such as MELAA. Such data is seen as useful only in this instance where there is a lack of comprehensive disaggregated follow-up data on the MELAA ethnic group.

The positive development of the TIMSS (1994) study was that in 1997 the Ministry of Education established the Mathematics and Science Taskforce as a deliberate effort to improve performance in these subjects and therefore in STEM careers. A variety of government funded initiatives followed the establishment of this Taskforce. In response to the poor performance of New Zealand students in the subsequent 1995 TIMSS assessment, support tools funded by the Ministry of Business, Innovation and Employment (MBIE) in the mid-2000s were introduced, which targeted student learning and skills for assessments. Some of them were the Assessment Resource Banks (ARBs), which were Science, Mathematics and English assessment resources for 7-14 year-olds, and the biotechnology and science learning (Buntting, 2012).

Indeed internationally it has been reported that, as a strategy, improvements and diversity in teaching strategies of sciences at both school level and university level can also contribute to the participation and retention of disadvantaged communities and women in STEM fields (George et al., 2001). Such programs that have provided students with an opportunity to engage in hands-on, real life projects have been successful in increasing female enrolment and retention.

Other initiatives include the Ministry of Education-funded resources such as the Laptops for Teachers Scheme, the School Network Upgrade Project and the Ultra-fast Broadband for Schools (Buntting, 2012). However, not much is known about the extent to which these resources were tailored to assist disadvantaged communities such as MELAA ethnic communities or MELAA women in particular. These resources could be beneficial if they had targeted low decile suburbs that have a lot of immigrants, such as Hillsborough and Mt Roskill schools in Auckland, which are known to have students from MELAA countries.

The justification for channelling such resources to regions or school zones with MELAA communities can be found in the fact that past research in New Zealand has linked being an immigrant and having English as second language (two conditions that describe MELAA communities) with lower scores in mathematics (Caygill & Kirkham, 2008). Students who always or almost always spoke English at home had, on average, higher mathematics achievement than those who sometimes or never spoke English at home. In terms of immigration status, it was found that on average mathematics scores were lower for immigrants than those born in New Zealand (Caygill & Kirkham, 2008).

In terms of simultaneity of disadvantage, it can be seen that if ethnic communities are disadvantaged overall, MELAA girls are more disadvantaged. This is taking into consideration the aggregate immigration disadvantage and other disadvantages associated with attitudes and gender gap margins in source countries. Source country disadvantages include the poor Gender Empowerment Measures (GEM) i.e. the efforts made by government and non-government organizations, individuals and private organizations in empowering women to participate in social, economic and political spheres of these countries. The GEM levels of these countries may be lower than that of New Zealand (Human Development Report, 2009).

Initiatives that Target Teachers' Professional Development

In New Zealand there are also initiatives that target teachers, as it was once reported that this country's teachers were having obstacles in innovative methods of teaching STEM subjects in the early 1990s (Buntting, Jones, McKinley, & Gan, 2013). This was followed by strategies targeting professional development of teachers, and these include the Growth and Innovation Framework (GIF) Technology Education Initiative funding for teachers' technology skills at all school levels. There are also the New Zealand Research and Researchers Teacher Fellowships of the Ministry of Business, Innovation and Employment (MBIE). However, a scrutiny of the documents of these programs does not show any evidence of targeting ethnic minority MELAA or the women in this group, which may indicate that this group is left out of important national programs. Staff development programs that target the MELAA group may be desirable, as such staff may end up acting as role models for ethnic minority students.

ORGANIZATIONAL AND GOVERNMENT INITIATIVES

It has been argued that government policy is important as an empowerment tool for women to participate in STEM careers, as policy determines what happens in the labor market through educational and labor legislation. Moreover, government campaigns are often aimed at influencing social attitudes (Dekkers, 1996). It has also been established that legal enablers such as creating structural, attitudinal, and gender egalitarianism are the first steps towards enabling participation of women in educational settings and in the job market (Jordan & Yeomans, 2003). Another direct government influence is the funding of educational reforms and research to promote interest in STEM subjects, as has already happened in New Zealand (Langen & Dekkers, 2005; Ministry of Education, 2014). Below is a review of what government ministries are doing in New Zealand which may directly or indirectly influence the participation of women in STEM careers.

Key Government Ministries

A search for initiatives targeting women in tertiary and workplaces was conducted as described in the section on search process. The selection of private organizations and key government ministries (shown in Table 2) was also based on regular citing in issues related to women in New Zealand literature (Buntting et al., 2013). Government ministries of direct interest to this issue included ones that were linked through STEM education and STEM careers and the economic development needs: the Tertiary Education section of the Ministry of Education, the Ministry of Education and the Ministry of Business, Innovation and Employment (MBIE). Other ministries and government departments were linked through general policy coordination (State Services commission), advocacy (Ministry for Women, Human Rights Commission, ARMS), and ethnic specific issues (Office of Internal Affairs).

A key issue observed is that within these ministries is that ethnic issues about STEM subjects and careers are not spelt out in key government policies or strategies. The Ministry for Women has strategies for the inclusion of women in STEM careers, but there are none that specifically target ethnic women (see table 2). Moreover, in the Ministry of Education and MBIE, although there is potential for inclusion of policies about the ethnic MELAA women in schools, universities and in employment sectors, there is not much evidence of the inclusion of MELAA women in this ministry's policies. Ethnic MELAA women policies do not exist despite the fact that in all these sectors there are clear policies regarding Maori and Pacific Island communities, justifiably because of past colonial injustices (for Maori) and documented low participation levels in tertiary institutions and in the economy (for Pacific peoples). The existence of Maori and Pacific policies is both a strength and model of evidence that the same could happen for ethnic communities, although it is acknowledged that research evidence of their needs already shows that equity policies are overdue for these two groups.

The State Services Commission

The State Services Commission administers the New Zealand Public Service Equal Employment Opportunities (EEO) Policy which has strong emphasis on diversity and equality under section 6 of the State Sector Act 1988 (SSA). One of its objective in implementing the policy is to promote, develop and monitor Equality and Diversity Policy across all departments in the Public Service. The State Services Commission specifies that equality will be achieved when one of the following conditions prevails in organisations:

- *inclusive, respectful and responsive organisational cultures which enable access to work, equitable career opportunities, and maximum participation for members of designated groups and all employees (Bowen, 2009)*

The above is to be achieved partly through basing appointments on merit, "while recognising the employment aspirations of Māori, *ethnic and minority groups, women,* [my emphasis] and people with disabilities (Bowen, 2009). Two things are of critical importance in the function and policy strategy of the State Services Commission. Firstly, this is one of the few government agencies with that has a policy and legislative platform that can be capitalised on in advocating the issues of MELAA ethnic women. Secondly, EEO has a focus on both ethnicity and gender. Despite such strengths one issue that has been observed is that in its documents there is no clear separate data on MLAA ethnic group or MLAA women. It is also surprising that there is lack of coordination in making consistent ethnic categorization between Statistics New Zealand and the State Services Commission although both are government agency.

The Office for Ethnic Communities

The office of ethnic affairs has a minister (Minister for Ethnic communities) whose portfolio is similar but separate from other ethnic ministries such as Ministry of Maori Affairs, and Ministry of Pacific Island Affairs. A critical analysis of the websites of these ministries shows that there are policy leverage potentials within these ministries that can be fully utilized for MELAA purposes. For example, both the Maori and the Pacific Island ministries seem to have a diverse range of programs and scholarships that are either specifically for their own women in STEM fields or that can be exploited by women to advance their STEM careers. On the other hand, STEM career scholarships from the Office for Ethnic Communities are scarce. The second observation about a comparison of initiatives of the Office for Ethnic Communities and the other two ethnic ministries is that the Maori and Pacific ministries have a strong tradition of advocacy for quotas in enrolments, such as the Maori and Pacific Admission Scheme for Health Professions at the University of Auckland and University of Otago. On the other hand, there seems to be no clear advocacy for quotas in tertiary institutions for MELAA ethnic communities despite the fact that this ministry has refugees and involuntary migrants who can be considered vulnerable.

The other issue about ethnic MELAA women being under the Office for Ethnic Communities is 'ethnic lumping' – a situation where a diverse range of ethnic groups are put under one umbrella despite evidence of diversity of disadvantage or needs. For example, as will be discussed later, people from continental Europe are under this group together with people from Africa and the Middle-East.

OTHER ORGANIZATIONS

In addition to government ministries described above recognized private, non-governmental and charitable trusts were scrutinized for statements that overtly mention ethnic MELAA women or ethnic MELAA group in general. The organizations in table 4 were searched with the key words 'science, technology, engineering and mathematics' (STEM) in combination with 'ethnic woman' or 'women of color' or 'refugee women'. It was found that although most of these did have statements on STEM or encouraging and advocating for women to participate in STEM none of them mentioned ethnic women as can be seen in table 4. From a critical race theory lens this situation is of concern as the women of color appear to be either left out in key policy instruments or have little or no mention in key government documents.

It was also seen that among these ministries one issue which can potentially affect the strength of policy leverage and research among ethnic MELAA women is that the definition of ethnic communities seems to be different between government agencies. For example, Statistics New Zealand shows Asians as a standalone category whereas the Office for Ethnic Communities includes Asians and those from Continental Europe in its definition. This issue of lack of consistency can pose problems in both advocacy and research.

In addition to government bodies, there are some recognized private, non-governmental and charitable trusts in New Zealand that have leverage to advocate for either women in general or for women in the MELAA group. For example, the RMS is a well-recognized charitable trust that carries out functions that are decentralized to it by central government, such as resettlement functions. This organization has the specific task of helping immigrant women into leadership. The strength of this organization is that it works on the premise that these are vulnerable communities. However, this organization that takes a broad definition of ethnic communities. This organization caters for individuals who fall under the broad umbrella 'migrant and refugee' category.

STEM INITIATIVES TARGETING WOMEN IN GENERAL

Among all the avenues of promoting STEM careers, those initiatives targeting women tend to be largely from government ministries, private organizations, charitable trusts and professional organizations. Examples include Women in Trades networks in Auckland, Wellington and Christchurch. These networks are run by tradeswomen for tradeswomen and aim to build support for women working in trades (Ministry for Women, 2015). Women's academic support groups are of special significance to note, as in the USA it has been found that the retention of Latinos in STEM subjects has reportedly been enhanced by faculty support and support groups (Arksey & O'Malley, 2005; United States Department of Health and Human Services, 2015).

Table 4. STEM initiatives in New Zealand and their race and gender focus

Initiative	Women in general	Ethnicities in general	Ethnic MELAA women in STEM
The Institution of Professional Engineers New Zealand (IPENZ): Women in Engineering Programme	Yes	Yes	No
Electricity Supply Industry Training Association (ESITO): Women in Power and the Ultimit initiative	Yes	Yes	Yes
Woman in Innovation	Yes	No	No
Edna Waddell Undergraduate Scholarships for Women in Technology and Engineering	Yes	No	No
NZ Women in Leadership (NZWIL) Programme	Yes	No	No
Zonta Women in Science Award	Yes	No	No
CPIT Woman in trades initiatives	Yes	No	No

Seven organizations (shown in Table 4) in New Zealand were selected and their websites were scrutinised for evidence of a focus on ethnic minority women. The search for a focus on ethnic minorities or ethnic minority women included reading all the displayed pages on the websites as well as searching the websites using the words 'ethnic minorities', 'other ethnicities', 'ethnic women'. The documents concerning application for scholarships on each website were also scrutinised for texts that suggested a focus on minority women.

None of the websites of these organizations showed anything to support ethnic MELAA women in STEM fields. Another criticism of these initiatives is about the limited number of awards. Although funding limitations are acknowledged, it is argued that scholarships that select one or two will be based on the Darwinian 'survival of the fittest' principles. Even if they are for disadvantaged groups, knowledge of access to funding resources is a privileged enjoyed by the dominant social groups. Those who know about these resources may tend to be the advantaged. This leaves ethnic women in marginal positions. Another criticism of these scholarships is that some of them are awarded based on excellence in academic achievement. Using a critical race theory lens, some of these scholarships can be seen to further inequalities, as high achievers tend to be Europeans and immigrant groups such as Asians.

Research Initiatives on STEM Studies and Careers

Research efforts in understanding factors that lead to success and persistence in STEM careers are some of the initiatives addressing issues in this area. As has been briefly stated before, in addition to organizational empowerment, in New Zealand there are some research centers that target STEM careers. A search was made for research activities in research centers using the terms 'ethnic' or 'minority' combined with any one of the following words: 'science', 'engineering', 'technology' or 'mathematics' or 'STEM'. The search included the descriptions of members of the institutes and also the publication hubs of the research centers. Members' names were also searched in academic databases such as Google

Table 5. Research activities of selected STEM centers in New Zealand

STEM-related research center	Women-related themes	MELAA Ethnic themes	Other ethnic themes	Research on enabling factors or curriculum and teaching issues
Community for Undergraduate Learning in the Mathematical Sciences (CULMS), University of Auckland	Yes	No	Yes	Yes
Technology, Environmental, Mathematics and Science Education Research Centre (TEMS), University of Waikato	Yes	No	Yes	Yes
Engineering Education Research Unit, University of Waikato	Yes	No	Yes	Yes
AUT-STEM CENTRE, AUT University	Yes	No	No	Yes

Scholar. These results were supplemented by manual searches for the publications of the members and their co-authors on the websites of these centers (Table 5).

Among the four research centers examined it was hard to find research that focused on the MELAA group from a New Zealand perspective. The results yielded tended to be researches on either Maori and Pacific students or generic about non-European students. Other themes of these research centers tended to be either women generic i.e. having women's themes but not women of the ethnic MELAA group or investigation of general student populations. Research about the participation of the MELAA group in these centers tended to be scarce. Although there is much scholarship about Latin-Americans in the USA (Chapa & De La Rosa, 2006; Ministry of Business Innovation and Employment & Ministry of Education, 2014) such scholarship may not be directly transferable to the New Zealand context because their experiences are different in New Zealand. Moreover, research about participation of Africans in STEM even in their countries of origin is scarce, except for one study (Banatvala & Doyal, 1998). It is therefore a concern that while there is such a paucity of literature about the MELAA group, not much research is happening to investigate factors that may inhibit or facilitate success among these communities. Admittedly this may be due to the small numbers of students of this group. Unless research efforts specifically target ethnic minority women such research may not yield noticeable benefits to women of the MELAA group.

A CRITICAL ANALYSIS OF THE STATUS QUO

Race and the Problem of Homogeneity

A critical analysis of the issues discussed above reveals that the major issue in enabling the participation of ethnic MELAA women in STEM careers is linked to the lack of consistent, disaggregated data for this specific group. There are a number of examples of key official surveys and documents that are said to be for either 'NZ major ethnic groups' or 'NZ ethnic groups' which do not say anything about the MELAA group. Some these overtly exclude this group, while others group them in the 'other' category or annex them with another ethnic group e.g. 'Asians and Other' (Statistics New Zealand, 2005b).

Consequently, issues for this group tend to be hidden in broad and inconsistent categories such as the 'other' group. It can therefore be seen that the marginal position of ethnic MELAA women is exacerbated by the lack of clear and consistent criteria for grouping these women. From the perspective of critical race theory, this could be interpreted as the tendency of racist practices to ignore nuances, identities, and difference among "like" groups (Chan-Malik, 2015; Whitley, 2005). Such a tendency to "lump" all people of color in one category can often create a situation in which social justice work is done for none of these groups.

In terms of the need for disaggregated data, another issue that warrants analysis using a critical race lens concerns the inclusion in the MELAA group of migrants from South Africa as a generic group. In fact, South African migrants may be dominated by races of European descent and this may skew the data on participation levels of the MELAA group. Due to their cultural similarities with western countries, they may not face the same obstacles as MELAA immigrants from African countries (Banatvala & Doyal, 1998).

Disadvantages from Traditions in Source Countries

The other issue relates to the diverse traditions of the countries of origin for ethnic MELAA women. It is expected that the values of countries of origin will have an influence on their beliefs about STEM subjects. Indeed it has been found that one issue of imbalance been women and men in STEM fields lies with attitudes, values and beliefs in different countries; for example the belief that men are physically superior to women and therefore destined for STEM fields is still held in some societies (Smyth & Nosek, 2015). Hanson, Schaub, & Baker (1996) argued that in the ideological climate in Finland, as well as in former East Germany, where gender equity has always been a priority, such egalitarian ideologies largely account for the high numbers of women in higher STEM professions. Moreover, Bradley and Charles (2004) also argue that attitudinal and gender egalitarianism have a stronger effect on women's participation in STEM fields (Langen & Dekkers, 2005). It is also expected that these notions could still be held in New Zealand given that this country consists of immigrants from different countries.

Furthermore, on the diversity of traditions in source countries such as the Middle East and some parts of Africa, it can be argued that these women bring with them disadvantages related to gender imbalance (World Economic Forum, 2014). In the Gender Gap Index (GGI) (inequality of achievement between women and men) of the Global Gender Gap Report (2014) the ranking of Latin American countries were Uruguay 82 out of 142 countries that participated, and Venezuela 86, while African countries such as Uganda and Algeria were ranked at 88 and 126 respectively. In the Middle East countries such as United Arab Emirates were ranked at 115 (World Economic Forum, 2014). New Zealand was ranked 13 out of the 142 countries. Comparable English-speaking countries such as the USA were ranked 20, United Kingdom (UK) 26, Canada 19, and Australia 24. The above rankings show that if the participation levels of women are worrying at national level, vulnerable groups may be at an even greater disadvantage due to the possible influence of the GGI in their countries.

Lack of Inclusion Policies Targeting MELAA Women

The issue of a lack of inclusion policies that encourage participation of ethnic MELAA minority women in particular is a concern. This situation raises issues about equity in access and equality of opportunities for contributing to the economy for these women. It is in this regard that a concern is expressed that lack of deliberate policies for monitoring the STEM participation issues of this community and its unique needs. In terms of the critical race theory, this situation makes the ethnic women suffer a double disadvantage because of their ethnicity and because of their gender.

Furthermore, in this country, most equity policies tend not to overtly include the MELAA communities and in cases where they are included, they tend to be annexed under statements that specifically target Maori and Pacific agendas. An example is the current Tertiary Education Strategy document, where ethnic communities are not overtly placed in a separate category. Instead they fall under 'priority 3: Boosting achievement of Maori and Pasifika' (Ministry of Education, 2014, p. 12). Moreover, there are some strategic areas for the Ministry of Education which include the achievement of Maori and Pasifika students (Ministry of Education, 2007) but there is no specific focus on ethnic minorities. Even in most influential education reports (See table 1) reviewed in this work there was little or no mention of ethnic MLAA group except in some instances where they fell under the "Asian and other group".

On this note, another issue worth discussing, which is a strength for advocacy for ethnic MELAA women's issues, is the existence of Maori and Pacific initiatives that aim to empower those groups and the women of those ethnicities. The structures and systems that exist within the organizations of those two ethnic groups can be used as models, as they are already established as opposed to the ethnic community ones. Support systems that often include minority or female role models and advisors, and advice from advanced Maori and Pacific students and staff, already exist for these groups. Admittedly in New Zealand these ethnic community initiatives are strengthened and supported by overt government policies where there are even centers in tertiary institutions as part of government policy strategies. Although such policies do not exist for MELAA ethnic minorities these others could be used as models for strategies for minority communities. On this note, studies (Bonous-Hammarth, 2000) have also emphasized the role of peers and mentors as well as faculty in generating interest in STEM careers (Bonous-Hammarth, 2000).

OTHER CONTEXTUAL ISSUES FACED BY WOMEN IN GENERAL

The discussion above has mainly been focused on contextual factors such as policy and support systems. It is therefore important to acknowledge that there is the biological/ cognitive dimension and even attitudinal issues that should be factored into the discussion. The knowledge about biological dispositions in STEM aptitudes can be useful in the New Zealand context, especially for advocacy for resources to be channelled towards girls and women's instructional resources. This is because research has shown that spatial skills can be improved dramatically and in a short time with any form of structured intervention or training (Ceci, Williams, & Barnett, 2009). On this note caution is taken as it has been argued that although female and male brains have been found to be physically distinct, the fact that these differences have an impact on cognitive strengths and weaknesses is still unclear.

Other issues inherent in this discussion include attitudinal factors. These include the issue of likeability and social approval of STEM occupations i.e. the compatibility between occupation and perceptions of appropriate female roles. This relates to social views and traditions about gender stereotypes. For example, Roger and Duffield (2000) contended that in Scotland the stereotypical image of STEM areas of study as being boring, masculine and remote from everyday life was an important factor in the general lack of interest in STEM careers for women (Roger & Duffield, 2000). Similarly there are contextual factors, such as the notion that women tend to lack interest in STEM fields mainly because they reflect male settings (Hill, Corbett, & St Rose, 2010). Furthermore, work-life balance issues seem not to accommodate the lifestyles of women, which include family responsibilities (George et al., 2001). Consequently, it has been found that there tend to be higher attrition rates for women than men in STEM fields. Such attrition rates may undermine any prior recruitment efforts made by tertiary institutions and employment organizations (Hewlett, 2008). In New Zealand there is a potential for empowerment initiatives since there is existing policy leverage of the State Services Commission and the Department of Labour. The two state agencies have statements in their strategic goals that relate to work life balance for a diverse workforce (including gender diversity).

THE WAY FORWARD: AVENUES FOR LEVERAGING EQUITY POLICIES TARGETING ETHNIC MELAA WOMEN

This chapter argues that although the participation levels of MELAA women is worrying there are policy and organizational advocacy platforms that can be exploited. These platforms include the equity and diversity strategies of the New Zealand Department of Internal affairs, the State Services Commission, and the Human Rights Commission. The mechanisms for addressing skills shortage and the organizational platforms such as the Office of Ethnic Communities, and the ministry of businesses and innovation are such other leverages. It is against this background that policy leverages, especially the 2014 policy on STEM careers are suggested as avenues for empowering these women, (Ministry of Business Innovation and Employment & Ministry of Education, 2014). The 2014 policy on STEM studies was undertaken by the government to increase funding in this area while reducing funding in arts. Such a leverage in this policy could be exploited through multi-stakeholder strategies involving different government departments/ministries. These may include Ministry of Women Affairs, the Ministry of Ethnic Affairs, the Ministry of Education-Tertiary Education Section and the Ministry of Education. The Ministry of Education is crucial in the sense that it has been claimed that achievement at school level is critical in later success or interest in STEM subjects. The Ministry of Woman's Affairs could make an additional strand of advocacy; that is, advocating for ethnic women within their policies of empowering woman in general.

Furthermore, inclusion of MELAA women in the skills shortage list (Long Term Skill Shortage List) can play a crucial role in getting ethnic MELAA women into STEM fields. This is because as New Zealand taps talent through local recruitment and immigration's Long Term Skill Shortage List Work Category- those qualified in an occupation for which New Zealand has a sustained and ongoing skill shortage (Banerjee, 2010). In that way newly migrated MELAA women in STEM fields will act as role models to other settled New Zealand MELAA women.

Similarly, a strong case is made that ethnic MELAA women should be included separately in equity policies alongside other disadvantaged groups such as Pacific and Maori communities if their issues are to be taken seriously. It is argued that the existence of an ethnic communities' ministry alongside other ministries representing ethnic groups is a strength that needs to be taken advantage of. Legislative platforms such as Pay and Employment Equity (about ensuring that pay, conditions, access to the full range of jobs at all levels, and experiences in the workplace, are not affected by gender) and the Equal Employment Opportunity one cannot provide enough leverage for the grounding of policies to address inequities faced by women from ethnic minority groups if used in isolation.

There is room for leverage of equity policies targeting ethnic MELAA women in the component of the Tertiary Education Strategy that focusses on vulnerable groups. It is here argued that universities are important mechanisms for assisting women to achieve in STEM careers and therefore should have separate policy instruments that target ethnic minority groups. In this way ethnic minority women can be empowered by tertiary education mechanisms too., In this regard, it is suggested that the recent announcement of a new government policy to increase funding for studies in STEM fields is an opportunity for ethnic women to ride the tide.

IMPLICATIONS

The literature contributing to issues of empowering women is important in the sense that any effort that is made in unpacking issues that lead to disparities in access to employment affecting ethnic women is important as it a step towards empowering women. This is due to the fact that although women still earn less than men, those in STEM fields tend to earn more than the workforce in other sectors (Dey & Hill, 2007). It has also been argued that the opportunity to pursue a career in STEM fields is also matter of pay equity as occupational segregation accounts for the majority of the gender wage gap (Dey & Hill, 2007). This chapter is therefore of significance to women not only to MELAA women but women in general as the equity initiatives proposed are part of strategies to improve their earning capacity and their economic independence.

Moreover, any effort that taps into the talent of ethnic MELAA women can also be seen as a diversity strategy in two senses: gender diversity and ethnic diversity. These two forms of diversity are critical to both innovation and economic development because STEM fields have a big influence on how products and services are designed. Thus, with no gender and ethnic diversity in STEM fields there is potential for needs unique to women or ethnic communities to be overlooked (George et al., 2001). Furthermore, this work is significant in that it is a step towards advocating for the needs of vulnerable immigrant women. The immigrant status of the MELAA group calls for responsibility on the part of the government to integrate them. Hence any effort made to empower this group is one step towards its successful integration.

RECOMMENDATIONS

In light of the discussion above the following recommendations are made:

- That STEM support projects should not only be initiated by the government but by relevant gender and ethnic specific support groups. These should involve broad strategies that aim at attraction, recruitment and encouraging persistence at universities and in workplaces.
- There is need for the introduction of diversity programs that encourage and support the MELAA ethnic minority students and continue to do so in collaboration with job market programs and MELAA community groups.

SUMMARY AND FUTURE RESEARCH

This research has raised issues about the lack of identification of ethnic MELAA minority groups and their inconsistent identification as either 'Asian and Other' or just 'Other' or MELAA. The discussion has demonstrated that these group should be earmarked as vulnerable because of their immigrant status and the wide gender gap margins in their source countries. Also statistics that show comparable representation with other groups might be misleading because STEM areas are one of the ways by which migrants come to New Zealand. There is also the issue of significance to women as the equity initiatives proposed are part of strategies to improve the earning capacity and the economic independence of New Zealand women. Recommendations are being made for government to frame national STEM policy and strategies in a way that will be inclusive for women in general and MELAA ethnic women specifically. It is also argued that efforts to sustain New Zealand productivity and economic competiveness both locally and internationally should be anchored on empowering women in general and ethnic minorities in particular into STEM careers. However, as has been argued in this chapter, the issue of MELAA women participation in STEM fields still has a lot of grey areas that need to be explored. In this regard the following areas are suggested as possible avenues for future research.

- There is need to better understand the contributions of known factors such as teaching methods, diversity policies and other programs in STEM success among disaggregated gender and ethnic variables.
- There is need for more gender and ethnic specific research to understand the decision making processes undertaken by students in selection of STEM careers, as well as the best methods of influencing or facilitating such decisions.
- There is need for collection of critical STEM education and STEM career data which is disaggregated to take into account gender, ethnicity and other forms of disadvantage, such as immigration and source-country factors.
- There is need for future research to aim to better understand gender and ethnic specific factors that facilitate or limit student entry and persistence in STEM programs in tertiary institutions and industry
- There is need for more gender and ethnic specific research to understand contextual factors such as government funding, cost of education, and the family, social, institutional and career factors that influence persistence in STEM education and careers by ethnic groups.

CONCLUSION

The MELAA ethnic group and MELAA women in particular can provide an untapped reservoir of talent that could be used to fill the engineering and technical workforce shortages that come as a result of the shift in the demographic composition of the New Zealand population structure. Efforts to further the interests of the MELAA group in general and MELAA women in particular direct steps in furthering empowerment of these women. This is because STEM participation issues cannot be separated from employment outcomes and also the social and economic well-being of this group. Under critical race theory any effort that targets women alone without having an additional strand that targets MELAA women may risk overlooking the interests of the disadvantaged. The aim should be to develop a policy that accounts for race and gender equity as part of an agenda to eliminate simultaneity of disadvantage. The 2014 policy on STEM subjects can provide both momentum and leverage for advancing the cause of ethnic MELAA women in STEM careers. It is assumed that this explanation of the problem is a step towards responsive policy-making. Evidence presented in this chapter may be of significance at international level too, as the ethnic minority women of New Zealand also constitute ethnic communities that are found in other English-speaking western countries, especially Canada, Australia, USA and the United Kingdom.

REFERENCES

Arksey, H., & O'Malley, L. (2005). Scoping studies: Towards a methodological framework. *International Journal of Social Research Methodology*, *8*(1), 19–32. doi:10.1080/1364557032000119616

Banatvala, N., & Doyal, L. (1998). Knowing when to say "no" on the student elective: Students going on electives abroad need clinical guidelines. *BMJ: British Medical Journal*, *316*(7142), 1404–1405. doi:10.1136/bmj.316.7142.1404 PMID:9572746

Banerjee, A. (2010). Medical electives: A chance for international health. *Journal of the Royal Society of Medicine*, *103*(1), 6–8. doi:10.1258/jrsm.2009.090266 PMID:20056663

Bonous-Hammarth, M. (2000). Pathways to success: Affirming opportunities for science, mathematics, and engineering majors. *The Journal of Negro Education*, 92–111.

Bowen, G. A. (2009). Document analysis as a qualitative research method. *Qualitative Research Journal*, *9*(2), 27-40.

Bradley, K., & Charles, M. (2004). Uneven roads: understandingw omen's status in higher education. In D. Baker, B. Fuller, E. Hannum, & R. Werum (Eds.), *Inequalitieasc rosss ocietiesf:a milies, schools and persisting stratification* (pp. 247–274). Amsterdam: Elsevier.

Buntting, C., Jones, A., McKinley, L., & Gan, M. (2013). *STEM initiatives and issues in New Zealand.* Retrieved from http://www.acola.org.au/PDF/SAF02Consultants/Consultant%20Report%20-%20New%20Zealand.pdf

Caygill, R., & Kirkham, S. (2008). *Year 5 Students' mathematics achievement in 2010/11: New Zealand results from the International Mathematics and Science Study (TIMSS).* Retrieved from http://www.educationcounts.govt.nz/publications/numeracy/TIMSS-200607/34160/9

Ceci, S. J., Williams, W. M., & Barnett, S. M. (2009). Women's underrepresentation in science: Sociocultural and biological considerations. *Psychological Bulletin*, *135*(2), 218–261. doi:10.1037/a0014412 PMID:19254079

Chan-Malik, S. (2015). Panethnic Religious Institutions. Asian American Religious Cultures, 98.

Chapa, J., & De La Rosa, B. (2006). The problematic pipeline demographic trends and Latino participation in graduate Science, Technology, Engineering, and Mathematics programs. *Journal of Hispanic Higher Education*, *5*(3), 203–221. doi:10.1177/1538192706288808

Chipman, S. F., & Thomas, V. G. (1987). The Participation of Women and Minorities in Mathematical, Scientific, and Technical Fields. *Review of Research in Education*, *14*, 387–430. doi:10.2307/1167316

Cole, D., & Espinoza, A. (2008). Examining the academic success of latino students in science technology engineering and mathematics (STEM) majors. *Journal of College Student Development*, *49*(4), 285–300. doi:10.1353/csd.0.0018

Day, J. C. (1992). *Population projections of the United States, by age, sex, race, and Hispanic origin: 1992 to 2050*. US Department of Commerce, Economics and Statistics Administration, Bureau of the Census.

Dey, J. G., & Hill, C. (2007). *Behind the pay gap*. Washington, DC: American Association of University Women Educational Foundation.

Durie, M. (1998). *Whaiora: Maori health development* (2nd ed.). Auckland: Oxford University Press.

Durie, M. (2003). *Nga kahui pou launching Maori futures*. Wellington: Huia publishers.

Education Review Office. (2012). *Science in the New Zealand Curriculum: Years 5 to 8*. Wellington, NZ: Ministry of Education.

Elit, L., Hunt, M., Redwood-Campbell, L., Ranford, J., Adelson, N., & Schwartz, L. (2011). Ethical issues encountered by medical students during international health electives. *Medical Education*, *45*(7), 704–711. doi:10.1111/j.1365-2923.2011.03936.x PMID:21649703

George, Y. S., Neale, D. S., Van Horne, V., & Malcolm, S. M. (2001). *In pursuit of a diverse science, technology, engineering, and mathematics workforce*. Paper presented at the American Association for the Advancement of Science.

Hanson, S. L., Schaub, M., & Baker, D. P. (1996). Gender stratification in the science pipeline: A Comparative Analysis of Seven Countries. *Gender & Society*, *10*(3), 271–290. doi:10.1177/089124396010003005

Hassan Al Marzouqi, A., & Forster, N. (2011). An exploratory study of the under-representation of emirate women in the United Arab Emirates' information technology sector. *Equality, Diversity and Inclusion. International Journal (Toronto, Ont.)*, *30*(7), 544–562.

Hewlett, S. A. (2008). *The Athena factor: Reversing the brain drain in science, engineering, and technology*. Harvard Business School Watertown.

Hill, C., Corbett, C., & St Rose, A. (2010). *Why So Few? Women in Science, Technology, Engineering, and Mathematics*. Retrieved from http://www.aauw.org/resource/why-so-few-women-in-science-technology-engineering-mathematics/

Human Development Report. (2009). *Overcoming barriers: Human mobility and development*. Retrieved from http://hdr.undp.org/en/search/node/GEM%202009

IPENZ. (2010). *Minority groups in engineering education*. Wellington: National Engineering Education Plan Project.

Langen, A. V., & Dekkers, H. (2005). Cross-national differences in participating in tertiary science, technology, engineering and mathematics education. *Comparative Education*, *41*(3), 329–350. doi:10.1080/03050060500211708

Marginson, S., Tytler, R., Freeman, B., & Roberts, K. (2013). *STEM: country comparisons: international comparisons of science, technology, engineering and mathematics (STEM) education. Final report*. Australian Council of Learned Academies.

Ministry for Women. (2014). *A snapshot of New Zealand women*. Retrieved from http://women.govt.nz/our-work#sthash.8mv6wmBO.dpuf

Ministry for Women. (2015). *Occupational segregation*. Retrieved from http://women.govt.nz/our-work/economic-independence/paid-and-unpaid-work/occupational-segregation

Ministry of Business Innovation and Employment, & Ministry of Education. (2014). *A nation of curious minds: A national strategic plan for science in society*. Retrieved from http://www.msi.govt.nz/assets/MSI/Update-me/Science-in-society-project/science-in-society-plan.pdf

Ministry of Education. (2014). *Tertiary Education Strategy 2014 - 2019*. Retrieved from http://www.minedu.govt.nz/NZEducation/EducationPolicies/TertiaryEducation/PolicyAndStrategy/TertiaryEducationStrategy2014-2019.aspx

Ministry of Pacific Island Affairs. (2010). *Education and Pacific Peoples in New Zealand*. Retrieved from http://www.stats.govt.nz/browse_for_stats/people_and_communities/pacific_peoples/pacific-progress-health.aspx

Ministry of Research Science and Technology. (1998). *Mathematics in New Zealand: Past, Present and Future*. Wellington: Ministry of Research, Science and Technology.

Mpofu, C. (2014). *The global dynamics of the medical workforce: A case study of flows from New Zealand to Australia*. Paper presented at the 10th Annual Association of American Medical Colleges Health Workforce Research Conference, Washington, DC.

National Center for Education Statistics. (2005). *Digest of education statistics*. Washington, DC: National Center for Education Statistics.

Ritchie, J., & Lewis, J. (Eds.). (2003). *Qualitative research practice: A guide for social science students and researchers*. London: Sage.

Roberts, J., Gardiner, B., Gilbert, J., & Vaughan, K. (2008). *Trading choices: young people's career decisions and gender segregation in the trades*. Retrieved from http://women.govt.nz/documents/trading-choices-young-people%E2%80%99s-career-decisions-and-gender-segregation-trades-2008

Roger, A., & Duffield, J. (2000). Factors underlying persistent gendered option choices in school science and technology in Scotland. *Gender and Education*, *12*(3), 367–383. doi:10.1080/713668300

Shettle, C., Roey, S., Mordica, J., Perkins, R., Nord, C., Teodorovic, J., . . . Brown, J. (2007). The Nation's Report Card [TM]: America's High School Graduates. NCES 2007-467. National Center for Education Statistics.

Smedley, A. (2007). *Race in North America: Evolution of a worldview*. Boulder, CO: Westview Press.

Smyth, F. L., & Nosek, B. A. (2015). On the gender–science stereotypes held by scientists: Explicit accord with gender-ratios, implicit accord with scientific identity. *Frontiers in Psychology*, 6. PMID:25964765

Statistics New Zealand. (2004). Degrees of difference: The employment of university qualified immigrants in New Zealand. Wellington, New Zealand: Author.

Statistics New Zealand. (2005a). *Focusing on women*. Wellington: Statistics New Zealand.

Statistics New Zealand. (2005b). *School achievement in maths and science*. Retrieved from http://www2.stats.govt.nz/domino/external/web/nzstories.nsf/092edeb76ed5aa6bcc256afe0081d84e/6b69fd321fb4557fcc256b1800067224?OpenDocument

Statistics New Zealand. (2006a). *Distribution of NZ population aged 15 and over by highest qualification and field of study 2006*. Wellington: Statistics New Zealand.

Statistics New Zealand. (2006b). *Distribution of the working age population by labour force status, highest qualification and ethnic group 2006 - percentages*. Wellington: Statistics New Zealand.

Statistics New Zealand. (2006c). *Domestic equivalent full-time students enrolled by field of study, level of study, age group and gender, 2013*. Wellington: Statistics New Zealand.

Statistics New Zealand. (2013). Historical population estimates tables. *Estimates and projections*. Retrieved from http://www.stats.govt.nz/browse_for_stats/population/estimates_and_projections/historical-population-tables.aspx

The Royal Society of New Zealand. (2014). *Women in Science – a Royal Society Overview*. Retrieved from http://www.royalsociety.org.nz/expert-advice/papers/yr2005/women-in-science/

United States Department of Health and Human Services. (2015). *Medically underserved areas/populations*. Retrieved from http://www.hrsa.gov/shortage/mua/index.html

Whitley, R. (2005). Book Review: Cultural Diversity, Mental Health and Psychiatry: The Struggle Against Racism. *Transcultural Psychiatry*, *42*(3), 507–509. doi:10.1177/136346150504200310

World Economic Forum. (2014). *The Global Gender Gap Report 2014*. Retrieved from http://www.weforum.org/reports/global-gender-gap-report-2014

KEY TERMS AND DEFINITIONS

Biculturalism: In ideological and socio-political spheres biculturalism is a deliberate confrontation and negotiation of two conflicting value systems both considered to be of equal value. In the New Zealand context biculturalism is about the relationship between the Maori and the Europeans whose cultures are considered to co-exist in a uniquely distinct manner.

Decile (School Decile): In New Zealand a school's decile is a measure of the proportion of students from low socio-economic communities, where decile 1 schools have the highest proportion of students from socio-economically disadvantaged communities, while decile 10 schools have the lowest proportion of students from these communities.

Gender Empowerment Measure (GEM): An index measuring the level of women's participation in political and economic decision-making.

Gender-Related Development Index (GDI): Inequality of achievement between women and men.

Gender Inequality Index (GII): An index for measurement of gender disparity that was introduced in the 2010 Human Development Report (20th anniversary edition) by the United Nations Development Programme (UNDP).

Maori: New Zealand's indigenous people.

Pacific Peoples/ Pasifika: The broad group of ethnic groups of peoples whose ancestry is from Pacific Island nations.

Retention: Is concerned with how long students persist in their studies, and specifically records what percentage of students stay in study until they have successfully completed.

Statistics New Zealand: A government department and New Zealand's national statistical office, a major source of official statistics, and leads the Official Statistics System.

Trends in International Mathematics and Science Study (TIMSS): Conducted by the International Association for the Evaluation of Educational Achievement (IEA).

Chapter 12
Prestigious College Courses and Race/Skin Color:
Some Thoughts about Higher Education in Brazil

Patricia Costa Pereira da Silva
Fluminense Federal University, Brazil

ABSTRACT

The discussion promoted in this text is part of the inquiry of PhD in Education in progress since 2012 at Fluminense Federal University (Niterói, Brazil). The above-mentioned doctoral research is dedicated to the study of educational trajectories of prestigious black professionals in Brazil. The main objective of this text is to bring up a reflection on the studies about the relation between prestigious college courses and race/skin color in Brazil carried out in the contemporaneousness. For this purpose, it is necessary to analyze the process of selection of the professional careers that demand the ownership of the bachelor's degree (higher education). This text is supported by studies about social stratification, such as Wright (1978), Goldthorpe (1987), Januzzi (2003) and Queiroz (2004). Bibliographical inquiry was the methodology elected for the development of the present text.

INTRODUCTION

How can we define what a prestigious profession is? This was one of the issues that appeared in the doctoral research of the author of this text. Based on studies that show that a larger number of blacks is inserted in the careers of lesser status and social prestige, the question raised has become more important in face of the need to discuss the prestige scale in which the claim is based. There are several ways to establish prestigious levels of hierarchy of professions, such as wages and position offers. Each study presents a particular type of hierarchy of courses and therefore the prestige scale.

The aim of this chapter is to bring up a reflection on contemporary studies about the participation of blacks in prestigious professions in Brazil. Therefore, it is necessary to analyze the process of selection/choice of the careers that require a college degree ownership. Recently, some academic studies show

DOI: 10.4018/978-1-5225-0174-9.ch012

that white men have predominance in the careers of higher prestige and social status; on the other hand, the number of blacks in the university is restricted to certain areas of knowledge that have less social prestige and, consequently, more easily accessed, such as Library and Social Work. It is noticed that the social hierarchy of professions reflects the provision of courses in the power structures of the university campuses.

Interest in this subject arose from the fact that both women and blacks are present in the university unevenly white men with respect to carriers. Everything indicates that white men dominate careers of higher prestige and social status. Thus, it aims to verify to what point this university market puts women and blacks in next levels of choice and possibilities.

BACKGROUND

The Prestigious Scale of Professions in Brazil

A priori, it is important to say that there is correspondence between the social inclusion of one in society and one's possession or not of means of production. The occupation exercised, the insertion in the production process, the position in the labor market and the possession or not of means of production would be some of the most classic criteria of the stratification of individuals (Januzzi, 2003).

According to Januzzi (2003), numerous methodologies have been dedicated to the construction of a socioeconomic index for occupations and their socio-occupational classifications. Often used in Anglo-Saxon studies, a way to classify occupations is to group them according to their salary level and required education. The occupations are grouped and prioritized in groups or categories of different socioeconomic status. It is called this methodology of socioeconomic scales occupations or socio-occupational scales. In this methodology, socioeconomic status matches

the social prestige conferred by population to the occupations or the relative position of employment in a qualifying ranking ruled by some simple socioeconomic indicator or compound as provided income, typical skill level occupations, education of the workforce allocated them or even a combination of these dimensions (Valle Silva, 1978, 1985; Jorrat And Acosta, 1992; Bukstein, 1997 apud Januzzi, 2003, p. 249).

Januzzi (Ibid) proposed the classification of occupations in five Brazilian strata. The author argues that the construction of socioeconomic status concept involves pragmatic choices and subjective preferences. The author of the proposal is supported by a wide Anglo-Saxon tradition, as Wright (1978) and Goldthorpe (1987), which reflected empirically the significant socioeconomic differentials in earnings, schooling, quality and safety in the workplace. Jannuzzi proposition (op. cit.) ranks occupations in strata through an index based on socioeconomic distance between the occupations, which can be measured by different methods. A methodology proposed by the author considers the differences and similarities of occupations in relation to its urban/rural character, manual/non-manual, to activity groups, skill level, control and autonomy. The author based his conclusions also on data from the National Survey of Household Sample (PNAD), to propose their socio-occupational scale aggregation of Brazilian occupations in five major groups:

1. **High:** Doctor, engineer, university professor, entrepreneurs, top managers and posts in public administration (judges, prosecutors, sheriffs, military officers, etc.);

2. **Medium-High:** Accounting and administration technicians, masters and foremen in the industry, primary and secondary school teachers, realtors, police inspectors, postmen, traders (owners) and farmers;
3. **Medium:** Mechanical turner, assemblers of electrical equipment, sellers, cashiers, self-employed dealers, pre-school teachers, drivers, students inspectors, nursing assistants, administrative and office assistants, police officers and other ranks of the armed forces;
4. **Medium-Low:** Occupations in the food industry, occupations in the textile industry, masons, painters, waiters, watchmen, porters, stevedores, street vendors; and
5. **Low:** Rural workers provided employees or self-employed (sharecropper producers or partners), in addition to low-status urban occupations, such as servants of masons, washerwomen, domestic workers and garbage collectors.

The first social group of the scale, termed as high, comprises individuals who have the condition of large landowners and/or dealing in command posts, direction or higher specialization. Therefore, it includes the occupations with higher income and education, that is, of higher socioeconomic status. As can be seen in the table above, the magistrates are typical representatives of this stratum, as well as doctors, engineers, businessmen and other leading posts in public administration (promoters, delegates, military officers, deans, etc.).

The second group, termed as medium-high, groups the small owners, managers and supervisors, and skilled employees and office-skilled technicians in industry and services. Examples of occupations framed in this group: accounting and administration technicians, basic education teachers, realtors, police inspectors and farmers. With less power to command or qualification, these occupations have a lower average income, however, when compared to the previous group, the differences between education indicators and precariousness are not so expressive.

The third group, known as the medium, embraces the majority of occupations in commerce, services and skilled jobs in the industry. Typical occupations of this group are: lathes, assemblers of electrical equipment, sellers, cashiers, administrative and clerical assistants, police officers and other ranks of the armed forces. Occupational insecurity indicators are significantly worse.

The fourth group, called medium-low, contains the workers of the area which provides services that require low-skilled people, construction and traditional industry.

Finally, the last socio-occupational group, called down, encompasses rural workers, provided employees or self-employed (sharecropper producers or partners), and urban occupations of low status, as mason servants, valets, maids and garbage collectors.

The social prestige is strongly connected to the hierarchy of occupations, ie, social stratification. Among top-level professionals, proportion of whites tends to increase as the social prestige of careers (and of course, the standard of remuneration) increases, while the opposite occurres with black. In careers as Social Work, Nursing and Letters, the presence black is generally higher than in others, such as Architecture, Law, Mechanical Engineering and Medicine, in which whites often have wide prevalence.

The Construction of the Prestigious Range of University Courses in Brazil

Course's social prestige is usually associated with socioeconomic status of their students, as shown by the literature on the subject (Braga, Peixoto, Bogutchi, 2001). The well defined categories served therefore as proxy of the socioeconomic status of the candidate. A study that can be referenced to understand the hierarchy of university courses in Brazil is Queiroz (2004). The author in the text entitled *The black and Brazilian society*, combines the applicant/opening ratio for college with the offerings of the labor market. In this study, the author presented relevant advances the understanding of how the hierarchy of prestige of university courses reflects the hierarchy of professions. Therefore, the author took as reference a research on the value of professions in the metropolitan region of Salvador labor market (RMS). That research was based on a survey of consulting human-resources firms which work in the RMS labor market and took as a reference the list of courses offered by the Bahia Federal University (UFBA). The author sent a questionnaire to more than 20 human-resources agencies in Bahia. The research resulted in the following prestigious scale of five positions:

1. **High:** Medicine, law, dentistry, administration, data processing, electrical engineering, psychology, civil engineering, mechanical engineering, architecture, chemical engineering;
2. **Medium-High:** Economics, journalism, veterinary medicine, sanitary engineering, nursing, education, accounting, nutrition, industrial chemical, mining engineering;
3. **Medium:** Secretariat, pharmacy, agronomy, bachelor and bachelor of chemistry, physical education, industrial design, social sciences, arts, literature, mathematics;
4. **Medium-Low:** Philosophy, life sciences, music, physical geology, geophysics, composition and conducting, statistics, performing arts, dance, history; and
5. **Low:** Design and plastic, instrument, library, corner, degree in the grade 1 science, geography, museology.

According to Queiroz (2004), at all the universities the following courses are among those with higher index candidate/vacancy and can therefore be considered prestigious: Medicine, law, dentistry, management, psychology, computer science, civil engineering, electrical engineering, mechanical engineering.

Also according to Queiroz (2004, p. 74),

(...) the blacks are reserved the least socially valued courses, such as those of teacher training, for example. Even in these courses they are, in general, minority. It can be said that the reputation of careers is formed by the aggregate of its symbolic and market values, reflecting more disputed access courses. That dispute, in turn, reveals strong character of social selection observed, for instance, in the categories income and student body color. The presence of blacks in prestigious courses can be configured as a state of emergency and is the subject of one of the sections of this chapter.

Through the analysis of sociocultural characteristics of the student body, Maria da Graça Setton (2002) reflects on a possible hierarchy of university courses of the Humanities at the São Paulo University (USP). From Bourdieu's contribution to the study of the cultural phenomena, the author analyzes the students' career choices through a number of provisions that represent the structure of different types of capital (economic capital, social capital and cultural capital).

This analysis led the author to create, based on the article *Democratization of higher education*, written by Aparecida Joly Gouveia (1968)[1], three classes of university courses: the select, intermediaries and popular. The select courses are represented by the high concentration of at least two forms of capital (economic, cultural and social); the intermediate, by the average concentration of one or both forms of capital; and popular, by the low concentration of one, two or all forms of capital.

The author classified the following courses as select[2]: law, management, architecture, radio/tv, publishing and public relations. Setton (2002) identified a match between these courses and those who direct their students to careers with good expectations of economic gains and acquisition of social status. In addition, the author also points out that such courses have great prominence in the social imaginary because they present, especially, great competitiveness in their entrance examinations. It is therefore possible to infer that the select courses are highly selective and intended for a tiny group of students.

Setton (2002) classified the following courses as popular: library science, education, geography, literature, history, social sciences and accounting. They are the ones that direct their students to undervalued careers in the labor market. With expectation of low wages and narrow employment opportunities in prestigious posts, students of these courses have a common characteristic: they head for teaching, career with low social status.

The intermediate courses have median rates in the variables income and parental occupation. They stand out for having a significant number of students with volume of economic and social capital in relatively distinctive levels or those with high levels only in cultural capital. The following courses can be classified as intermediate: tourism, journalism, advertising and film.

Regarding the gender question, Setton (2002) found a clear tendency of men to choose the select courses. Only at the Law School balance between the number of men and women appeared; however, it was still possible to see a small male advantage. As for popular courses, the author noticed a reverse trend: there is a feminization of the students of the courses. Teaching courses, social services and library can be cited as examples of this fact.

An article by Kaizô Iwakami Beltran and Moema Poli Teixeira (2005) titled *Color and gender selectivity of university careers* brings up a basic assumption about the construction of the prestigious range of professions that somehow converge with Setton (op. cit.) thought on the select and popular courses: the more feminine for the student body, the lower the social prestige of the course. This hypothesis was inspired by Bourdieu studies. According to Beltran and Teixeira, "(...) the most female professions have lower market (and social) value", a natural choice "for the less privileged groups would be to access (for example, a degree) less masculine careers" (Ibid, p. 145). That is, the courses with female majority would tend to reduce his prestige before society.

According to Bourdieu (1999), the sexual division of labor is guided by the assumption that men should dominate the fields of power and public space; while women dominate the private space. From a list of 335 careers, Bourdieu identifies what he termed as female careers. Elementary school teacher, nurse, nutritionist, maid, secretary and receptionist may be cited as the main female careers, since their functions appear as a supplement to domestic functions (education, care and services). In the Bourdieu's words (1999, pp. 112-113)

If the old structures of the sexual division still seem to determine the direction and shape of the changes, it is because, besides being targeted levels, in careers, in positions more or less strongly sexed, they act through three practical principles that not only women, but also their own environment, put into action in your choices: According to the first of these principles, the functions that suit women are extension of domestic functions: education, care, service; second, that a woman cannot have authority over men and therefore has every possibility of being all things to all alike, find themselves sidelined by a man to a position of authority or be relegated to subordinate roles, assist; the third gives the man a monopoly on maintenance of technical objects and machines.

Through studies of Limongi et al. (2002) that were carried out based on data from USP's entrance examination, Beltran and Teixeira attest "(...) the idea of the socioeconomic gap, considering the color group, family income or sex". Data also suggest that the socially less privileged groups have a higher difference of the most privileged groups in the most competitive careers.

THE BLACKS IN THE BRAZILIAN HIGHER EDUCATION

The Choice of Career and Success in Vestibular

The Vestibular[3] is the main mechanism of entrance in Brazilian universities. For decades, it became an object of controversy in academia. Using old newspapers as source of information, it is possible to observe some publications about the entrance of college exam. An edition of *Folha Dirigida* – a Brazilian newspaper – which was published at October 15th, 1991, brings the headline in one of his sections: *Vestibular: portrait of poverty*. The text makes notes on the top of vacancies in private higher education in relation to the public. It also points that the Brazilian university population was small: about 10% of the population between 20 and 24 years. In 2004, according Census of Higher Education, the rate remained the same.

The article of *Folha Dirigida* brings a rich source of information that is obviously linked to the social context of the early nineties. In addition to the data above, the following items are noteworthy: in some courses, almost all students were from the private universities; the richest occupied the free places; Brazilian Ministery of Education wanted to reduce social injustice with the creation of evening classes; and the entrance exam was revealing of social injustice. This last item makes me refer to another report, published in 1973, on *O Globo* newspaper. The text provides a discussion about the format entrance exam. For some, the model established for the test cannot vouch knowledge student, harmed the poor, and should be reviewed. It was at that belly which was born examination of multiple choices, as we know it today. For Guimaraes (2003), one of the causes for low absorption of black in higher education today is that "(...) examination Vestibular leaves no room for other qualities and potential of students are evaluated" (p. 78).

One aspect that draws attention in educational research on the production of racial inequality in Brazil is the choice made by the subjects in registration times in the university entrance exams. The option for a course is commonly interpreted as a rational decision process that sometimes is criticized. Hodkinson and Sparkes (1997), in the text entitled *Careership: The sociological theory of career decision making*, argue that the assessment of the costs and benefits associated with each choice is less consciously and systematically than it is suggested by the theories of rational choice. One of the alternatives is eventually discarded by the individual. In their social and school life, people can learn to visualize certain alternatives as inappropriate or unreachable, and others as appropriate and desirable. The American sociologist Jon Elster (1994) believes that human actions move according to the opportunities that the medium offers. According to Elster (Ibid, p. 51), "actions are evaluated and chosen by themselves, but as more or less efficient means to a further end". The information and representations that individuals hold relating to the labor market, depend on the social environment of origin. The authors point out that individuals do not react directly and uniquely to an opportunity structure objectively defined.

For Bourdieu (1983), it is unlikely that individuals can act freely in society, taking precedence over governing structures. If, as stated by Bourdieu (Ibid), the proximity of individuals in the social space is revealed by the similarity of habitus, the social condition should approach, by practices, the different racial groups. Because of the high selectivity, as said above, it can be concluded that the choice for a course is geared towards the achievement of a certain social status and the

(...) The avoidance of discriminatory situations and symbolic violence perpetrated against non-whites are mutually reinforcing, causing normally blacks and mulattos to regulate their aspirations in accordance with what is culturally imposed and defined as the 'proper place' for color people (Hasenbalg And Valle Silva, 1988, p. 167).

In addition to the experiences of discrimination, economic needs are a considerable burden upon the choices of careers. Black students mostly have the urgency about employment to support themselves (Teixeira, 1998). Soon, they choose to join university courses of low competition and having the willingness to reconcile studies with work, as in the case of undergraduate courses and Human Services. A common sense prevails when choosing in order to "ensure a minimum adaptation to the likely course of this world through reasonable anticipations, adjusted broadly (outside any calculation), the objective possibilities" (Bourdieu, 1999, pp. 284-285).

Social and economical arrangements interfere in the choice of course. For Zago (2008, p. 163), this condition has had "[...] influence on access to the most prestigious careers, because it is associated school background and a series investments that become credentials with considerable weight in selection processes [...]".

These studies prove, therefore, the existence of significant racial inequalities among those who managed to enter higher education, showing that whites have the most privileged situation in public universities, both as it regards to the ratio that are represented, as to in connection to the areas and courses they have access to.

Blacks in Prestigious College Courses

(...) Missing blacks in our universities because the educational gap between blacks and whites remains for more than a century. The most subtle Brazilian racist speech is silence, the transformation of black hindrance in a society that ideally would not have race prejudice. (Ricardo Henriques in an interview with journalist Elio Gaspari)

The trajectory of Brazilian students in public school is marked by tragic repeated failure stories; but the school failure mainly affects the black students. They are those who bear the weight of large number of exclusion. Children and young black are affected in different ways by subtly racist the Brazilian education system. Racism is expressed, for example, in different access racial segments: higher proportions of non-literate people are among the blacks; the participation of blacks in educational system decreases as the years of schooling increases. Another face of racism in this Brazilian education system is the silence of curricula about the black history and culture. At school, practices also reach negatively black children when they reinforce a stereotyped and depreciated image of black, with heavy consequences for their self-esteem and the construction of their identity. All these phenomena contribute to reduce the black student opportunities that may result from a successful scholar trajectory.

Brazil is considered the country with the highest proportion of blacks outside Africa. Salvador, Bahia, is the second largest city of black population in the world. Despite the significant presence of black among Brazil population, considerable distances separate blacks from whites in various sectors of society, and it also happens in labor market and education. Slavery heritage has been invoked as an argument to justify the present situation of disadvantage of the black population. This argument proves to be fragile when considered the long time since the abolition of slave labor. The weakness of this explanation is evident when one notes that the precarious economic conditions of blacks in the post-abolition do not much differ from that foreigner worker groups, constituted by whites, who arrived in Brazil, since the late nineteenth century. Significant part of economic, political and intellectual elites from Brazil comes from these groups of poor immigrants, while the position of majority of black population remained almost unchanged. Thus, there is no explain for the precarious living conditions of blacks in Brazil today, except for the effect devastating racism.

According to Queiroz (2004), 63% of students of Letters are black and brown. In the Librarianship course, they are 62%. In the Visual Arts, 58%, and in Accounting, 57%. The author researched 3,100 students from Bahia Federal University (UFBA, in Portuguese acronym), which had, at the time of the survey, 3,600 students. By the response of university students, 41.7% of students who entered UFBA are black and brown; the white are totaled 49.8%. The difference (8.1%) could be ignored if the Bahian black population was not almost 80% of inhabitants. According to the survey, 65% of Medicine students who began studying in UFBA last year are white. In the entrance exam for Law, 60% of approved are white. In Dentistry, the whites are also 60%.

Search carried out by the social department of UFBA (Federal University of Bahia) in five public universities in Brazil reveals that blacks attend courses considered of "low social prestige". Coordinated by Professor Delcele Mascarenhas Queiroz, the research involved 12,278 students enrolled for all courses in the first half of 2000 in UFBA, at Rio de Janeiro Federal University (UFRJ, in Portuguese acronym), at Brasília University (UnB, in Portuguese acronym), at Maranhão Federal University (UFMA, in Portuguese acronym) and Paraná Federal University (UFPR, in Portuguese acronym). According to the data, only 1.9% of the students enrolled in the Medical School at UFBA are black. Whites who attend Medicine are majority at UFBA, with 65.4%. The research showed, in all universities, a strong racial selectivity in the access to high social prestige courses, showing that this privilege is also predominantly white.

At UFMA, blacks are also minority at Medicine School: only 5.9%, against 55.9% of white students. Even with the sum of browns (35.3%), whites still represent more than half of the students.

Bahia and Maranhão are the two Brazilian states that record the highest concentration of black and brown, according to IBGE (Brazilian Institute of Geography and Statistics). According to the data of the institute, 79.1% of the population of Bahia are black and brown. In Maranhão, the percentage is slightly lower: 78.7% of the inhabitants are black and brown. In Paraná, the difference is even higher. According to the survey, 84.9% of Medicine students enrolled in the first semester at Paraná Federal University are white. The researchers found no black registered in the first semester of this Medicine course.

In the surveyed universities, the number of blacks enrolled in courses of low social prestige is much higher. At UFBA, for example, 17.4% of students enrolled in the first semester of History are black. Other 28.3% are brown. In Maranhão, 11.1% of students enrolled in the same course are black. The mulattos account for 25%.

In the distribution of data by color, students located in highly prestigious courses show the privileged position of whites in relation to access to high social prestige courses in all investigated universities. The biggest distance between racial segments is at UFPR. There, the white/black ratio is 217.5 to one;

for the white/brown ratio is 13 to one; and the white/yellow ratio is 17.5 to one. At UFRJ, for each black enrolled at high prestigious college courses, there are 48.5 whites.At the universities from the Northeast and Federal District, the relationship is less unequal, but distances remain quite significant. At UnB, despite high family income of blacks, they are absent from the high prestige courses. For each brown in this prestigious level courses, there are 2.8 white. At UFBA, for each black, there are 13 whites and two browns. The shorter distances are in UFMA: for each black are ten whites and 1.7 brown.

According the infographic *Retrato dos Negros no Brasil*, made by *Rede Angola*[4], more than half of the Brazilian population is self-declared as black or brown in the census conducted by IBGE in 2010, but only 26 of 100 students from universities in Brazil are black. Although still far below, the black population access to higher education increased by 232% between 2000 and 2010; of every hundred students, less than three, or 2.66%, are brown or black.

The publication of the results of the Socioeconomic Survey of the last bone of Fuvest[5] reveals that of almost 12,000 approved at USP on 2013, 78.7% are white and 2.4% are black. The survey also shows that 62.9% had studied at private schools, and 22% attended High School in public schools. Almost half of the students enrolled in the ten most competitive courses at USP belongs to A- (income above US$ 12,440) or B- (above R $ 6,220) class families. Data from the socioeconomic questionnaire applied by Fuvest show that the ones that entered in these careers in 2013 are predominantly white, former private-school students and with family income above ten monthly minimum wages.

Study on the profile of students enrolled for Vestibular 2012 at São Paulo University (USP, in Portuguese acronym), the Brazil largest university, shows that among the 774 enrolled in the 10 most competitive courses, only four declared themselves as black. In the three most competitive careers – Medicine, Engineering and Law School – no one said black, according to the questionnaire applied by Fuvest, responsible for the entrance exam (Vestibular)

The data show that the elitist profile and white largest university, which has resisted the implementation of affirmative actions for black and indigenous, has been increasing. The year before, in 2011, 10 people had declared black in the 10 most prestigious courses.

In São Paulo, black and brown correspond to 34.6% (29.1% browns and 5.5% of black) of the 42 million of people. São Paulo is the state with the largest black population in Brazil, in absolute numbers: about 14.5 million of black people. According to Census 2010, conducted by IBGE, 63.9% of São Paulo declares white; 1.4, yellow 1.4%, and 0.1%, Indians.

To scholars of the field, the information that blacks are under-represented in social spaces of prestige are not new. Several studies devoted to questioning Brazilian educational inequality show that it is the privilege of white participation in high social prestige courses, such as Law, Engineering and Medicine.

Inequalities in the Brazilian Academic-Scientific System

The research conducted by Applied Economic Research Institute (IPEA, in Portuguese acronym) – entitled *Portrait of gender and race inequalities*, which was published in 2008, reveals that it is in the educational field where high inequalities in access and especially the permanence of different population groups. Racial differences are very striking and expand the higher the level of education.

In the educational field, racial inequalities are evident. According to Plan National Household Sampling (PNAD, in Portuguese acronym) in 2007, 62% of white youths between 15 and 17 years were attending school; among black population, this rate is reduced by half. In higher education, the differences are even greater. According to the Census of Higher Education 2007, 19.9% of white youths aged from 18 to 24 are in higher education; 7% is the index of black population in this educational segment.

In line with the policies of the Federal Government developed in recent decades, in order to increase the diversity in higher education, in 2013, the National Council for Scientific and Technological Development (CNPq, in Portuguese acronym) included the item color/race, according to the classification of the Brazilian Institute of Geography and Statistics (IBGE), the Lattes Platform meet the growing demand for this information. This way, in order to update all Lattes resumes, information about race color of students, scholars and researchers from Brazil is required.

At this section, it will be presented the first analysis of the participation of black men and women (here black and mixed are considered) in the training system of grants and research CNPq. From collecting data on race/color of researchers and scholars, you can now perform studies to map the participation of ethnic and racial groups in Brazilian science and technology, as well as follow racial inclusion policies that have currently existed in the country, just like other structured programs and policies to specific niches, if required.

The exchanges in Brazil totaled 91,103 in January 2015. The participation of white fellows is very significant, about 58%. Black fellows are about a quarter of the total stock (26%). The percentage of yellow and indigenous is very small, and the percentage of indigenous does not reach 1%. It should be noted that those who do not wish to declare their color/race represent almost 11% of the total (9,918).

The stock abroad totaled 12,780 stock in December 2014. The percentage of whites is higher than the same group of scholars at home and reaches 64.8% of total. The percentage of yellow is also superior outside, and almost reaches 3%. However, the percentage of black men and women is lower, accounting for 18.8%, whether from brown or black: 16.4% and 2.4%, respectively. 91,303 of stock in the country, 46,232 are women, and 45,069 men. The largest share of women occurs in all races/color, except for the indians. Indigenous men total 195 individuals and indigenous women, 129, among fellows of CNPq in Brazil. It is interesting to note that the number of men who do not wish to declare their race/color is higher than women's one, and the number of those with the curriculum outdated.

White women represent 59% of female scholars, and black (brown and black), 26.8%. However, the participation of black is small: 4.8%. Among men, 56.3% represent whites, 24.3%, blacks, and 4.7% are black. Among the scholars abroad, the majority of female scholars that was observed in Country of bags is not repeated. In this mode, men outnumber except between yellow. Blacks (black and mixed) represent 19.6%, and black women, 17.7%. Numerically 120 are black women, and 184, black men.

Many researches and studies show the existence of inequalities between whites and blacks in Brazil. For Carlos Hasenbalg (1979), the reasons for the social subordination of black population in Brazil are found in discriminatory practices that are a direct result of inequalities between blacks and whites in various sectors of social life – health, education, labor market, etc.

SOLUTIONS AND RECOMMENDATIONS

The diagnosis of the participation of racial segments in top education can therefore support the debate on the need to adopt public policies to correct inequalities between whites and blacks that begins to take shape today in Brazilian society.

It is necessary to formulate policies in the educational field, as that reveals the reality of participation of black population in higher education:

(...) All social indicators show that black population is the one that suffers most from inequalities in the country. Therefore, specific policies focusing on the race are necessary. Quotas are a measure of inclusion in a society where racism is reflected in the educational system and deepens the social divide. Innovative and courageously, this discussion could draw paths for a change and a transformation in Brazilian society, paying off two effects: to show that change is possible and should not be delayed and reaffirm that the construction of new means bother as it leads to questions and reactions mainly of those in comfortable place reserved for small elites ever (...). (Jornal da Tarde, São Paulo, September 17th, 2009)

The low participation of blacks in highly prestigious university courses puts the challenge of seeking to establish mechanisms to overcome this situation to the whole society. It is from this context that affirmative action policies assume a political dimension necessary in Brazil:

Until contrary is proven, affirmative action policies have, in other contexts, contributed to the promotion of groups historically deprived of essential opportunities for their development and strengthening of their skills as human beings.

(...) It is undeniable that the experiences that culminated in action programs, global in scope, have been responsible for a more collective elaboration of the concept of development and renovation of the understanding of equality, equity and social participation. (Sant'anna, 2001, p. 363)

In contemporary times, to have a higher level course completed is an important differential in the labor market, because it implies a greater chance of upward mobility. Affirmative actions aim to achieve a double objective:

- To permit equal opportunity and treatment – which, in the medium term, allows for economic materialization of equality; and - The work one reverse pedagogy that not only rebuild the black esteem, but also educate all against racism. (Santos apud Siss, 2003, p.131)

Hence, the plausible hypothesis that affirmative action, to the extent that it provides blacks access to universities, may be responsible for a greater number of student graduations and consequently that they constitute an important factor of increase in income upward social mobility. The presentation of this data, even though it is brief, proves to be essential to discuss the promotion of policies that contribute to the implementation of social equity.

CONCLUSION

It can be seen that, with regard to academic careers, both women and blacks are present in the university unequally compared to white men. Previous studies on the selection exams for admission in universities found that the selection process to higher education is strongly conditioned by the social structure, which makes entry into the university a selection of pre-selected (Klein, 1982). Although in the book *Male dominance* Bourdieu make greater reference to the gender issue, the author mentions that blacks have, through the skin color, the stigma that affects negatively all that are or what they do:

On one hand, regardless of their position in the social space, women have in common the fact that they are separated from men by a negative symbolic coefficient, such as skin color for black, or any other sign of belonging to a stigmatized social group, which negatively affects all they are and do and is in the basis of a systematic set of homologous differences. (Bourdieu, 1999, p. 111)

The analysis showed significant inequalities between racial segments in higher education, indicating that Brazilian university is a space of white predominance. At almost all universities, whites accounted for rates higher than half of the students. The comparison between the representativeness of racial segments in universities and their expression in the population of each state, revealed significant distances, pointing to the over-representation of whites and the under-representation of blacks, even in states in which blacks represent the majority of population, such as Bahia and Maranhão.

The survey also showed a strong racial selectivity in access to higher courses social prestige, showing that it is also predominantly white that privilege. These results therefore show that the Brazilian federal universities investigated have strongly selective areas, particularly marked by inequality racial.

Although the number of university students from popular spaces increases year to year, Brazilian university remains a busy space, mostly by the middle classes and by elites, both white. The opening to the working class and the populations of different sociocultural origins in its faculties, especially in its student table, is an ongoing process that,

(...) In terms of access to higher education, starting from a selection between very few, predominant in Brazilian higher education until the mid-twentieth century, and selection between few, reached with the changes from the 60s, we have, in the late 90s, a selection between many. However, even though it is less elitist, we are far from a Selection between all, which characterizes an effective democratization of access to higher education in Brazil (see Santos, 1998). It is noteworthy, however, that defining higher education as a right for all does not mean necessarily the universalization of this level of education, but rather the possibility of all segments for seeing themselves represented there. (Moehlecke & Gabriel, 2006, p. 3)

In Brazil, although university system has never officially excluded blacks, general access to higher education has always been highly selective. There is an elimination mechanism that makes it clear that – a young upper layer is eighty times more likely to enter university that the son of an agricultural wage and forty times more than a worker's son, whose chances are still twice as high those of a young median class (Bourdieu, 1998, p. 41). It can also be said that the history of education shows us that whites had a much greater chance of admission in universities in the busiest courses or higher social standing – for to blacks. If the system is open to few,

it dramatically enhances the case of black students. Although comprising 45% of Brazilians, the black and brown population (according to the classification of the IBGE) that have completed tertiary education is only 2% and 12% of that total, respectively, compared with 83% of the white population. (Moehlecke, 2004,p. 757)

Brazil has a long history of denied racial inequalities in those, despite the deep distances between whites and blacks, the representations of race relations were influenced by the idea of 'racial democracy'. This self-image has hindered the emergence of a critical view about the reality of race relations in Brazil.

REFERENCES

Beltrão, K., & Teixeira, M. (2005). Cor e gênero na seletividade das carreiras universitárias. In *Os mecanismos de discriminação racial nas escolas brasileiras*. Rio de Janeiro: IPEA.

Bourdieu, P. (1999). *A dominação masculina*. Rio de Janeiro: Bertrand Brasil.

Brasil, I. P. E. A. (2008). *Retrato das desigualdades de gênero e raça*. Brasília: IPEA.

Elster, J. (1994). *Peças e engrenagens das ciências sociais*. Rio de Janeiro: Relume-Dumará.

Gouveia, A. J. (1968). Democratização do ensino superior. *Revista Brasileira de Estudos Pedagógicos, 50*(112), 232-244.

Guimarães, A., & Sérgio, A. (1995). "Raça", racismo e grupos de cor no Brasil. *Estudos Afroasiáticos, 27*, 45–63.

Hasenbalg, C., & Valle Silva, N. (1991). Raça e oportunidades educacionais no Brasil. In *Desigualdades raciais no Brasil contemporâneo*. Belo Horizonte: UFMG/Cedeplar.

Hasenbalg, C., & Valle Silva, N. (1998). *Estrutura social, mobilidade e raça*. Rio de Janeiro: Iuperj/ Vértice.

Hasenbalg, C., Valle Silva, N., & Lima, M. (1999). *Cor e estratificação social*. Rio de Janeiro: Contra Capa.

Jannuzzi, P. M. (2003). Estratificação sócio-ocupacional para estudos de mercado e pesquisa social no Brasil. *Perspectiva, 17*(3-4). Accessed on May 4, 2014, at http://www.scielo.br/Scielo.Php?Script=Sci_Arttext&Pid=S0102-88392003000300024&Lng=En&Nrm=Iso

Klein, R. & Ribeiro, S. C. A. (1982). Divisão interna da universidade: posição social das carreiras. *Educação e Seleção*, 5.

Limongi, F., Carneiro, L., Silva, L., & Mancuso, W. (n.d.). *Acesso à Universidade de São Paulo: atributos socioeconômicos dos excluídos e dos ingressantes no exame vestibular*. Work Document 03/02. Accessed on Dec 1, 2013, at http://nupps.usp.br/Downloads/Docs/Dt0203.Pdf

Moehlecke, S. (2002). Ação afirmativa: História e debates no Brasil. *Cadernos de Pesquisa, São Paulo, 117*(117), 197–217. doi:10.1590/S0100-15742002000300011

Moehlecke, S., & Gabriel, C. T. (2006). Conexões de saberes: uma outra visão sobre o ensino superior. *Revista Contemporânea de Educação da UFRJ*. Retrieved from http://www.educacao.ufrj.br/artigos/n2/numero2-smoehlecke.pdf

Negros na UFBA. (n.d.). Accessed on Dec 1, 2013, at http://www1.folha.uol.com.br/fsp/cotidian/ff23089813.htm#_=_

Queiroz, D. M. (2001). *Raça, gênero e educação superior*. (Doctoral Thesis). Faculty of Education, Universidade Federal da Bahia, Brazil.

Queiroz, D. M. (2004). O negro e a universidade brasileira. *Historia Actual Online, 1*(3), 73-82, 2004. Accessed on Aug 28, 2006, at http://www.historia-actual.com/hao/volumes/volume1/issue3/esp/v1i3c8.pdf

Queiroz, D. M. (2014). *Quem são os negros? Classificação racial no Brasil: aproximações e divergências*. Accessed on Sept 13, 2014, at http://www.anped.org.br/24/t0344239895038.doc

Sant'Anna, W. (2001). Novos marcos para as relações étnico/raciais no Brasil: uma responsabilidade coletiva. In Anais de Seminários Regionais Preparatórios Para Conferência Mundial Contra Racismo, Discriminação Racial, Xenofobia e Intolerância Correlata. Brasília: Ministério da Justiça e Secretaria de Estado dos Direitos Humanos.

Santos, J. C. (1998). *Por uma teoria da redemocratização do ensino superior: perspectivas e implicações*. Rio de Janeiro: Fórum Educacional.

Setton, M. G. J. (2002). A divisão interna do campo universitário: Uma tentativa de classificação. *Revista Brasileira de Estudos Pedagógicos, Brasília, 80*(196), 451–471.

Siss, A. (2003). *Afro-brasileiros, cotas e ações afirmativas: razões históricas.* Niterói: Penesb.

Tavares, I., Braga, M. L. S., & Lima, B. S. (2015). *Análise sobre a participação de negras e negros no sistema científico*. Retrieved from http://www.cnpq.br/documents/10157/1f95db49-f382-4e22-9df7-933608de9e8d

USP tem 78,7% de calouros brancos e 2,4% de pretos, diz pesquisa. (n.d.). Accessed on Dec 1, 2013, at http://g1.globo.com/educacao/noticia/2013/05/usp-tem-787-de-calouros-brancos-e-24-de-pretos-diz-pesquisa.html

ENDNOTES

1 Gouveia classifies some courses offered by three institutions of higher education (University Mackenzie, USP and Catholic University – PUC – São Paulo), according to the social background of their students. To build such classification, the author was based on differences in social background of students and father's occupation as the main criterion classification

2 The author made the ranking of the capital of pupils volume. In his analysis tool, the economic capital has a greater weight when compared to other capitals. In addition, the author used some categories of analysis, which generated the repetition of some courses in different classes, such as psychology, which appears as select course and as an intermediate course.

3 The word *vestibular* comes from the Latin *vestibulum*, meaning entry. Formerly the expression vestibular-examination (entrance exam) was used. Over time, we started to use only vestibular to designate this type of evidence. By the early twentieth century, Brazilian universities were occupied by students from traditional schools, such as Dom Pedro II, in Rio de Janeiro. With the increase in demand exceeding the number of places available, the former Minister of Justice and Business, da Cunha Rivadávia Correa, established the college entrance in Brazil in 1911. The evidence was written and oral, contained language issues, such as English, foreign language, Science (Mathematics, Physics and Chemistry) and contents of the first year of college, where students resorted to special classes to study specific subjects. Hence, the emergence of cramming. Source: http://www.vestibular.brasilescola.com/especial/a-origem-vestibular-no-brasil.htm. Access November 03 2010.

4 See: http://www.redeangola.info/multimedia/retrato-dos-negros-no-brasil/

5 Universitary Fundation for Vestibular, in Portuguese acronomy.

Chapter 13
Becoming a Trauma Surgeon:
How Race Impacted My Journey

Ruby Skinner
Kern Medical Center, USA

ABSTRACT

Surgery is a medical specialty that has a rich history of rigorous training, and the development of a young surgeon requires both individual study and external reinforcement to ensure competency. Although, women are entering medical school at increasing numbers, they only make up to 19% of American surgeons. Minority women represent an even smaller number. Recent studies document that minorities report challenges during surgical residency that may inhibit successful surgical training. These challenges are explored as they relate to limited mentoring and collegial isolation while training in trauma surgery.

I felt like a rock star when I read the article entitled "The Bullet and the Damage Done", which was published in the Rolling Stone Magazine in the spring of 2003. I was a trauma surgery fellow at the University of Pennsylvania and was featured in the article, which focused on the growing public health threat of penetrating trauma in the inner cities of America· (Wilkinson 2003)· Despite the 18 years of school and training with an associated impressive academic pedigree, it seemed very superficial in a way that an article in a non-medical magazine seemed to solidify my accomplishments. At that time, I was about to embark on a career in trauma surgery, and my journey was defined by endless sacrifice and development of a superb surgical skillset that in many ways was seemingly ignored by my medical colleagues, yet I was passed along because I successfully put in the work.

Thinking back on my experience in general surgery residency training, there is a certain amount of expected lonely drudgery that one must endure to meet the demands of learning the field of surgery. Mastering thc pathophysiology of surgical disease, the anatomy and techniques related to numerous operations, and the proper peri-operative management of patients requires 5 years of training. The progression from the first year resident intern to becoming a chief fifth year resident is nothing short of a miracle. There are objective criteria that must be met every year during the training period that allows for progression, and it is based upon technical precision demonstrated the operating room,

DOI: 10.4018/978-1-5225-0174-9.ch013

clinical knowledge demonstrated daily in clinical rounds, conferences, and exams, and competence in working with teams.

The mastery of the aforementioned criteria allowing for advancement in training is also met with the acknowledgment of surgical expertise from peers and hospital staff. This acknowledgement portends to the development of self-confidence and a strong surgical identity that is vital to developing the surgical leadership skills required to direct an operating room staff. This acknowledgment is as subtle as supporting staff taking complete direction from the senior surgical resident in a non-verbal manner, or as obvious as surgical junior residents looking to the senior most resident for guidance with blind trust. The leadership trait arising from this acknowledgment gradually becomes an inherent skill-set that facilitates the surgeon to make critical decisions, allowing for safe and skilled surgical intervention. In my training, this type specific acknowledgement was not always evident and was a source of internal struggle and stresses that I had to overcome to maintain the confidence and surgical identify that contributed to my successful matriculation in surgical training. I became a surgical leader not so much by external cues as described above, but from a rich internal foundation that I developed by trusting the knowledge, and instincts that allowed for the correct surgical judgment that was required of all surgeons.

The issue of minorities being underrepresented in surgery is not a new concept. Racial disparities in surgery are rooted in a long history of discrimination. In 1868 the department of surgery was established at Howard University, but it wasn't until 1928 that the department had an African American chief of surgery. Black surgeons were denied admittance to major surgical societies, and in the mid-1940's Dr. Charles Drew refused membership with the American College of Surgeons, due to discriminatory practices of not accepting well qualified African American surgeons. My former mentor, the late Dr. Claude Organ Jr., was a prominent African American surgeon, and served as the second African American president of the American College of Surgeons. He had enormous influence in the surgical world as an international leader, and was a living legend to many young African American surgeons (Townsend 2012). My success is directly linked to his prominence which allowed for my access to elite research and fellowship training. Dr. Organ was one of the original founders of the Society of Black Academic Surgeons, which is an active organization with members that are high ranking academic surgeons at diverse institutions. Despite these major accomplishments and advances, as a whole, there are still major impediments that exist for blacks to thrive in the field of surgery. Recent data shows that nationally minorities, in particular, black surgery residents don't feel like they fit in as surgery residents which may impact successful matriculation (Butler 2008). This is not surprising since African-Americans comprise less than 3% of academic surgical faculty (Wong 2013). There is very little data specifically on minority women in surgery, and one survey study from 2012 documented that minority women in surgery reported unequal pay compared to white women and their male counterparts. They also reported discriminatory work environments (Frohman 2015).

There is more data on women as minorities in surgery and the disparities they face. The surgical field traditionally has been male-dominated and is associated with a strong tradition of exclusivity. Women in medicine account for 30 percent according to a US Census Bureau in 2009. This number although growing, is not reflective of women pursuing careers in surgery, as women constitute about 19 percent of all U.S. surgeons (Kirk 2014). The work life balance is a common theme that women seem to struggle with when considering a surgical career. The demands of surgery are both physically and mentally rigorous and the training is the longest of all of the medical specialties. There are numerous accounts of women surgeon, who describe many of the isolation issues that minorities face (Cassell

1998). Early in my career, I was resistant to embrace the concept that women and minorities faced some of the same discriminatory issues, as many of mentors were Caucasian males. My experiences with other women surgeons were not always positive, but it was perhaps erroneous on my part to make assumptions about women surgeons as whole, given the low overall percentage of women in the field. The plight of the female surgeon regardless of race has many similarities to the plight of African American surgeons and there is a lot to be gained from understanding the history of both groups.

Recently I was listening to a lecture on the history of trauma surgery given by a nationally recognized surgeon, who talked about a book titled "The Making of a Surgeon", published in 1968 (Nolen 1968). The book was written by a white male surgeon who described his journey as a surgical trainee from medical school, thorough the residency process. The book exemplified the rich tradition of the surgical training process being that of developing surgical skill, clinical acumen and high moral character. I read the book with this project in mind and found that it accurately described the residency process and characterized the external and internal struggles of a young surgeon in training. The struggles of high performance under strict scrutiny, particularly during the early years of training completely resonated with my experience. This scrutiny however, later changed to acceptance and acknowledgment of expertise as the author progressed successfully through the training process. The acknowledged expertise then facilitated the development of confidence, leadership and grace under pressure allowing for the emergence of a competent surgeon.

Despite my training over four decades later after the book was published, I identified with the experiences and clinical situations that the author cleverly described. His poignant narrative regarding situations during his early years of training, were quite comical and most modern surgeons can relate to similar experiences which may been pivotal in their growth. The rite of passage of the junior surgical trainee to the senior trainee and then to the independent surgeon is built on a rich tradition of processes that mimic the military. This type of training in theory is important for the young surgeon to develop the stamina and fortitude that is required to take on the risk of surgical procedures in a successful manner. The major difference between my experiences and what the author described, however, was that along the way he had constant external reinforcement of his progress both good and bad, and this external reinforcement allowed for the development of his confidence and demeanor as a budding young surgeon. The lack of persistent reinforcement in my case for the most part not malicious in nature, but may have originated from the lack of societal perceptions of women of color being leaders. The good clinical outcomes that I was able to achieve and at times outstanding saves were obviously good for the patients, but perhaps were unfathomable to many, as being due to my expertise. This could have been detrimental to my growth as a surgeon because there were significant experiences in the rich tradition of surgical training that I missed. I was able to reconcile that discrepancy, however, based upon limited interactions with a few mentors that provided the feedback about my work when it was deemed necessary.

A prime example of the lack of appreciation of my abilities as a surgeon was when I was in training and on call one evening for trauma. A young African American patient came in through the emergency room after being shot in the neck. This type of injury pattern can be life threatening, and he was bleeding profusely from the wound. I rushed him to the operating room and controlled the bleeding from the carotid artery in an area of the neck, where the artery is very difficult to gain access to. This type of injury pattern is rare, and is associated with high mortality rates. My management of this injury resulted in the patient surviving, with no complications. The next morning, I was so proud of my case, and expected that the other surgeons would at least acknowledge that it was a great save. To my surprise, only the senior most surgeon in the group voiced that it was a great case, and the more junior surgeons only had multiple harsh critiques of my management, despite the patient being in good condition, and despite that fact that they had never been faced with a similar injury. I was devastated at the response that day, and it took me a long time to recover from what I considered to be a major professional slight. I had witnessed on several occasions where my non-African American colleagues had managed cases that weren't even as technically challenging as my case, and were quickly commended on their good judgment and surgical expertise. The quick decisions I made that evening, came from an internal confidence that I had developed up to then, and it was exhilarating to control such a life threatening and chaotic situation with a few quick decisions that resulted in a patient surviving a rare life threatening injury. It was at that point on the morning when my colleagues seemed to be annoyed with the save, that I realized that in order for me to not become consumed with anger and resentment, that I needed to view their behavior from more of a sociological standpoint, and turn the negative energy into my own internal positive reinforcement about my progress. These justifications was my coping strategy which enabled me move forward and start my career with an open and fresh outlook on the possibilities that were my future.

The unspoken truth and the lack of external validation in many ways influenced how I naturally became attracted to trauma surgery as a specialty of general surgery. In contrast to general surgery, in which disease patterns are predicable, and the surgical work up and operative interventions are equally predictable, trauma surgery is chaotic, unpredictable and requires creative rigor in order to bring patients from near death to life in short period of time. The results of trauma are dramatic, and there are no words to necessarily describe the miracles that can occur. The lack of predictability that comes along with trauma surgery and in many instances the lack of external cues that drive diagnoses seemed to have mimicked my residency experience. In other words, the aforementioned lack of external validation of my skill set although subtle was in many ways similar to the chaotic presentation of trauma patients who arrived to the hospital in extremis. I was able to manage the chaos of my training environment by internal forces that developed and drove me to excellence, and these same forces were seamlessly applied in the trauma field, as the ability to control chaos, and make the right decisions, with minimal objective information, came natural to me, and were essential for success in trauma surgery.

Trauma surgery was a relatively new sub-specialty of general surgery when I decided to pursue the extra training. I seemed to inherently understand how the body worked when faced with injury, and spent three years doing bench research related to the cellular changes associated with injury and shock. My identity as a surgeon was based on my ability to easily recognize injury patterns, even with limited objective information, and make crucial decisions that resulted in the prevention of early deaths. From a physiologic standpoint, I understood what happened to the body on a cellular level in response to injury, and from a surgical standpoint, I was trained and understood how to change the course of injury to the restoration of health and recovery. I came to understand this about myself early in my training, and there were times when it was difficult for my more experienced colleagues to comprehend this level of talent. There were subtle ways in which the acknowledgment came, such as, when the surgeons who were my teachers would indirectly ask for my advice when faced with tough surgical decisions. I understood the complexity of those dynamic situations, and would subtly provide assistance by showing up and doing procedures, or by privately suggesting alternative approaches to their management. Those situations would largely go unnoticed by many, but were small victories for me, and allowed for the development of the confidence that I would need for the rest of my career.

As previously discussed, I had a few mentors who were national leaders in surgery and were sources of protection providing structured guidance during my training and early years in practice. Without that source of support, I don't think that the internal drive that I developed would have been enough to allow me to progress through training. There were always negative forces that were directed my way, particularly if I performed exceptionally well and my mentors would generally disperse those forces if they became too volatile. Despite that protection, to which I will always remain grateful, my daily dealings with colleagues and medical staff was always limited, and the natural comradery between surgical colleagues and staff associated with mutual admiration and respect was definitely lacking. My impact was most apparent and most appreciated from the patients that I helped. Ultimately, good patient outcomes and satisfaction are all that matter and that source of validation enhanced my drive in trauma surgery.

Moving forward to the present, my practice is well established and I have taken on leadership positions related to trauma surgery program development, surgical intensive care program development, surgical research and teaching. In many ways, despite these accomplishments, the natural acknowledgement from peers and staff, and comradery that develops with most well established surgeons is still for the most part absent. The difference is that is has less of an internal impact of stress or anxiety, because my accomplishments have not been limited by the lack of validation. I find myself reveling in the isolation associated with my gender and race when it comes to major achievements. I have had great mentorship, perhaps not on a consistent basis, but enough guidance to be able to rise to a level of undeniable expertise, despite the lack of collegial camaraderie along the way. The fact that I have been placed in positions of leadership, is recognition enough, and it hopefully will impact perceptions that will be prohibitive to the isolation that I experienced, allowing the experiences of women of color in surgery, to be more in par with what the author of aforementioned book. Surgical talent and expertise should be recognized and groomed during the early years of training. With the proper guidance most young surgeons can develop their specific talents and gain the expertise needed to be efficient competent practitioners.

Although there is no empirical evidence that characters on television have an impact on perceptions of women of color in medicine, it is intriguing to witness the growth of women of color in surgery being portrayed on television shows and to imagine that these images may impart some normalcy for young female surgeons as they move up the surgical ladder to leadership positions. The impact of media imagery is undeniable, and may impart some useful images that could facilitate acceptance of women of color as being natural leaders in all areas of medicine and science. I have also witnessed a substantial increase in minority women training in surgery at major institutions compared to when I trained over two decades ago. I often refer to my experience in surgery training as what the author Ralph Ellison described in the Invisible Man (Ellison 1947). The lack of eternal validation of my skillset made me at times feel invisible, but the internal fortitude that I mastered that propelled me forward in the field of trauma surgery, allowed my work to become visible. Ultimately one's body of work should be enough to overcome professional isolation, and the fortitude to continue to produce and to provide guidance to those who seek it, is all that matters and should lead to both personal and professional fulfillment. The barriers that existed in the past by far are becoming less impactful for women of color to thrive in surgery and medicine. My desire is that the future generations of surgeons who happen to be women of color, will be visible in all accounts and that their work and experience can be reflective of the natural surgical evolution of the American surgeon, which is based upon a rich tradition of training and surgical expertise.

REFERENCES

Butler, P. D., Longaker, M. T., & Britt, L. D. (2008). Major deficit in the number of underrepresented minority academic surgeons persists. *Annals of Surgery*, *248*(5), 704–711. doi:10.1097/SLA.0b013e31817f2c30 PMID:18948795

Cassell, J. (1998). *The woman in the surgeon's body*. Cambridge, MD: Harvard University Press.

Ellison, R. (1947). *The invisible man*. New York, NY: Random House, Inc.

Frohman, H. A. (2015). The nonwhite women surgeon: A rare species. *The Journal of Surgical Research*, *72*(6), 1266–1271. PMID:26160130

Kirk, N. (2014). Fostering more women surgeons. UConn Today.

Nolen, W. (1968). *The making of a surgeon*. Minneapolis, MN: Mid-List Press.

Townsend, C. M. (2012). *Sabiston Textbook of Surgery*. Philadelphia, PA: WB Saunders Co.

Wilkinson, P. (2003). *The damage the bullet has done. Rolling Stone*.

Wong, R. L., Sullivan, M. C., Yeo, H. L., Roman, S. A., Bell, R. H. Jr, & Sosa, J. A. (2013). Race and surgical residency: Results from a national survey of 4339 US general surgery residents. *Annals of Surgery*, *257*(4), 782–787. doi:10.1097/SLA.0b013e318269d2d0 PMID:23001076

Related References

To continue our tradition of advancing information science and technology research, we have compiled a list of recommended IGI Global readings. These references will provide additional information and guidance to further enrich your knowledge and assist you with your own research and future publications.

Aas, I. H. (2013). Improving patient safety with telemedicine: Exploring organizational factors. In A. Moumtzoglou & A. Kastania (Eds.), *E-health technologies and improving patient safety: Exploring organizational factors* (pp. 56–70). Hershey, PA: Medical Information Science Reference; doi:10.4018/978-1-4666-2657-7.ch004

Aboelfotoh, M. H., Martin, P., & Hassanein, H. (2013). Ubiquitous multimedia data access in electronic health care systems. In D. Tjondronegoro (Ed.), *Tools for mobile multimedia programming and development* (pp. 191–227). Hershey, PA: Information Science Reference; doi:10.4018/978-1-4666-4054-2.ch011

Abreu, J. F., Almeida, P., & Silva, T. (2013). iNeighbour TV: A social TV application to promote wellness of senior citizens. In R. Martinho, R. Rijo, M. Cruz-Cunha, & J. Varajão (Eds.), Information systems and technologies for enhancing health and social care (pp. 1-19). Hershey, PA: Medical Information Science Reference. doi:10.4018/978-1-4666-3667-5.ch001

Aceti, V., & Luppicini, R. (2013). Exploring the effect of mhealth technologies on communication and information sharing in a pediatric critical care unit: A case study. In J. Tan (Ed.), *Healthcare information technology innovation and sustainability: Frontiers and adoption* (pp. 88–108). Hershey, PA: Medical Information Science Reference; doi:10.4018/978-1-4666-2797-0.ch006

Adami, I., Antona, M., & Stephanidis, C. (2014). Ambient assisted living for people with motor impairments. In G. Kouroupetroglou (Ed.), *Disability informatics and web accessibility for motor limitations* (pp. 76–104). Hershey, PA: Medical Information Science Reference; doi:10.4018/978-1-4666-4442-7.ch003

Adeyemo, O. (2013). The nationwide health information network: A biometric approach to prevent medical identity theft. In I. Management Association (Ed.), User-driven healthcare: Concepts, methodologies, tools, and applications (pp. 1636-1649). Hershey, PA: Medical Information Science Reference. doi:10.4018/978-1-4666-2770-3.ch081

Åkerberg, A., Lindén, M., & Folke, M. (2013). Pedometer cell phone applications and future trends in measuring physical activity. In R. Martinho, R. Rijo, M. Cruz-Cunha, & J. Varajão (Eds.), *Information systems and technologies for enhancing health and social care* (pp. 324–339). Hershey, PA: Medical Information Science Reference; doi:10.4018/978-1-4666-3667-5.ch021

Al Hamouche, V. (2014). Making quality control decisions in radiology department: A decision support system for radiographers' performance appraisal using PACS. In C. El Morr (Ed.), *Research perspectives on the role of informatics in health policy and management* (pp. 48–61). Hershey, PA: Medical Information Science Reference; doi:10.4018/978-1-4666-4321-5.ch004

Al-Khudairy, S. (2014). Caring for our aging population: Using CPOE and telehomecare systems as a response to health policy concerns. In C. El Morr (Ed.), *Research perspectives on the role of informatics in health policy and management* (pp. 153–166). Hershey, PA: Medical Information Science Reference; doi:10.4018/978-1-4666-4321-5.ch010

Albuquerque, C. (2013). The study of social needs as a strategic tool for the innovation of the social care sector: The contribution of new technologies. In M. Cruz-Cunha, I. Miranda, & P. Gonçalves (Eds.), *Handbook of research on ICTs for human-centered healthcare and social care services* (pp. 347–365). Hershey, PA: Medical Information Science Reference; doi:10.4018/978-1-4666-3986-7.ch018

Alexandrou, D. A., & Pardalis, K. V. (2013). SEMantic PATHways: Modeling, executing, and monitoring intra-organizational healthcare business processes towards personalized treatment. In A. Moumtzoglou & A. Kastania (Eds.), *E-health technologies and improving patient safety: Exploring organizational factors* (pp. 98–123). Hershey, PA: Medical Information Science Reference; doi:10.4018/978-1-4666-2657-7.ch007

Algarín, A. D., Demurjian, S. A., Ziminski, T. B., Sánchez, Y. K., & Kuykendall, R. (2014). Securing XML with role-based access control: Case study in health care. In A. Ruiz-Martinez, R. Marin-Lopez, & F. Pereniguez-Garcia (Eds.), *Architectures and protocols for secure information technology infrastructures* (pp. 334–365). Hershey, PA: Information Science Reference; doi:10.4018/978-1-4666-4514-1.ch013

Alhaqbani, B., & Fidge, C. (2013). A medical data trustworthiness assessment model. In I. Management Association (Ed.), User-driven healthcare: Concepts, methodologies, tools, and applications (pp. 1425-1445). Hershey, PA: Medical Information Science Reference. doi:10.4018/978-1-4666-2770-3.ch071

Almeida, L., Menezes, P., & Dias, J. (2013). Augmented reality framework for the socialization between elderly people. In M. Cruz-Cunha, I. Miranda, & P. Gonçalves (Eds.), *Handbook of research on ICTs for human-centered healthcare and social care services* (pp. 430–448). Hershey, PA: Medical Information Science Reference; doi:10.4018/978-1-4666-3986-7.ch023

Amer, M. B., Amawi, M., & El-Khatib, H. (2013). A novel neural fuzzy approach for diagnosis of potassium disturbances. In J. Tan (Ed.), *Healthcare information technology innovation and sustainability: Frontiers and adoption* (pp. 208–218). Hershey, PA: Medical Information Science Reference; doi:10.4018/978-1-4666-2797-0.ch013

Andersen, S. T., & Jansen, A. (2013). Innovation in ICT-based health care provision. In J. Tan (Ed.), *Healthcare information technology innovation and sustainability: Frontiers and adoption* (pp. 58–72). Hershey, PA: Medical Information Science Reference; doi:10.4018/978-1-4666-2797-0.ch004

Angjellari-Dajci, F., Lawless, W. F., Stachura, M. E., Wood, E. A., & DiBattisto, C. (2013). Economic evaluations for service delivery in autism spectrum disorders: benefit-cost analysis for emerging telehealth systems. In M. Cruz-Cunha, I. Miranda, & P. Gonçalves (Eds.), *Handbook of research on ICTs and management systems for improving efficiency in healthcare and social care* (pp. 16–42). Hershey, PA: Medical Information Science Reference; doi:10.4018/978-1-4666-3990-4.ch002

Archibald, D., MacDonald, C. J., Hogue, R., & Mercer, J. (2013). Accessing knowledge from the bedside: Introducing the tablet computer to clinical teaching. In C. Rückemann (Ed.), *Integrated information and computing systems for natural, spatial, and social sciences* (pp. 96–109). Hershey, PA: Information Science Reference; doi:10.4018/978-1-4666-2190-9.ch005

Archondakis, S. (2013). Static telecytological applications for proficiency testing. In V. Gulla, A. Mori, F. Gabbrielli, & P. Lanzafame (Eds.), *Telehealth networks for hospital services: New methodologies* (pp. 228–239). Hershey, PA: Medical Information Science Reference; doi:10.4018/978-1-4666-2979-0.ch015

Arling, P. A., Doebbeling, B. N., & Fox, R. L. (2013). Improving the implementation of evidence-based practice and information systems in healthcare: A social network approach. In J. Tan (Ed.), *Healthcare information technology innovation and sustainability: Frontiers and adoption* (pp. 247–270). Hershey, PA: Medical Information Science Reference; doi:10.4018/978-1-4666-2797-0.ch016

Arriaga, P., Esteves, F., & Fernandes, S. (2013). Playing for better or for worse?: Health and social outcomes with electronic gaming. In M. Cruz-Cunha, I. Miranda, & P. Gonçalves (Eds.), *Handbook of research on ICTs for human-centered healthcare and social care services* (pp. 48–69). Hershey, PA: Medical Information Science Reference; doi:10.4018/978-1-4666-3986-7.ch003

Arslan, P. (2014). Collaborative participation in personalized health through mobile diaries. In K. Rızvanoğlu & G. Çetin (Eds.), *Research and design innovations for mobile user experience* (pp. 150–181). Hershey, PA: Information Science Reference; doi:10.4018/978-1-4666-4446-5.ch009

Assis-Hassid, S., Reychav, I., Pliskin, J. S., & Heart, T. H. (2013). The effects of electronic medical record (EMR) use in primary care on the physician-patient relationship. In M. Cruz-Cunha, I. Miranda, & P. Gonçalves (Eds.), *Handbook of research on ICTs for human-centered healthcare and social care services* (pp. 130–150). Hershey, PA: Medical Information Science Reference; doi:10.4018/978-1-4666-3986-7.ch007

Azar, A. T. (2013). Overview of biomedical engineering. In I. Management Association (Ed.), Bioinformatics: Concepts, methodologies, tools, and applications (pp. 1-28). Hershey, PA: Medical Information Science Reference. doi:10.4018/978-1-4666-3604-0.ch001

Azar, A. T., & Eljamel, M. S. (2014). Medical robotics. In I. Management Association (Ed.), Robotics: Concepts, methodologies, tools, and applications (pp. 1116-1147). Hershey, PA: Information Science Reference. doi:10.4018/978-1-4666-4607-0.ch054

Baladrón, C., Aguiar, J. M., Calavia, L., Carro, B., & Sánchez-Esguevillas, A. (2014). Learning on the move in the web 2.0: New initiatives in m-learning. In *K-12 education: Concepts, methodologies, tools, and applications* (pp. 1693–1714). Hershey, PA: Information Science Reference; doi:10.4018/978-1-4666-4502-8.ch097

Bauer, K. A. (2013). Caught in the web: The internet and the demise of medical privacy. In I. Management Association (Ed.), User-driven healthcare: Concepts, methodologies, tools, and applications (pp. 1252-1272). Hershey, PA: Medical Information Science Reference. doi:10.4018/978-1-4666-2770-3.ch063

Belsis, P., Skourlas, C., & Gritzalis, S. (2013). Secure electronic healthcare records management in wireless environments. In M. Khosrow-Pour (Ed.), *Interdisciplinary advances in information technology research* (pp. 202–219). Hershey, PA: Information Science Reference; doi:10.4018/978-1-4666-3625-5.ch015

Benito, R. V., Vega-Colado, C., Coco, M. B., Cuadrado, R., Torres-Zafra, J. C., Sánchez-Pena, J. M., & López-Miguel, A. et al. (2013). New electro-optic and display technology for visually disabled people. In M. Cruz-Cunha, I. Miranda, & P. Gonçalves (Eds.), *Handbook of research on ICTs for human-centered healthcare and social care services* (pp. 687–718). Hershey, PA: Medical Information Science Reference; doi:10.4018/978-1-4666-3986-7.ch036

Bergenti, F., Poggi, A., & Tomaiuolo, M. (2013). Using multi-agent systems to support e-health services. In M. Cruz-Cunha, I. Miranda, & P. Gonçalves (Eds.), *Handbook of research on ICTs for human-centered healthcare and social care services* (pp. 549–567). Hershey, PA: Medical Information Science Reference; doi:10.4018/978-1-4666-3986-7.ch029

Berler, A., & Apostolakis, I. (2014). Normalizing cross-border healthcare in Europe via new e-prescription paradigms. In C. El Morr (Ed.), *Research perspectives on the role of informatics in health policy and management* (pp. 168–208). Hershey, PA: Medical Information Science Reference; doi:10.4018/978-1-4666-4321-5.ch011

Bernardi, R. (2013). Information technology and resistance to public sector reforms: A case study in Kenya. In I. Management Association (Ed.), User-driven healthcare: Concepts, methodologies, tools, and applications (pp. 14-33). Hershey, PA: Medical Information Science Reference. doi:10.4018/978-1-4666-2770-3.ch002

Bernhard, S., Al Zoukra, K., & Schtte, C. (2013). From non-invasive hemodynamic measurements towards patient-specific cardiovascular diagnosis. In *Data mining: Concepts, methodologies, tools, and applications* (pp. 2069–2093). Hershey, PA: Information Science Reference; doi:10.4018/978-1-4666-2455-9.ch106

Best, C., O'Neill, B., & Gillespie, A. (2013). Assistive technology for cognition: Enabling activities of daily living. In M. Cruz-Cunha, I. Miranda, & P. Gonçalves (Eds.), *Handbook of research on ICTs for human-centered healthcare and social care services* (pp. 112–129). Hershey, PA: Medical Information Science Reference; doi:10.4018/978-1-4666-3986-7.ch006

Bhattacharya, P., Asanga, A. P., & Biswas, R. (2013). Stomodeum to proctodeum: Email narratives on clinical problem solving in gastroenterology. In I. Management Association (Ed.), User-driven healthcare: Concepts, methodologies, tools, and applications (pp. 1073-1091). Hershey, PA: Medical Information Science Reference. doi:10.4018/978-1-4666-2770-3.ch054

Boboc, C., & Ţiţan, E. (2014). Migration of medical doctors, health, medical education, and employment in eastern and central Europe. In A. Driouchi (Ed.), *Labor and health economics in the Mediterranean region: Migration and mobility of medical doctors* (pp. 158–191). Hershey, PA: Medical Information Science Reference; doi:10.4018/978-1-4666-4723-7.ch007

Botella, C., Baños, R. M., Etchemendy, E., Castilla, D., García-Palacios, A., & Alcañiz, M. (2013). An e-health system for promoting wellbeing in the elderly: The butler system. In I. Management Association (Ed.), User-driven healthcare: Concepts, methodologies, tools, and applications (pp. 838-852). Hershey, PA: Medical Information Science Reference. doi:10.4018/978-1-4666-2770-3.ch042

Bradai, N., Chaari, L., & Kamoun, L. (2013). A comprehensive overview of wireless body area networks (WBAN). In J. Rodrigues (Ed.), *Digital advances in medicine, e-health, and communication technologies* (pp. 1–32). Hershey, PA: Medical Information Science Reference; doi:10.4018/978-1-4666-2794-9.ch001

Brown, K. E., Bayley, J., & Newby, K. (2013). Serious game for relationships and sex education: Application of an intervention mapping approach to development. In S. Arnab, I. Dunwell, & K. Debattista (Eds.), *Serious games for healthcare: Applications and implications* (pp. 135–166). Hershey, PA: Medical Information Science Reference; doi:10.4018/978-1-4666-1903-6.ch007

Brown, M. (2013). Will comparative effectiveness research lead to healthcare rationing? In I. Management Association (Ed.), User-driven healthcare: Concepts, methodologies, tools, and applications (pp. 1487-1507). Hershey, PA: Medical Information Science Reference. doi:10.4018/978-1-4666-2770-3.ch074

Burlamaqui, A. M., Azevedo, S. O., Silva, I. R., Silva, S. R., Souto da Silva, G. H., & Benjamim, X. C. … Felix de Castro, A. (2013). Low cost t-health and t-social with ginga: Experience with mime TV, ImFine, and iFunnyCube interactive TV programs. In M. Cruz-Cunha, I. Miranda, & P. Gonçalves (Eds.), Handbook of research on ICTs for human-centered healthcare and social care services (pp. 303-318). Hershey, PA: Medical Information Science Reference. doi:10.4018/978-1-4666-3986-7.ch015

Burns, J., Blanchard, M., & Metcalf, A. (2013). Bridging the digital divide in Australia: The potential implications for the mental health of young people experiencing marginalisation. In *Digital literacy: Concepts, methodologies, tools, and applications* (pp. 772–793). Hershey, PA: Information Science Reference; doi:10.4018/978-1-4666-1852-7.ch040

Buyurgan, N., Rardin, R. L., Jayaraman, R., Varghese, V. M., & Burbano, A. (2013). A novel GS1 data standard adoption roadmap for healthcare providers. In J. Tan (Ed.), *Healthcare information technology innovation and sustainability: Frontiers and adoption* (pp. 41–57). Hershey, PA: Medical Information Science Reference; doi:10.4018/978-1-4666-2797-0.ch003

Cabrita, M. D., & Cabrita, M. (2014). Applying social marketing to healthcare: Challenges and opportunities. In A. Kapoor & C. Kulshrestha (Eds.), *Dynamics of competitive advantage and consumer perception in social marketing* (pp. 78–97). Hershey, PA: Business Science Reference; doi:10.4018/978-1-4666-4430-4.ch004

Carroll, J. M., & Rosson, M. B. (2013). Community networks: Infrastructure and models for therapeutic support. In M. Cruz-Cunha, I. Miranda, & P. Gonçalves (Eds.), *Handbook of research on ICTs for human-centered healthcare and social care services* (pp. 187–207). Hershey, PA: Medical Information Science Reference; doi:10.4018/978-1-4666-3986-7.ch010

Castiglioni, I., Gilardi, M. C., & Gallivanone, F. (2013). E-health decision support systems for the diagnosis of dementia diseases. In A. Moumtzoglou & A. Kastania (Eds.), *E-health technologies and improving patient safety: Exploring organizational factors* (pp. 84–97). Hershey, PA: Medical Information Science Reference; doi:10.4018/978-1-4666-2657-7.ch006

Catley, C., Smith, K., McGregor, C., James, A., & Eklund, J. M. (2013). A framework for multidimensional real-time data analysis: A case study for the detection of apnoea of prematurity. In A. Gangopadhyay (Ed.), *Methods, models, and computation for medical informatics* (pp. 16–35). Hershey, PA: Medical Information Science Reference; doi:10.4018/978-1-4666-2653-9.ch002

Chaari, L., & Kamoun, L. (2013). QoS concepts and architecture over wireless body area networks for healthcare applications. In J. Rodrigues (Ed.), *Digital advances in medicine, e-health, and communication technologies* (pp. 114–130). Hershey, PA: Medical Information Science Reference; doi:10.4018/978-1-4666-2794-9.ch007

Chamberlin, B., Maloney, A., Gallagher, R. R., & Garza, M. L. (2013). Active video games: Potential for increased activity, suggestions for use, and guidelines for implementation. In S. Arnab, I. Dunwell, & K. Debattista (Eds.), *Serious games for healthcare: Applications and implications* (pp. 191–212). Hershey, PA: Medical Information Science Reference; doi:10.4018/978-1-4666-1903-6.ch009

Charisi, A., Korvesis, P., & Megalooikonomou, V. (2013). Similarity searching of medical image data in distributed systems: Facilitating telemedicine applications. In A. Gangopadhyay (Ed.), *Methods, models, and computation for medical informatics* (pp. 58–77). Hershey, PA: Medical Information Science Reference; doi:10.4018/978-1-4666-2653-9.ch004

Chattopadhyay, A., Malhotra, K., & Chatterjee, S. (2013). Ethical guidelines for the quality assessment of healthcare. In I. Management Association (Ed.), User-driven healthcare: Concepts, methodologies, tools, and applications (pp. 1332-1347). Hershey, PA: Medical Information Science Reference. doi:10.4018/978-1-4666-2770-3.ch066

Chen, J. Y., Xu, H., Shi, P., Culbertson, A., & Meslin, E. M. (2013). Ethics and privacy considerations for systems biology applications in predictive and personalized medicine. In I. Management Association (Ed.), Bioinformatics: Concepts, methodologies, tools, and applications (pp. 1378-1404). Hershey, PA: Medical Information Science Reference. doi:10.4018/978-1-4666-3604-0.ch071

Chen, Y. (2013). Construction of digital statistical atlases of the liver and their applications to computer-aided diagnosis. In J. Wu (Ed.), *Technological advancements in biomedicine for healthcare applications* (pp. 68–79). Hershey, PA: Medical Information Science Reference; doi:10.4018/978-1-4666-2196-1.ch008

Cheng, B., Stanley, R. J., De, S., Antani, S., & Thoma, G. R. (2013). Automatic detection of arrow annotation overlays in biomedical images. In J. Tan (Ed.), *Healthcare information technology innovation and sustainability: Frontiers and adoption* (pp. 219–236). Hershey, PA: Medical Information Science Reference; doi:10.4018/978-1-4666-2797-0.ch014

Cherian, E. J., & Ryan, T. W. (2014). Incongruent needs: Why differences in the iron-triangle of priorities make health information technology adoption and use difficult. In C. El Morr (Ed.), *Research perspectives on the role of informatics in health policy and management* (pp. 209–221). Hershey, PA: Medical Information Science Reference; doi:10.4018/978-1-4666-4321-5.ch012

Chowdhry, N., Ashraf, A., Chowdhry, B., Baloch, A., Ansari, A., & De Meer, H. (2013). Grid for post operative care through wireless sensor networks. In I. Management Association (Ed.), User-driven healthcare: Concepts, methodologies, tools, and applications (pp. 514-531). Hershey, PA: Medical Information Science Reference. doi:10.4018/978-1-4666-2770-3.ch026

Chun, S. A., Kwon, J. H., & Lee, H. (2013). Social credential-based role recommendation and patient privacy control in medical emergency. In A. Gangopadhyay (Ed.), *Methods, models, and computation for medical informatics* (pp. 215–237). Hershey, PA: Medical Information Science Reference; doi:10.4018/978-1-4666-2653-9.ch013

Cohen, J. F., Bancilhon, J., & Sergay, S. (2013). An empirical study of patient willingness to use self-service technologies in the healthcare context. In M. Cruz-Cunha, I. Miranda, & P. Gonçalves (Eds.), *Handbook of research on ICTs and management systems for improving efficiency in healthcare and social care* (pp. 378–395). Hershey, PA: Medical Information Science Reference; doi:10.4018/978-1-4666-3990-4.ch019

Constantinides, P. (2013). The development of the English national health information infrastructure. In I. Management Association (Ed.), User-driven healthcare: Concepts, methodologies, tools, and applications (pp. 968-991). Hershey, PA: Medical Information Science Reference. doi:10.4018/978-1-4666-2770-3.ch049

Constantinides, P. (2013). The development of a regional health information infrastructure in Greece. In I. Management Association (Ed.), User-driven healthcare: Concepts, methodologies, tools, and applications (pp. 992-1017). Hershey, PA: Medical Information Science Reference. doi:10.4018/978-1-4666-2770-3.ch050

Corrigan, D., Hederman, L., Khan, H., Taweel, A., Kostopoulou, O., & Delaney, B. (2013). An ontology-driven approach to clinical evidence modelling implementing clinical prediction rules. In A. Moumtzoglou & A. Kastania (Eds.), *E-health technologies and improving patient safety: Exploring organizational factors* (pp. 257–284). Hershey, PA: Medical Information Science Reference; doi:10.4018/978-1-4666-2657-7.ch016

Corritore, C. L., Kracher, B., Wiedenbeck, S., & Marble, R. (2013). Foundations of trust for e-health. In I. Management Association (Ed.), User-driven healthcare: Concepts, methodologies, tools, and applications (pp. 1167-1193). Hershey, PA: Medical Information Science Reference. doi:10.4018/978-1-4666-2770-3.ch059

Costa, Â., Andrade, F., & Novais, P. (2013). Privacy and data protection towards elderly healthcare. In M. Cruz-Cunha, I. Miranda, & P. Gonçalves (Eds.), *Handbook of research on ICTs for human-centered healthcare and social care services* (pp. 330–346). Hershey, PA: Medical Information Science Reference; doi:10.4018/978-1-4666-3986-7.ch017

Coughlan, J., & Brinkman, W. (2013). Design considerations for delivering e-learning to surgical trainees. In J. Rodrigues (Ed.), *Digital advances in medicine, e-health, and communication technologies* (pp. 341–350). Hershey, PA: Medical Information Science Reference; doi:10.4018/978-1-4666-2794-9.ch020

D'Abundo, M. L. (2013). Electronic health record implementation in the United States healthcare industry: Making the process of change manageable. In V. Wang (Ed.), *Handbook of research on technologies for improving the 21st century workforce: Tools for lifelong learning* (pp. 272–286). Hershey, PA: Information Science Publishing; doi:10.4018/978-1-4666-2181-7.ch018

D'Abundo, M. L. (2013). Electronic health record implementation in the United States healthcare industry: Making the process of change manageable. In I. Management Association (Ed.), User-driven healthcare: Concepts, methodologies, tools, and applications (pp. 382-395). Hershey, PA: Medical Information Science Reference. doi:10.4018/978-1-4666-2770-3.ch019

D'Andrea, A., Ferri, F., & Grifoni, P. (2013). RFID technologies in the health sector. In A. Moumtzoglou & A. Kastania (Eds.), *E-health technologies and improving patient safety: Exploring organizational factors* (pp. 140–147). Hershey, PA: Medical Information Science Reference; doi:10.4018/978-1-4666-2657-7.ch009

D'Astolfo, C. (2014). Investing in a "rehabilitation model" to improve the decision-making process in long-term care. In C. El Morr (Ed.), *Research perspectives on the role of informatics in health policy and management* (pp. 37–47). Hershey, PA: Medical Information Science Reference; doi:10.4018/978-1-4666-4321-5.ch003

Daniel, V. M. (2013). Genomics and genetic engineering: Playing God? In I. Management Association (Ed.), Bioinformatics: Concepts, methodologies, tools, and applications (pp. 249-267). Hershey, PA: Medical Information Science Reference. doi:10.4018/978-1-4666-3604-0.ch013

Dasgupta, S. S. (2013). Cyber capability framework: A tool to evaluate ICT for development projects. In M. Cruz-Cunha, I. Miranda, & P. Gonçalves (Eds.), *Handbook of research on ICTs for human-centered healthcare and social care services* (pp. 399–406). Hershey, PA: Medical Information Science Reference; doi:10.4018/978-1-4666-3986-7.ch021

Davis, S. A. (2013). Global telemedicine and ehealth: Advances for future healthcare – Using a systems approach to integrate healthcare functions. In V. Gulla, A. Mori, F. Gabbrielli, & P. Lanzafame (Eds.), *Telehealth networks for hospital services: New methodologies* (pp. 15–32). Hershey, PA: Medical Information Science Reference; doi:10.4018/978-1-4666-2979-0.ch002

Deng, M., Petkovic, M., Nalin, M., & Baroni, I. (2013). Home healthcare in cloud computing. In M. Cruz-Cunha, I. Miranda, & P. Gonçalves (Eds.), *Handbook of research on ICTs and management systems for improving efficiency in healthcare and social care* (pp. 614–634). Hershey, PA: Medical Information Science Reference; doi:10.4018/978-1-4666-3990-4.ch032

DeVany, M., Knobloch-Ludwig, K., Penticoff, M., Assimacopoulos, A., & Speedie, S. (2013). Telehealth implementation: The voice of experience. In S. Sarnikar, D. Bennett, & M. Gaynor (Eds.), *Cases on healthcare information technology for patient care management* (pp. 126–139). Hershey, PA: Medical Information Science Reference; doi:10.4018/978-1-4666-2671-3.ch008

Ding, W., Qiu, P., Liu, Y., & Feng, W. (2013). Current omics technologies in biomarker discovery. In I. Management Association (Ed.), Bioinformatics: Concepts, methodologies, tools, and applications (pp. 465-497). Hershey, PA: Medical Information Science Reference. doi:10.4018/978-1-4666-3604-0.ch027

Dingli, A., Abela, C., & D'Ambrogio, I. (2013). PINATA: Taking e-health a step forward. In A. Moumtzoglou & A. Kastania (Eds.), *E-health technologies and improving patient safety: Exploring organizational factors* (pp. 173–195). Hershey, PA: Medical Information Science Reference; doi:10.4018/978-1-4666-2657-7.ch012

Dogra, D. P., Nandam, K., Majumdar, A. K., Sural, S., Mukhopadhyay, J., Majumdar, B., & Mukherjee, S. et al. (2013). A tool for automatic hammersmith infant neurological examination. In J. Rodrigues (Ed.), *Digital advances in medicine, e-health, and communication technologies* (pp. 301–311). Hershey, PA: Medical Information Science Reference; doi:10.4018/978-1-4666-2794-9.ch017

Doyle, J., Bertolotto, M., & Wilson, D. (2013). Towards multimodal mobile GIS for the elderly. In *Digital literacy: Concepts, methodologies, tools, and applications* (pp. 590–609). Hershey, PA: Information Science Reference; doi:10.4018/978-1-4666-1852-7.ch031

Dragoni, A. F. (2013). Health services through digital terrestrial television. In V. Gulla, A. Mori, F. Gabbrielli, & P. Lanzafame (Eds.), *Telehealth networks for hospital services: New methodologies* (pp. 207–227). Hershey, PA: Medical Information Science Reference; doi:10.4018/978-1-4666-2979-0.ch014

Dragoni, A. F. (2013). Virtual carer: A first prototype. In V. Gulla, A. Mori, F. Gabbrielli, & P. Lanzafame (Eds.), *Telehealth networks for hospital services: New methodologies* (pp. 290–299). Hershey, PA: Medical Information Science Reference; doi:10.4018/978-1-4666-2979-0.ch019

Driouchi, A. (2014). Introduction to labor and health economics: Mobility of medical doctors in the Mediterranean region. In A. Driouchi (Ed.), *Labor and health economics in the Mediterranean region: Migration and mobility of medical doctors* (pp. 1–22). Hershey, PA: Medical Information Science Reference; doi:10.4018/978-1-4666-4723-7.ch001

Driouchi, A. (2014). Medical knowledge, north-south cooperation, and mobility of medical doctors. In A. Driouchi (Ed.), *Labor and health economics in the Mediterranean region: Migration and mobility of medical doctors* (pp. 376–395). Hershey, PA: Medical Information Science Reference; doi:10.4018/978-1-4666-4723-7.ch015

Driouchi, A. (2014). Skilled human resources in the health sectors and impacts of new health technologies on health workforce in developing economies. In A. Driouchi (Ed.), *Labor and health economics in the Mediterranean region: Migration and mobility of medical doctors* (pp. 23–50). Hershey, PA: Medical Information Science Reference; doi:10.4018/978-1-4666-4723-7.ch002

Duan, X., Wang, X., & Huang, Q. (2013). Medical manipulators for surgical applications. In J. Wu (Ed.), *Technological advancements in biomedicine for healthcare applications* (pp. 111–122). Hershey, PA: Medical Information Science Reference; doi:10.4018/978-1-4666-2196-1.ch012

Dugdale, A. (2013). Australian patient organizations: Using digital technologies to engage health citizen communities in health policy. In I. Management Association (Ed.), User-driven healthcare: Concepts, methodologies, tools, and applications (pp. 853-869). Hershey, PA: Medical Information Science Reference. doi:10.4018/978-1-4666-2770-3.ch043

Dunwell, I., & Jarvis, S. (2013). A serious game for on-the-ward infection control awareness training: Ward off infection. In S. Arnab, I. Dunwell, & K. Debattista (Eds.), *Serious games for healthcare: Applications and implications* (pp. 233–246). Hershey, PA: Medical Information Science Reference; doi:10.4018/978-1-4666-1903-6.ch011

Durant, K. T., McCray, A. T., & Safran, C. (2013). Identifying temporal changes and topics that promote growth within online communities: A prospective study of six online cancer forums. In A. Gangopadhyay (Ed.), *Methods, models, and computation for medical informatics* (pp. 78–97). Hershey, PA: Medical Information Science Reference; doi:10.4018/978-1-4666-2653-9.ch005

El-Farargy, N. (2013). Refresher training in clinical psychology supervision: A blended learning approach. In A. Benson, J. Moore, & S. Williams van Rooij (Eds.), *Cases on educational technology planning, design, and implementation: A project management perspective* (pp. 295–317). Hershey, PA: Information Science Reference; doi:10.4018/978-1-4666-4237-9.ch016

Enquist, H. (2013). From idea to use: lessons learned from a participatory ICT healthcare case study. In I. Management Association (Ed.), User-driven healthcare: Concepts, methodologies, tools, and applications (pp. 1037-1053). Hershey, PA: Medical Information Science Reference. doi:10.4018/978-1-4666-2770-3.ch052

Ervin, K. (2014). Legal and ethical considerations in the implementation of electronic health records. In J. Krueger (Ed.), *Cases on electronic records and resource management implementation in diverse environments* (pp. 193–210). Hershey, PA: Information Science Reference; doi:10.4018/978-1-4666-4466-3.ch012

Escayola, J., Trigo, J., Martínez, I., Martínez-Espronceda, M., Aragüés, A., Sancho, D., . . . García, J. (2013). Overview of the ISO/IEEE11073 family of standards and their applications to health monitoring. In I. Management Association (Ed.), User-driven healthcare: Concepts, methodologies, tools, and applications (pp. 357-381). Hershey, PA: Medical Information Science Reference. doi:10.4018/978-1-4666-2770-3.ch018

Facelli, J. C., Hurdle, J. F., & Mitchell, J. A. (2013). Medical informatics and bioinformatics. In I. Management Association (Ed.), Bioinformatics: Concepts, methodologies, tools, and applications (pp. 195-221). Hershey, PA: Medical Information Science Reference. doi:10.4018/978-1-4666-3604-0.ch010

Fakhar, A. (2014). Beyond brain drain: A case study of the benefits of cooperation on medical immigration. In A. Driouchi (Ed.), *Labor and health economics in the Mediterranean region: Migration and mobility of medical doctors* (pp. 294–313). Hershey, PA: Medical Information Science Reference; doi:10.4018/978-1-4666-4723-7.ch012

Falan, S. L., Han, B., Zoeller, L. H., Tarn, J. M., & Roach, D. M. (2013). Sustaining healthcare through waste elimination: A taxonomic analysis with case illustrations. In J. Tan (Ed.), *Healthcare information technology innovation and sustainability: Frontiers and adoption* (pp. 18–39). Hershey, PA: Medical Information Science Reference; doi:10.4018/978-1-4666-2797-0.ch002

Farinha, C., & Mira da Silva, M. (2013). Identifying requirements for healthcare information systems with focus groups. In M. Cruz-Cunha, I. Miranda, & P. Gonçalves (Eds.), *Handbook of research on ICTs for human-centered healthcare and social care services* (pp. 491–510). Hershey, PA: Medical Information Science Reference; doi:10.4018/978-1-4666-3986-7.ch026

Fernando, J. (2013). The protocols of privileged information handling in an e-health context: Australia. In I. Management Association (Ed.), User-driven healthcare: Concepts, methodologies, tools, and applications (pp. 737-759). Hershey, PA: Medical Information Science Reference. doi:10.4018/978-1-4666-2770-3.ch036

Ferreira, L., Teixeira, A., & Cunha, J. P. (2013). Medical information extraction in European Portuguese. In M. Cruz-Cunha, I. Miranda, & P. Gonçalves (Eds.), *Handbook of research on ICTs for human-centered healthcare and social care services* (pp. 607–626). Hershey, PA: Medical Information Science Reference; doi:10.4018/978-1-4666-3986-7.ch032

Fitzpatrick, J. (2013). An exploration of the experiences of migrant women: Implications for policy development of effective user driven health care delivery systems. In I. Management Association (Ed.), User-driven healthcare: Concepts, methodologies, tools, and applications (pp. 954-967). Hershey, PA: Medical Information Science Reference. doi:10.4018/978-1-4666-2770-3.ch048

Fitzpatrick, J., & Ako, W. (2013). Developing community ontologies in user driven healthcare. In I. Management Association (Ed.), User-driven healthcare: Concepts, methodologies, tools, and applications (pp. 655-672). Hershey, PA: Medical Information Science Reference. doi:10.4018/978-1-4666-2770-3.ch032

Flores, A. E., Win, K. T., & Susilo, W. (2013). Secure exchange of electronic health records. In I. Management Association (Ed.), IT policy and ethics: Concepts, methodologies, tools, and applications (pp. 1059-1079). Hershey, PA: Information Science Reference. doi:10.4018/978-1-4666-2919-6.ch048

Flores, A. E., Win, K. T., & Susilo, W. (2013). Secure exchange of electronic health records. In I. Management Association (Ed.), User-driven healthcare: Concepts, methodologies, tools, and applications (pp. 1403-1424). Hershey, PA: Medical Information Science Reference. doi:10.4018/978-1-4666-2770-3.ch070

Forrestal, E. J., Cellucci, L. W., Zeng, X., Kennedy, M. H., & Smith, D. (2013). Health information technology collaboration in community health centers: The community partners HealthNet, Inc. In S. Sarnikar, D. Bennett, & M. Gaynor (Eds.), *Cases on healthcare information technology for patient care management* (pp. 171–196). Hershey, PA: Medical Information Science Reference; doi:10.4018/978-1-4666-2671-3.ch011

Frederico, C. (2013). Nutrition games. In S. Arnab, I. Dunwell, & K. Debattista (Eds.), *Serious games for healthcare: Applications and implications* (pp. 167–190). Hershey, PA: Medical Information Science Reference; doi:10.4018/978-1-4666-1903-6.ch008

Freitas, L., Pereira, R. T., Pereira, H. G., Martini, R., Mozzaquatro, B. A., Kasper, J., & Librelotto, G. (2013). Ontological representation and an architecture for homecare pervasive systems. In R. Martinho, R. Rijo, M. Cruz-Cunha, & J. Varajão (Eds.), *Information systems and technologies for enhancing health and social care* (pp. 215–234). Hershey, PA: Medical Information Science Reference; doi:10.4018/978-1-4666-3667-5.ch015

Frigo, C. A., & Pavan, E. E. (2014). Prosthetic and orthotic devices. In *Assistive technologies: Concepts, methodologies, tools, and applications* (pp. 549–613). Hershey, PA: Information Science Reference; doi:10.4018/978-1-4666-4422-9.ch028

Fukami, T., & Wu, J. (2013). Image fusion method and the efficacy of multimodal cardiac images. In J. Wu (Ed.), *Technological advancements in biomedicine for healthcare applications* (pp. 47–54). Hershey, PA: Medical Information Science Reference; doi:10.4018/978-1-4666-2196-1.ch006

Furdu, I., & Patrut, B. (2013). ICT applications and solutions in healthcare: Present and perspectives. In M. Cruz-Cunha, I. Miranda, & P. Gonçalves (Eds.), *Handbook of research on ICTs and management systems for improving efficiency in healthcare and social care* (pp. 559–576). Hershey, PA: Medical Information Science Reference; doi:10.4018/978-1-4666-3990-4.ch029

Gabbrielli, F. (2013). Telemedicine R&D influencing incoming strategies and organization models. In V. Gulla, A. Mori, F. Gabbrielli, & P. Lanzafame (Eds.), *Telehealth networks for hospital services: New methodologies* (pp. 250–264). Hershey, PA: Medical Information Science Reference; doi:10.4018/978-1-4666-2979-0.ch017

Gaivéo, J. M. (2013). Security of ICTs supporting healthcare activities. In M. Cruz-Cunha, I. Miranda, & P. Gonçalves (Eds.), *Handbook of research on ICTs for human-centered healthcare and social care services* (pp. 208–228). Hershey, PA: Medical Information Science Reference; doi:10.4018/978-1-4666-3986-7.ch011

Gallo, C. (2013). A multichannel framework for multimedia content deployment in e-health environments. In M. Cruz-Cunha, I. Miranda, & P. Gonçalves (Eds.), *Handbook of research on ICTs and management systems for improving efficiency in healthcare and social care* (pp. 872–891). Hershey, PA: Medical Information Science Reference; doi:10.4018/978-1-4666-3990-4.ch045

Gao, Y., Yang, W., Yang, J., Satoshi, T., & Wu, J. (2013). Neural mechanisms of audiovisual integration in integrated processing for verbal perception and spatial factors. In J. Wu (Ed.), *Biomedical engineering and cognitive neuroscience for healthcare: Interdisciplinary applications* (pp. 327–336). Hershey, PA: Medical Information Science Reference; doi:10.4018/978-1-4666-2113-8.ch034

Garzo, A., Carmien, S. P., & Madina, X. (2013). Mapping input technology to ability. In I. Management Association (Ed.), User-driven healthcare: Concepts, methodologies, tools, and applications (pp. 480-501). Hershey, PA: Medical Information Science Reference. doi:10.4018/978-1-4666-2770-3.ch024

Gatzidis, C. (2013). First-person shooter game engines and healthcare: An examination of the current state of the art and future potential. In S. Arnab, I. Dunwell, & K. Debattista (Eds.), *Serious Games for Healthcare: Applications and Implications* (pp. 76–89). Hershey, PA: Medical Information Science Reference; doi:10.4018/978-1-4666-1903-6.ch004

Gaughan, M. (2013). The hybridized nature of America's health care system: Medicare as a case of both market and public failure. In M. Merviö (Ed.), *Healthcare management and economics: Perspectives on public and private administration* (pp. 154–164). Hershey, PA: Medical Information Science Reference; doi:10.4018/978-1-4666-3982-9.ch012

Gavgani, V. Z. (2013). Information therapy (ix) service and patients' preference. In A. Gangopadhyay (Ed.), *Methods, models, and computation for medical informatics* (pp. 117–125). Hershey, PA: Medical Information Science Reference; doi:10.4018/978-1-4666-2653-9.ch007

Gavgani, V. Z. (2013). Ubiquitous information therapy service through social networking libraries: An operational web 2.0 service model. In I. Management Association (Ed.), User-driven healthcare: Concepts, methodologies, tools, and applications (pp. 673-688). Hershey, PA: Medical Information Science Reference. doi:10.4018/978-1-4666-2770-3.ch033

Germaine-McDaniel, N. S. (2013). The emerging hispanic use of online health information in the United States: Cultural convergence or dissociation? In I. Management Association (Ed.), User-driven healthcare: Concepts, methodologies, tools, and applications (pp. 1607-1621). Hershey, PA: Medical Information Science Reference. doi:10.4018/978-1-4666-2770-3.ch079

Gholson, J., & Tennyson, H. (2013). One system of care, one electronic chart. In S. Sarnikar, D. Bennett, & M. Gaynor (Eds.), *Cases on healthcare information technology for patient care management* (pp. 55–69). Hershey, PA: Medical Information Science Reference; doi:10.4018/978-1-4666-2671-3.ch003

Ghosh, B. (2013). Healthcare systems using clinical data: Addressing data interoperability challenges. In S. Sarnikar, D. Bennett, & M. Gaynor (Eds.), *Cases on healthcare information technology for patient care management* (pp. 208–223). Hershey, PA: Medical Information Science Reference; doi:10.4018/978-1-4666-2671-3.ch013

Gibson, L., Sloan, D., & Moncur, W. (2013). E-health and digital inclusion. In I. Management Association (Ed.), User-driven healthcare: Concepts, methodologies, tools, and applications (pp. 197-210). Hershey, PA: Medical Information Science Reference. doi:10.4018/978-1-4666-2770-3.ch011

Gillies, A. C., & Howard, J. (2013). Information as change agent or barrier in health care reform? In J. Tan (Ed.), *Healthcare information technology innovation and sustainability: Frontiers and adoption* (pp. 1–17). Hershey, PA: Medical Information Science Reference; doi:10.4018/978-1-4666-2797-0.ch001

Glusker, A., & Hoelscher, E. (2014). Transitioning from print to online-only resources: The experience of a medium-sized healthcare organization library. In J. Krueger (Ed.), *Cases on electronic records and resource management implementation in diverse environments* (pp. 226–242). Hershey, PA: Information Science Reference; doi:10.4018/978-1-4666-4466-3.ch014

Gofuku, A., Fukumori, S., & Sato, K. (2013). A mirror visual feedback therapy system applying virtual reality technology. In J. Wu (Ed.), *Biomedical engineering and cognitive neuroscience for healthcare: Interdisciplinary applications* (pp. 73–80). Hershey, PA: Medical Information Science Reference; doi:10.4018/978-1-4666-2113-8.ch008

Gomes, C., Sperandio, F., Peles, A., Borges, J., Brito, A. C., & Almada-Lobo, B. (2013). An operating theater planning decision support system. In R. Martinho, R. Rijo, M. Cruz-Cunha, & J. Varajão (Eds.), *Information systems and technologies for enhancing health and social care* (pp. 69–86). Hershey, PA: Medical Information Science Reference; doi:10.4018/978-1-4666-3667-5.ch005

Grissom, T. E., DuKatz, A., Kordylewski, H. A., & Dutton, R. P. (2013). Bring out your data: The evolution of the national anesthesia clinical outcomes registry (NACOR). In A. Gangopadhyay (Ed.), *Methods, models, and computation for medical informatics* (pp. 126–145). Hershey, PA: Medical Information Science Reference; doi:10.4018/978-1-4666-2653-9.ch008

Gudes, O., Kendall, E., Yigitcanlar, T., Han, J. H., & Pathak, V. (2013). Developing a competitive city through healthy decision-making. In I. Management Association (Ed.), User-driven healthcare: Concepts, methodologies, tools, and applications (pp. 808-822). Hershey, PA: Medical Information Science Reference. doi:10.4018/978-1-4666-2770-3.ch040

Gugerty, B., & Maranda, M. J. (2013). The promises and challenges of health information technology. In I. Management Association (Ed.), User-driven healthcare: Concepts, methodologies, tools, and applications (pp. 57-76). Hershey, PA: Medical Information Science Reference. doi:10.4018/978-1-4666-2770-3.ch004

Guimarães, C., Antunes, D. R., García, L. S., & Fernandes, S. (2013). Information challenges of the deaf in their health and social care needs. In M. Cruz-Cunha, I. Miranda, & P. Gonçalves (Eds.), *Handbook of research on ICTs for human-centered healthcare and social care services* (pp. 93–111). Hershey, PA: Medical Information Science Reference; doi:10.4018/978-1-4666-3986-7.ch005

Guinalíu, M., Marta, J., & Subero, J. M. (2013). Social networks as a tool to improve the life quality of chronic patients and their relatives. In M. Cruz-Cunha, I. Miranda, & P. Gonçalves (Eds.), *Handbook of research on ICTs for human-centered healthcare and social care services* (pp. 172–186). Hershey, PA: Medical Information Science Reference; doi:10.4018/978-1-4666-3986-7.ch009

Gullà, V. (2013). Leading the technological innovation in healthcare systems: The telematic medicine approach. In V. Gulla, A. Mori, F. Gabbrielli, & P. Lanzafame (Eds.), *Telehealth networks for hospital services: New methodologies* (pp. 134–153). Hershey, PA: Medical Information Science Reference; doi:10.4018/978-1-4666-2979-0.ch009

Gullà, V., & Cancellotti, C. (2013). Telemedicine in emergency: A first aid hospital network experience. In V. Gulla, A. Mori, F. Gabbrielli, & P. Lanzafame (Eds.), *Telehealth networks for hospital services: New methodologies* (pp. 240–248). Hershey, PA: Medical Information Science Reference; doi:10.4018/978-1-4666-2979-0.ch016

Guo, M., Yu, Y., Yang, J., & Wu, J. (2013). The crossmodal between the visual and tactile for motion perception. In J. Wu (Ed.), *Biomedical engineering and cognitive neuroscience for healthcare: Interdisciplinary applications* (pp. 99–108). Hershey, PA: Medical Information Science Reference; doi:10.4018/978-1-4666-2113-8.ch011

Habib, M. K. (2014). Web-based multi-user distributed and collaborative environment supporting emergency and relief activities. In I. Management Association (Ed.), Crisis management: Concepts, methodologies, tools and applications (pp. 425-445). Hershey, PA: Information Science Reference. doi:10.4018/978-1-4666-4707-7.ch019

Hafeez-Baig, A., & Gururajan, R. (2013). Phenomena of adoption of wireless handheld devices: A case of healthcare setting. In H. Muga & K. Thomas (Eds.), *Cases on the diffusion and adoption of sustainable development practices* (pp. 20–43). Hershey, PA: Information Science Reference; doi:10.4018/978-1-4666-2842-7.ch002

Haida, M. (2013). Implications of NIRS brain signals. In J. Wu (Ed.), *Biomedical engineering and cognitive neuroscience for healthcare: Interdisciplinary applications* (pp. 120–128). Hershey, PA: Medical Information Science Reference; doi:10.4018/978-1-4666-2113-8.ch013

Han, D., & Braun, K. L. (2013). Promoting active ageing through technology training in Korea. In *Digital literacy: Concepts, methodologies, tools, and applications* (pp. 572–589). Hershey, PA: Information Science Reference; doi:10.4018/978-1-4666-1852-7.ch030

Haniff, D. (2013). Usability engineering and e-health. In I. Management Association (Ed.), User-driven healthcare: Concepts, methodologies, tools, and applications (pp. 1446-1468). Hershey, PA: Medical Information Science Reference. doi:10.4018/978-1-4666-2770-3.ch072

Hareva, D. H., Okada, H., & Oka, H. (2013). Ecological momentary assessment using a mobile phone. In J. Wu (Ed.), *Technological advancements in biomedicine for healthcare applications* (pp. 398–407). Hershey, PA: Medical Information Science Reference; doi:10.4018/978-1-4666-2196-1.ch039

Harnett, B. (2013). Patient centered medicine and technology adaptation. In I. Management Association (Ed.), User-driven healthcare: Concepts, methodologies, tools, and applications (pp. 77-98). Hershey, PA: Medical Information Science Reference. doi:10.4018/978-1-4666-2770-3.ch005

Hatton, J. D., Schmidt, T. M., & Jelen, J. (2013). Adoption of electronic health care records: Physician heuristics and hesitancy. In R. Martinho, R. Rijo, M. Cruz-Cunha, & J. Varajão (Eds.), *Information systems and technologies for enhancing health and social care* (pp. 148–165). Hershey, PA: Medical Information Science Reference; doi:10.4018/978-1-4666-3667-5.ch010

Heinrichs, L., Fellander-Tsai, L., & Davies, D. (2013). Clinical virtual worlds: The wider implications for professional development in healthcare. In K. Bredl & W. Bösche (Eds.), *Serious games and virtual worlds in education, professional development, and healthcare* (pp. 221–240). Hershey, PA: Information Science Reference; doi:10.4018/978-1-4666-3673-6.ch014

Heinrichs, W. L., Davies, D., & Davies, J. (2013). Virtual worlds in healthcare: Applications and implications. In S. Arnab, I. Dunwell, & K. Debattista (Eds.), *Serious games for healthcare: Applications and implications* (pp. 1–22). Hershey, PA: Medical Information Science Reference; doi:10.4018/978-1-4666-1903-6.ch001

Hirata, M., Yanagisawa, T., Matsushita, K., Sugata, H., Kamitani, Y., & Suzuki, T. … Yoshimine, T. (2013). Brain–machine interface using brain surface electrodes: Real-time robotic control and a fully implantable wireless system. In J. Wu (Ed.), Technological advancements in biomedicine for healthcare applications (pp. 362-374). Hershey, PA: Medical Information Science Reference. doi:10.4018/978-1-4666-2196-1.ch036

Hocine, N., & Gouaïch, A. (2013). Difficulty and scenario adaptation: An approach to customize therapeutic games. In S. Arnab, I. Dunwell, & K. Debattista (Eds.), *Serious games for healthcare: Applications and implications* (pp. 107–134). Hershey, PA: Medical Information Science Reference; doi:10.4018/978-1-4666-1903-6.ch006

Hoonakker, P., Cartmill, R. S., Carayon, P., & Walker, J. M. (2013). Development and psychometric qualities of the SEIPS survey to evaluate CPOE/EHR implementation in ICUs. In J. Tan (Ed.), *Healthcare information technology innovation and sustainability: Frontiers and adoption* (pp. 161–179). Hershey, PA: Medical Information Science Reference; doi:10.4018/978-1-4666-2797-0.ch010

Hoshi, K. (2013). Reframing dichotomies: Human experiential design of healthcare technologies. In I. Management Association (Ed.), User-driven healthcare: Concepts, methodologies, tools, and applications (pp. 1303-1331). Hershey, PA: Medical Information Science Reference. doi:10.4018/978-1-4666-2770-3.ch065

Høstgaard, A. M. (2013). End-user participation in health IT development: The EUPHIT method. In I. Management Association (Ed.), User-driven healthcare: Concepts, methodologies, tools, and applications (pp. 608-629). Hershey, PA: Medical Information Science Reference. doi:10.4018/978-1-4666-2770-3.ch030

Igrejas, G., Amaral, J. S., & Rodrigues, P. J. (2013). Fall detection systems to be used by elderly people. In M. Cruz-Cunha, I. Miranda, & P. Gonçalves (Eds.), *Handbook of research on ICTs for human-centered healthcare and social care services* (pp. 449–473). Hershey, PA: Medical Information Science Reference; doi:10.4018/978-1-4666-3986-7.ch024

Inomata, C., & Nitta, S. (2013). Nursing in integrative medicine and nurses' engagement in caring-healing: A discussion based on the practice and study of music therapy and nursing care for patients with neurodegenerative disorders. In J. Wu (Ed.), *Technological advancements in biomedicine for healthcare applications* (pp. 235–239). Hershey, PA: Medical Information Science Reference; doi:10.4018/978-1-4666-2196-1.ch025

Ishaq, G. M., Hussain, P. T., Iqbal, M. J., & Mushtaq, M. B. (2013). Risk-benefit analysis of combination vs. unopposed HRT in post-menopausal women. In I. Management Association (Ed.), Bioinformatics: Concepts, methodologies, tools, and applications (pp. 1424-1440). Hershey, PA: Medical Information Science Reference. doi:10.4018/978-1-4666-3604-0.ch073

Iwaki, S. (2013). Multimodal neuroimaging to visualize human visual processing. In J. Wu (Ed.), *Biomedical engineering and cognitive neuroscience for healthcare: Interdisciplinary applications* (pp. 274–282). Hershey, PA: Medical Information Science Reference; doi:10.4018/978-1-4666-2113-8.ch028

Ji, Z., Sugi, T., Goto, S., Wang, X., & Nakamura, M. (2013). Multi-channel template extraction for automatic EEG spike detection. In J. Wu (Ed.), *Biomedical engineering and cognitive neuroscience for healthcare: Interdisciplinary applications* (pp. 255–265). Hershey, PA: Medical Information Science Reference; doi:10.4018/978-1-4666-2113-8.ch026

Jiang, Y., Wang, S., Tan, R., Ishida, K., Ando, T., & Fujie, M. G. (2013). Motor cortex activation during mental imagery of walking: An fNIRS study. In J. Wu (Ed.), *Biomedical engineering and cognitive neuroscience for healthcare: Interdisciplinary applications* (pp. 29–37). Hershey, PA: Medical Information Science Reference; doi:10.4018/978-1-4666-2113-8.ch004

Johnson, K., & Tashiro, J. (2013). Interprofessional care and health care complexity: Factors shaping human resources effectiveness in health information management. In I. Management Association (Ed.), User-Driven Healthcare: Concepts, Methodologies, Tools, and Applications (pp. 1273-1302). Hershey, PA: Medical Information Science Reference. doi:10.4018/978-1-4666-2770-3.ch064

Jose, J. (2013). Pharmacovigilance: Basic concepts and applications of pharmacoinformatics. In I. Management Association (Ed.), Bioinformatics: Concepts, methodologies, tools, and applications (pp. 1453-1473). Hershey, PA: Medical Information Science Reference. doi:10.4018/978-1-4666-3604-0.ch075

Kalina, J., & Zvárová, J. (2013). Decision support systems in the process of improving patient safety. In A. Moumtzoglou & A. Kastania (Eds.), *E-health technologies and improving patient safety: Exploring organizational factors* (pp. 71–83). Hershey, PA: Medical Information Science Reference; doi:10.4018/978-1-4666-2657-7.ch005

Kalina, J., & Zvárová, J. (2013). Decision support systems in the process of improving patient safety. In I. Management Association (Ed.), Bioinformatics: Concepts, methodologies, tools, and applications (pp. 1113-1125). Hershey, PA: Medical Information Science Reference. doi:10.4018/978-1-4666-3604-0.ch057

Kamath, J. R., & Donahoe-Anshus, A. L. (2013). Electronic health record: Adoption, considerations and future direction. In I. Management Association (Ed.), User-driven healthcare: Concepts, methodologies, tools, and applications (pp. 34-56). Hershey, PA: Medical Information Science Reference. doi:10.4018/978-1-4666-2770-3.ch003

Karamagkioli, K. Z., & Karamagioli, E. (2013). European e-health framework: Towards more "patient-friendly" healthcare services? In I. Management Association (Ed.), User-driven healthcare: Concepts, methodologies, tools, and applications (pp. 760-775). Hershey, PA: Medical Information Science Reference. doi:10.4018/978-1-4666-2770-3.ch037

Kardaras, D. K., & Karakostas, B. (2013). Case studies in customization of e-health services. In I. Management Association (Ed.), User-driven healthcare: Concepts, methodologies, tools, and applications (pp. 1018-1036). Hershey, PA: Medical Information Science Reference. doi:10.4018/978-1-4666-2770-3.ch051

Kastania, A. N. (2013). Evaluation considerations for e-health systems. In I. Management Association (Ed.), User-driven healthcare: Concepts, methodologies, tools, and applications (pp. 1126-1140). Hershey, PA: Medical Information Science Reference. doi:10.4018/978-1-4666-2770-3.ch057

Kato, P. M. (2013). The role of the researcher in making serious games for health. In S. Arnab, I. Dunwell, & K. Debattista (Eds.), *Serious games for healthcare: Applications and implications* (pp. 213–231). Hershey, PA: Medical Information Science Reference; doi:10.4018/978-1-4666-1903-6.ch010

Katsura, S. (2013). Preservation and reproduction of human motion based on a motion-copying system. In J. Wu (Ed.), *Technological advancements in biomedicine for healthcare applications* (pp. 375–384). Hershey, PA: Medical Information Science Reference; doi:10.4018/978-1-4666-2196-1.ch037

Kikuchi, T. (2013). Human-friendly mechatronics systems with functional fluids and elastomers. In J. Wu (Ed.), *Technological advancements in biomedicine for healthcare applications* (pp. 94–101). Hershey, PA: Medical Information Science Reference; doi:10.4018/978-1-4666-2196-1.ch010

Kimura, T., Miura, T., Shinohara, K., & Doi, S. (2013). Visual attention in 3-D space while moving forward. In J. Wu (Ed.), *Biomedical engineering and cognitive neuroscience for healthcare: Interdisciplinary applications* (pp. 81–88). Hershey, PA: Medical Information Science Reference; doi:10.4018/978-1-4666-2113-8.ch009

Kldiashvili, E. (2013). Implementation of telecytology in Georgia. In V. Gulla, A. Mori, F. Gabbrielli, & P. Lanzafame (Eds.), *Telehealth networks for hospital services: New methodologies* (pp. 341–361). Hershey, PA: Medical Information Science Reference; doi:10.4018/978-1-4666-2979-0.ch022

Kolovou, L. T., & Lymberopoulos, D. K. (2013). The concept of interoperability for AAL systems. In I. Management Association (Ed.), User-driven healthcare: Concepts, methodologies, tools, and applications (pp. 1364-1385). Hershey, PA: Medical Information Science Reference. doi:10.4018/978-1-4666-2770-3.ch068

Koppar, A. R., & Sridhar, V. (2013). Malaria parasite detection: Automated method using microscope color image. In J. Rodrigues (Ed.), *Digital advances in medicine, e-health, and communication technologies* (pp. 289–300). Hershey, PA: Medical Information Science Reference; doi:10.4018/978-1-4666-2794-9.ch016

Kotsonis, E., & Eliakis, S. (2013). Information security standards for health information systems: The implementer's approach. In I. Management Association (Ed.), User-driven healthcare: Concepts, methodologies, tools, and applications (pp. 225-257). Hershey, PA: Medical Information Science Reference. doi:10.4018/978-1-4666-2770-3.ch013

Krishnaswamy, K., & Oates, T. (2014). Pathway to independence: Past, present, and beyond via robotics. In G. Kouroupetroglou (Ed.), *Disability informatics and web accessibility for motor limitations* (pp. 153–201). Hershey, PA: Medical Information Science Reference; doi:10.4018/978-1-4666-4442-7.ch005

Kuehler, M., Schimke, N., & Hale, J. (2013). Privacy considerations for electronic health records. In I. Management Association (Ed.), User-driven healthcare: Concepts, methodologies, tools, and applications (pp. 1387-1402). Hershey, PA: Medical Information Science Reference. doi:10.4018/978-1-4666-2770-3.ch069

Kurita, Y., Ikeda, A., Nagata, K., Okajima, M., & Ogasawara, T. (2013). Biomedical robotics for healthcare. In J. Wu (Ed.), *Technological advancements in biomedicine for healthcare applications* (pp. 160–169). Hershey, PA: Medical Information Science Reference; doi:10.4018/978-1-4666-2196-1.ch017

Lakkaraju, S., & Lakkaraju, S. (2013). Mobile device application in healthcare. In S. Sarnikar, D. Bennett, & M. Gaynor (Eds.), *Cases on healthcare information technology for patient care management* (pp. 276–307). Hershey, PA: Medical Information Science Reference; doi:10.4018/978-1-4666-2671-3.ch016

Laskowski, M. (2013). A prototype agent based model and machine learning hybrid system for healthcare decision support. In J. Rodrigues (Ed.), *Digital advances in medicine, e-health, and communication technologies* (pp. 230–253). Hershey, PA: Medical Information Science Reference; doi:10.4018/978-1-4666-2794-9.ch013

Lee, T. (2014). Mobile healthcare computing in the cloud. In J. Rodrigues, K. Lin, & J. Lloret (Eds.), *Mobile networks and cloud computing convergence for progressive services and applications* (pp. 275–294). Hershey, PA: Information Science Reference; doi:10.4018/978-1-4666-4781-7.ch015

Lelardeux, C., Alvarez, J., Montaut, T., Galaup, M., & Lagarrigue, P. (2013). Healthcare games and the metaphoric approach. In S. Arnab, I. Dunwell, & K. Debattista (Eds.), *Serious games for healthcare: Applications and implications* (pp. 23–49). Hershey, PA: Medical Information Science Reference; doi:10.4018/978-1-4666-1903-6.ch002

Lhotska, L., Bursa, M., Huptych, M., Chudacek, V., & Havlik, J. (2013). Interoperability of medical devices and information systems. In M. Cruz-Cunha, I. Miranda, & P. Gonçalves (Eds.), *Handbook of research on ICTs for human-centered healthcare and social care services* (pp. 749–762). Hershey, PA: Medical Information Science Reference; doi:10.4018/978-1-4666-3986-7.ch039

Li, X., Lin, Z., & Wu, J. (2013). Language processing in the human brain of literate and illiterate subjects. In J. Wu (Ed.), *Biomedical engineering and cognitive neuroscience for healthcare: Interdisciplinary applications* (pp. 201–209). Hershey, PA: Medical Information Science Reference; doi:10.4018/978-1-4666-2113-8.ch021

Liao, H. (2013). Biomedical information processing and visualization for minimally invasive neurosurgery. In J. Wu (Ed.), *Technological advancements in biomedicine for healthcare applications* (pp. 36–46). Hershey, PA: Medical Information Science Reference; doi:10.4018/978-1-4666-2196-1.ch005

Libreri, C., & Graffigna, G. (2014). How web 2.0 shapes patient knowledge sharing: The case of diabetes in Italy. In C. El Morr (Ed.), *Research perspectives on the role of informatics in health policy and management* (pp. 238–260). Hershey, PA: Medical Information Science Reference; doi:10.4018/978-1-4666-4321-5.ch014

Lin, C., Huang, Y., Li, C., & Jalleh, G. (2013). Key health information systems outsourcing issues from six hospital cases. In I. Management Association (Ed.), User-driven healthcare: Concepts, methodologies, tools, and applications (pp. 824-837). Hershey, PA: Medical Information Science Reference. doi:10.4018/978-1-4666-2770-3.ch041

Lin, L., Chen, Y., Wu, J., & Tennyson, R. D. (2013). What skill/knowledge is important to a nursing professional? In P. Ordóñez de Pablos (Ed.), *Business, technology, and knowledge management in Asia: Trends and innovations* (pp. 234–249). Hershey, PA: Business Science Reference; doi:10.4018/978-1-4666-2652-2.ch018

Liu, G. (2014). Using case costing data and case mix for funding and benchmarking in rehabilitation hospitals. In C. El Morr (Ed.), *Research perspectives on the role of informatics in health policy and management* (pp. 62–78). Hershey, PA: Medical Information Science Reference; doi:10.4018/978-1-4666-4321-5.ch005

Lixun, Z., Dapeng, B., & Lei, Y. (2013). Design of and experimentation with a walking assistance robot. In J. Wu (Ed.), *Technological advancements in biomedicine for healthcare applications* (pp. 123–127). Hershey, PA: Medical Information Science Reference; doi:10.4018/978-1-4666-2196-1.ch013

Lui, K. (2013). The health informatics professional. In I. Management Association (Ed.), User-driven healthcare: Concepts, methodologies, tools, and applications (pp. 120-141). Hershey, PA: Medical Information Science Reference. doi:10.4018/978-1-4666-2770-3.ch007

Ma, X., Chen, G., & Xiao, J. (2013). Understanding weight change behaviors through online social networks. In A. Gangopadhyay (Ed.), *Methods, models, and computation for medical informatics* (pp. 189–214). Hershey, PA: Medical Information Science Reference; doi:10.4018/978-1-4666-2653-9.ch012

Makikawa, M., Okada, S., Fujiwara, Y., & Esaki, M. (2013). Sleep monitoring system equipped with a flexible non-contact ECG, respiration, and body motion sensor. In J. Wu (Ed.), *Technological advancements in biomedicine for healthcare applications* (pp. 287–297). Hershey, PA: Medical Information Science Reference; doi:10.4018/978-1-4666-2196-1.ch030

Malindretos, G. (2013). Distribution and logistics outsourcing in the pharmaceutical sector. In D. Folinas (Ed.), *Outsourcing management for supply chain operations and logistics service* (pp. 202–222). Hershey, PA: Business Science Reference; doi:10.4018/978-1-4666-2008-7.ch012

Manvi, S. S., & B., M. R. (2013). A survey on health care services using wireless sensor networks. In M. Cruz-Cunha, I. Miranda, & P. Gonçalves (Eds.), *Handbook of research on ICTs for human-centered healthcare and social care services* (pp. 587-606). Hershey, PA: Medical Information Science Reference. doi:10.4018/978-1-4666-3986-7.ch031

Marshall, S., & Hogan, J. (2013). The use of medical simulation to improve patient safety. In A. Moumtzoglou & A. Kastania (Eds.), *E-health technologies and improving patient safety: Exploring organizational factors* (pp. 155–172). Hershey, PA: Medical Information Science Reference; doi:10.4018/978-1-4666-2657-7.ch011

Martin, J., & McKay, E. (2013). Mental health, post-secondary education, and information communications technology. In J. Lannon & E. Halpin (Eds.), *Human rights and information communication technologies: Trends and consequences of use* (pp. 196–213). Hershey, PA: Information Science Reference; doi:10.4018/978-1-4666-1918-0.ch012

Martin, J., & McKay, E. (2014). Mental health, post-secondary education, and information communications technology. In *Assistive technologies: Concepts, methodologies, tools, and applications* (pp. 1209–1226). Hershey, PA: Information Science Reference; doi:10.4018/978-1-4666-4422-9.ch063

Martin, V. (2013). Developing a library collection in bioinformatics: Support for an evolving profession. In I. Management Association (Ed.), Bioinformatics: Concepts, methodologies, tools, and applications (pp. 130-150). Hershey, PA: Medical Information Science Reference. doi:10.4018/978-1-4666-3604-0.ch007

Masayuki, K., Eiji, K., Tetsuo, T., & Nozomu, M. (2013). Evaluation of olfactory impairment in Parkinson's disease using near-infrared spectroscopy. In J. Wu (Ed.), *Biomedical engineering and cognitive neuroscience for healthcare: Interdisciplinary applications* (pp. 293–302). Hershey, PA: Medical Information Science Reference; doi:10.4018/978-1-4666-2113-8.ch030

Mazzanti, I., Maolo, A., & Antonicelli, R. (2013). E-health and telemedicine in the elderly: State of the art. In V. Gulla, A. Mori, F. Gabbrielli, & P. Lanzafame (Eds.), *Telehealth Networks for Hospital Services: New Methodologies* (pp. 33–43). Hershey, PA: Medical Information Science Reference; doi:10.4018/978-1-4666-2979-0.ch003

Mazzanti, I., Maolo, A., & Antonicelli, R. (2014). E-health and telemedicine in the elderly: State of the art. In *Assistive technologies: Concepts, methodologies, tools, and applications* (pp. 693–704). Hershey, PA: Information Science Reference; doi:10.4018/978-1-4666-4422-9.ch034

McCrossan, B. A., & Casey, F. A. (2013). The role of telemedicine in paediatric cardiology. In V. Gulla, A. Mori, F. Gabbrielli, & P. Lanzafame (Eds.), *Telehealth networks for hospital services: New methodologies* (pp. 44–88). Hershey, PA: Medical Information Science Reference; doi:10.4018/978-1-4666-2979-0.ch004

McGaha, J. (2013). Implementation issues on a national electronic health record network. In I. Management Association (Ed.), User-driven healthcare: Concepts, methodologies, tools, and applications (pp. 1236-1251). Hershey, PA: Medical Information Science Reference. doi:10.4018/978-1-4666-2770-3.ch062

McGinnes, S., & Burke, M. (2013). Hoping for the best: A qualitative study of information technology in primary care. In M. Cruz-Cunha, I. Miranda, & P. Gonçalves (Eds.), *Handbook of research on ICTs and management systems for improving efficiency in healthcare and social care* (pp. 1088–1108). Hershey, PA: Medical Information Science Reference; doi:10.4018/978-1-4666-3990-4.ch057

Medhekar, A., Wong, H. Y., & Hall, J. (2014). Innovation in medical tourism service marketing: A case of India. In A. Goyal (Ed.), *Innovations in services marketing and management: Strategies for emerging economies* (pp. 49–66). Hershey, PA: Business Science Reference; doi:10.4018/978-1-4666-4671-1.ch003

Medhekar, A., Wong, H. Y., & Hall, J. (2014). Medical tourism: A conceptual framework for an innovation in global healthcare provision. In A. Goyal (Ed.), *Innovations in services marketing and management: Strategies for emerging economies* (pp. 148–169). Hershey, PA: Business Science Reference; doi:10.4018/978-1-4666-4671-1.ch009

Menciassi, A., & Laschi, C. (2014). Biorobotics. In I. Management Association (Ed.), Robotics: Concepts, methodologies, tools, and applications (pp. 1613-1643). Hershey, PA: Information Science Reference. doi:10.4018/978-1-4666-4607-0.ch079

Mendes, D., & Rodrigues, I. P. (2013). A semantic web pragmatic approach to develop clinical ontologies, and thus semantic interoperability, based in HL7 v2.xml messaging. In R. Martinho, R. Rijo, M. Cruz-Cunha, & J. Varajão (Eds.), *Information systems and technologies for enhancing health and social care* (pp. 205–214). Hershey, PA: Medical Information Science Reference; doi:10.4018/978-1-4666-3667-5.ch014

Mendoza-González, R., Martin, M. V., & Rodríguez-Martínez, L. C. (2013). Identifying the essential design requirements for usable e-health communities in mobile devices. In I. Management Association (Ed.), User-driven healthcare: Concepts, methodologies, tools, and applications (pp. 533-552). Hershey, PA: Medical Information Science Reference. doi:10.4018/978-1-4666-2770-3.ch027

Mettler, T. (2013). Transformation of the hospital supply chain: How to measure the maturity of supplier relationship management systems in hospitals? In J. Tan (Ed.), *Healthcare information technology innovation and sustainability: Frontiers and adoption* (pp. 180–192). Hershey, PA: Medical Information Science Reference; doi:10.4018/978-1-4666-2797-0.ch011

Miesenberger, K., Nussbaum, G., & Ossmann, R. (2014). AsTeRICS: A framework for including sensor technology into AT solutions for people with motor disabilities. In G. Kouroupetroglou (Ed.), *Assistive technologies and computer access for motor disabilities* (pp. 154–179). Hershey, PA: Medical Information Science Reference; doi:10.4018/978-1-4666-4438-0.ch006

Miller, K., & Sankaranarayanan, S. (2013). Applications of policy based agents in wireless body sensor mesh networks for patient health monitoring. In J. Rodrigues (Ed.), *Digital advances in medicine, e-health, and communication technologies* (pp. 85–101). Hershey, PA: Medical Information Science Reference; doi:10.4018/978-1-4666-2794-9.ch005

Mohammadian, M., & Jentzsch, R. (2013). User and data classification for a secure and practical approach for patient-doctor profiling using an RFID framework in hospital. In S. Sarnikar, D. Bennett, & M. Gaynor (Eds.), *Cases on healthcare information technology for patient care management* (pp. 224–253). Hershey, PA: Medical Information Science Reference; doi:10.4018/978-1-4666-2671-3.ch014

Monguet, J. M., Huerta, E., Fernández, J., Ferruzca, M., & Badillo, S. (2013). E-health business models prototyping by incremental design. In I. Management Association (Ed.), User-driven healthcare: Concepts, methodologies, tools, and applications (pp. 776-790). Hershey, PA: Medical Information Science Reference. doi:10.4018/978-1-4666-2770-3.ch038

Montazemi, A. R., Pittaway, J. J., & Keshavjee, K. (2013). State of IS integration in the context of patient-centered care: A network analysis and research directions. In J. Tan (Ed.), *Healthcare information technology innovation and sustainability: Frontiers and adoption* (pp. 127–144). Hershey, PA: Medical Information Science Reference; doi:10.4018/978-1-4666-2797-0.ch008

Mori, A. R., Contenti, M., & Verbicaro, R. (2013). Policies on telemedicine-enhanced hospital services: Prioritization criteria for the interventions at regional level. In V. Gulla, A. Mori, F. Gabbrielli, & P. Lanzafame (Eds.), *Telehealth networks for hospital services: New methodologies* (pp. 1–14). Hershey, PA: Medical Information Science Reference; doi:10.4018/978-1-4666-2979-0.ch001

Morita, A. (2013). The quantitative EEG change in Parkinson's disease. In J. Wu (Ed.), *Biomedical engineering and cognitive neuroscience for healthcare: Interdisciplinary applications* (pp. 225–234). Hershey, PA: Medical Information Science Reference; doi:10.4018/978-1-4666-2113-8.ch023

Moromugi, S., & Ishimatsu, T. (2013). A strength training machine with a dynamic resistance control function based on muscle activity level. In J. Wu (Ed.), *Technological advancements in biomedicine for healthcare applications* (pp. 102–110). Hershey, PA: Medical Information Science Reference; doi:10.4018/978-1-4666-2196-1.ch011

Morrow, D., & Chin, J. (2013). Technology as a bridge between health care systems and older adults. In R. Zheng, R. Hill, & M. Gardner (Eds.), *Engaging older adults with modern technology: Internet use and information access needs* (pp. 59–79). Hershey, PA: Information Science Reference; doi:10.4018/978-1-4666-1966-1.ch004

Morrow, D., & Chin, J. (2013). Technology as a bridge between health care systems and older adults. In I. Management Association (Ed.), User-driven healthcare: Concepts, methodologies, tools, and applications (pp. 99-119). Hershey, PA: Medical Information Science Reference. doi:10.4018/978-1-4666-2770-3.ch006

Mostafa, R., Hasan, G. M., Kabir, A. A., & Rahman, M. A. (2013). Proposed framework for the deployment of telemedicine centers in rural Bangladesh. In J. Rodrigues (Ed.), *Digital advances in medicine, e-health, and communication technologies* (pp. 254–270). Hershey, PA: Medical Information Science Reference; doi:10.4018/978-1-4666-2794-9.ch014

Motoi, K., Ogawa, M., Yamakoshi, T., & Yamakoshi, K. (2013). Fusion physiological sensing system for healthcare. In J. Wu (Ed.), *Technological advancements in biomedicine for healthcare applications* (pp. 298–313). Hershey, PA: Medical Information Science Reference; doi:10.4018/978-1-4666-2196-1.ch031

Motorny, S. P. (2013). Big information technology bet of a small community hospital. In S. Sarnikar, D. Bennett, & M. Gaynor (Eds.), *Cases on healthcare information technology for patient care management* (pp. 70–94). Hershey, PA: Medical Information Science Reference; doi:10.4018/978-1-4666-2671-3.ch004

Moumtzoglou, A. (2013). Health 2.0 and medicine 2.0: Safety, ownership and privacy issues. In I. Management Association (Ed.), User-driven healthcare: Concepts, methodologies, tools, and applications (pp. 1508-1522). Hershey, PA: Medical Information Science Reference. doi:10.4018/978-1-4666-2770-3.ch075

Moumtzoglou, A. (2013). Risk perception as a patient safety dimension. In A. Moumtzoglou & A. Kastania (Eds.), *E-health technologies and improving patient safety: Exploring organizational factors* (pp. 285–299). Hershey, PA: Medical Information Science Reference; doi:10.4018/978-1-4666-2657-7.ch017

Msanjila, S. S. (2013). Emerging ICT challenges on provision of online HIV/AIDS advisory services. In M. Cruz-Cunha, I. Miranda, & P. Gonçalves (Eds.), *Handbook of research on ICTs for human-centered healthcare and social care services* (pp. 248–269). Hershey, PA: Medical Information Science Reference; doi:10.4018/978-1-4666-3986-7.ch013

Müller, A. (2013). Improving the identification of medication names by increasing phonological awareness via a language-teaching computer game (medicina). In S. Arnab, I. Dunwell, & K. Debattista (Eds.), *Serious games for healthcare: Applications and implications* (pp. 283–295). Hershey, PA: Medical Information Science Reference; doi:10.4018/978-1-4666-1903-6.ch014

Murakami, S., Kim, H., Tan, J. K., Ishikawa, S., & Aoki, T. (2013). The development of a quantitative method for the detection of periarticular osteoporosis using density features within ROIs from computed radiography images of the hand. In J. Wu (Ed.), *Technological advancements in biomedicine for healthcare applications* (pp. 55–67). Hershey, PA: Medical Information Science Reference; doi:10.4018/978-1-4666-2196-1.ch007

Nadathur, S. G., & Warren, J. R. (2013). Formal-transfer in and out of stroke care units: An analysis using Bayesian networks. In J. Tan (Ed.), *Healthcare information technology innovation and sustainability: Frontiers and adoption* (pp. 193–207). Hershey, PA: Medical Information Science Reference; doi:10.4018/978-1-4666-2797-0.ch012

Naidoo, V., & Naidoo, Y. (2014). Home telecare, medical implant, and mobile technology: Evolutions in geriatric care. In C. El Morr (Ed.), *Research perspectives on the role of informatics in health policy and management* (pp. 222–237). Hershey, PA: Medical Information Science Reference; doi:10.4018/978-1-4666-4321-5.ch013

Nakagawa, S. (2013). Bone-conducted ultrasonic perception: An elucidation of perception mechanisms and the development of a novel hearing aid for the profoundly deaf. In J. Wu (Ed.), *Technological advancements in biomedicine for healthcare applications* (pp. 148–159). Hershey, PA: Medical Information Science Reference; doi:10.4018/978-1-4666-2196-1.ch016

Nakai, M., & Niinomi, M. (2013). Recent progress in mechanically biocompatible titanium-based materials. In J. Wu (Ed.), *Technological advancements in biomedicine for healthcare applications* (pp. 206–212). Hershey, PA: Medical Information Science Reference; doi:10.4018/978-1-4666-2196-1.ch022

Nakakuki, T., & Okada-Hatakeyama, M. (2013). Methods for the analysis of intracellular signal transduction systems. In J. Wu (Ed.), *Technological advancements in biomedicine for healthcare applications* (pp. 347–353). Hershey, PA: Medical Information Science Reference; doi:10.4018/978-1-4666-2196-1.ch034

Nalin, M., Verga, M., Sanna, A., & Saranummi, N. (2013). Directions for ICT research in disease prevention. In M. Cruz-Cunha, I. Miranda, & P. Gonçalves (Eds.), *Handbook of research on ICTs for human-centered healthcare and social care services* (pp. 229–247). Hershey, PA: Medical Information Science Reference; doi:10.4018/978-1-4666-3986-7.ch012

Nap, H. H., & Diaz-Orueta, U. (2013). Rehabilitation gaming. In S. Arnab, I. Dunwell, & K. Debattista (Eds.), *Serious games for healthcare: Applications and implications* (pp. 50–75). Hershey, PA: Medical Information Science Reference; doi:10.4018/978-1-4666-1903-6.ch003

Narushima, T., & Ueda, K. (2013). Calcium phosphate coating on titanium by RF magnetron sputtering. In J. Wu (Ed.), *Technological advancements in biomedicine for healthcare applications* (pp. 223–233). Hershey, PA: Medical Information Science Reference; doi:10.4018/978-1-4666-2196-1.ch024

Nava-Muñoz, S., & Morán, A. L. (2013). A review of notifications systems in elder care environments: Challenges and opportunities. In M. Cruz-Cunha, I. Miranda, & P. Gonçalves (Eds.), *Handbook of research on ICTs for human-centered healthcare and social care services* (pp. 407–429). Hershey, PA: Medical Information Science Reference; doi:10.4018/978-1-4666-3986-7.ch022

Nganji, J. T., & Nggada, S. H. (2014). Adoption of blended learning technologies in selected secondary schools in Cameroon and Nigeria: Challenges in disability inclusion. In N. Ololube (Ed.), *Advancing technology and educational development through blended learning in emerging economies* (pp. 159–173). Hershey, PA: Information Science Reference; doi:10.4018/978-1-4666-4574-5.ch009

Nikolov, S., Vera, J., & Wolkenhauer, O. (2013). Bifurcation analysis of a model accounting for the 14-3-3s signalling compartmentalisation. In I. Management Association (Ed.), Bioinformatics: Concepts, methodologies, tools, and applications (pp. 851-859). Hershey, PA: Medical Information Science Reference. doi:10.4018/978-1-4666-3604-0.ch046

Nishiguchi, H. (2013). Description of and applications for a motion analysis method for upper limbs. In J. Wu (Ed.), *Technological advancements in biomedicine for healthcare applications* (pp. 1–10). Hershey, PA: Medical Information Science Reference; doi:10.4018/978-1-4666-2196-1.ch001

Nokata, M. (2013). Small medical robot. In J. Wu (Ed.), *Technological advancements in biomedicine for healthcare applications* (pp. 170–179). Hershey, PA: Medical Information Science Reference; doi:10.4018/978-1-4666-2196-1.ch018

Noritsugu, T. (2013). Wearable power assist robot driven with pneumatic rubber artificial muscles. In J. Wu (Ed.), *Technological advancements in biomedicine for healthcare applications* (pp. 139–147). Hershey, PA: Medical Information Science Reference; doi:10.4018/978-1-4666-2196-1.ch015

Noteboom, C. (2013). Physician interaction with EHR: The importance of stakeholder identification and change management. In S. Sarnikar, D. Bennett, & M. Gaynor (Eds.), *Cases on healthcare information technology for patient care management* (pp. 95–112). Hershey, PA: Medical Information Science Reference; doi:10.4018/978-1-4666-2671-3.ch005

O'Hanlon, S. (2013). Avoiding adverse consequences of e-health. In A. Moumtzoglou & A. Kastania (Eds.), *E-health technologies and improving patient safety: Exploring organizational factors* (pp. 13–26). Hershey, PA: Medical Information Science Reference; doi:10.4018/978-1-4666-2657-7.ch002

Ogawa, K., Nishio, S., Minato, T., & Ishiguro, H. (2013). Android robots as telepresence media. In J. Wu (Ed.), *Biomedical engineering and cognitive neuroscience for healthcare: Interdisciplinary applications* (pp. 54–63). Hershey, PA: Medical Information Science Reference; doi:10.4018/978-1-4666-2113-8.ch006

Ohashi, M., Sakimura, N., Fujimoto, M., Hori, M., & Kurata, N. (2013). Technical perspective of authentication policy extension for the adaptive social services and e-health care management. In M. Cruz-Cunha, I. Miranda, & P. Gonçalves (Eds.), *Handbook of research on ICTs for human-centered healthcare and social care services* (pp. 719–726). Hershey, PA: Medical Information Science Reference; doi:10.4018/978-1-4666-3986-7.ch037

Ohta, Y., & Uchida, M. (2013). Non-contact pulse monitoring using live imaging. In J. Wu (Ed.), *Technological advancements in biomedicine for healthcare applications* (pp. 240–246). Hershey, PA: Medical Information Science Reference; doi:10.4018/978-1-4666-2196-1.ch026

Ohuchida, K., & Hashizume, M. (2013). Biomedical robotics for healthcare. In J. Wu (Ed.), *Technological advancements in biomedicine for healthcare applications* (pp. 200–205). Hershey, PA: Medical Information Science Reference; doi:10.4018/978-1-4666-2196-1.ch021

Oikonomou, D., Moulianitis, V., Lekkas, D., & Koutsabasis, P. (2013). DSS for health emergency response: A contextual, user-centred approach. In R. Biswas (Ed.), *Clinical solutions and medical progress through user-driven healthcare* (pp. 51–69). Hershey, PA: Medical Information Science Reference; doi:10.4018/978-1-4666-1876-3.ch006

Okamoto, S., Hirotomi, T., Aoki, K., & Hosomi, Y. (2013). Evaluation of walking motions with the aid of walkers using acceleration sensors. In J. Wu (Ed.), *Biomedical engineering and cognitive neuroscience for healthcare: Interdisciplinary applications* (pp. 346–354). Hershey, PA: Medical Information Science Reference; doi:10.4018/978-1-4666-2113-8.ch036

Okamura, H. (2013). Rehabilitation of elderly people with dementia. In J. Wu (Ed.), *Biomedical engineering and cognitive neuroscience for healthcare: Interdisciplinary applications* (pp. 235–242). Hershey, PA: Medical Information Science Reference; doi:10.4018/978-1-4666-2113-8.ch024

Olvera-Lobo, M., & Gutiérrez-Artacho, J. (2013). Searching health information in question-answering systems. In M. Cruz-Cunha, I. Miranda, & P. Gonçalves (Eds.), *Handbook of research on ICTs for human-centered healthcare and social care services* (pp. 474–490). Hershey, PA: Medical Information Science Reference; doi:10.4018/978-1-4666-3986-7.ch025

Padovani, E., Orelli, R. L., Agnoletti, V., & Buccioli, M. (2013). Low cost and human-centered innovations in healthcare services: A case of excellence in Italy. In S. Saeed & C. Reddick (Eds.), *Human-centered system design for electronic governance* (pp. 239–252). Hershey, PA: Information Science Reference; doi:10.4018/978-1-4666-3640-8.ch014

Paninchukunnath, A. (2014). Healthcare services delivery in India: special reference to mother and child health. In A. Goyal (Ed.), *Innovations in services marketing and management: Strategies for emerging economies* (pp. 170–189). Hershey, PA: Business Science Reference; doi:10.4018/978-1-4666-4671-1.ch010

Paparountas, T., Nikolaidou-Katsaridou, M. N., Rustici, G., & Aidinis, V. (2013). Data mining and meta-analysis on DNA microarray data. In I. Management Association (Ed.), Bioinformatics: Concepts, methodologies, tools, and applications (pp. 1196-1236). Hershey, PA: Medical Information Science Reference. doi:10.4018/978-1-4666-3604-0.ch062

Parentela, G., Mancini, P., Naccarella, F., Feng, Z., & Rinaldi, G. (2013). Telemedicine, the European space agency, and the support to the african population for infectious disease problems: Potentiality and perspectives for Asia countries and China. In V. Gulla, A. Mori, F. Gabbrielli, & P. Lanzafame (Eds.), *Telehealth networks for hospital services: New methodologies* (pp. 89–96). Hershey, PA: Medical Information Science Reference; doi:10.4018/978-1-4666-2979-0.ch005

Paterson, G. I., MacDonald, J. M., & Mensink, N. N. (2014). The administrative policy quandary in Canada's health service organizations. In C. El Morr (Ed.), *Research perspectives on the role of informatics in health policy and management* (pp. 116–134). Hershey, PA: Medical Information Science Reference; doi:10.4018/978-1-4666-4321-5.ch008

Penchovsky, R. (2013). Engineering gene control circuits with allosteric ribozymes in human cells as a medicine of the future. In I. Management Association (Ed.), Bioinformatics: Concepts, methodologies, tools, and applications (pp. 860-883). Hershey, PA: Medical Information Science Reference. doi:10.4018/978-1-4666-3604-0.ch047

Pereira, O. R., Caldeira, J. M., & Rodrigues, J. J. (2013). An advanced and secure symbian-based mobile approach for body sensor networks interaction. In J. Rodrigues (Ed.), *Digital advances in medicine, e-health, and communication technologies* (pp. 33–48). Hershey, PA: Medical Information Science Reference; doi:10.4018/978-1-4666-2794-9.ch002

Petkovic, M., & Ibraimi, L. (2013). Privacy and security in e-health applications. In I. Management Association (Ed.), User-driven healthcare: Concepts, methodologies, tools, and applications (pp. 1141-1166). Hershey, PA: Medical Information Science Reference. doi:10.4018/978-1-4666-2770-3.ch058

Peyton, L., & Hu, J. (2013). Identity management and audit trail support for privacy protection in e-health networks. In I. Management Association (Ed.), User-driven healthcare: Concepts, methodologies, tools, and applications (pp. 1112-1125). Hershey, PA: Medical Information Science Reference. doi:10.4018/978-1-4666-2770-3.ch056

Pham, D. V., Halgamuge, M. N., Nirmalathas, T., & Moran, B. (2013). A centralized real-time e-healthcare system for remote detection and prediction of epileptic seizures. In I. Management Association (Ed.), User-driven healthcare: Concepts, methodologies, tools, and applications (pp. 326-356). Hershey, PA: Medical Information Science Reference. doi:10.4018/978-1-4666-2770-3.ch017

Phua, C., Roy, P. C., Aloulou, H., Biswas, J., Tolstikov, A., & Foo, V. S. … Xu, D. (2014). State-of-the-art assistive technology for people with dementia. In Assistive technologies: Concepts, methodologies, tools, and applications (pp. 1606-1625). Hershey, PA: Information Science Reference. doi:10.4018/978-1-4666-4422-9.ch085

Pino, A. (2014). Augmentative and alternative communication systems for the motor disabled. In G. Kouroupetroglou (Ed.), *Disability informatics and web accessibility for motor limitations* (pp. 105–152). Hershey, PA: Medical Information Science Reference; doi:10.4018/978-1-4666-4442-7.ch004

Portela, F., Cabral, A., Abelha, A., Salazar, M., Quintas, C., Machado, J., & Santos, M. F. et al. (2013). Knowledge acquisition process for intelligent decision support in critical health care. In R. Martinho, R. Rijo, M. Cruz-Cunha, & J. Varajão (Eds.), *Information systems and technologies for enhancing health and social care* (pp. 55–68). Hershey, PA: Medical Information Science Reference; doi:10.4018/978-1-4666-3667-5.ch004

Previtali, P. (2013). Grid technology for archive solutions in health care organizations. In A. Moumtzoglou & A. Kastania (Eds.), *E-health technologies and improving patient safety: Exploring organizational factors* (pp. 148–154). Hershey, PA: Medical Information Science Reference; doi:10.4018/978-1-4666-2657-7.ch010

Purkayastha, S. (2013). Design and implementation of mobile-based technology in strengthening health information system: Aligning mhealth solutions to infrastructures. In I. Management Association (Ed.), User-driven healthcare: Concepts, methodologies, tools, and applications (pp. 689-713). Hershey, PA: Medical Information Science Reference. doi:10.4018/978-1-4666-2770-3.ch034

Qi, G., & Wu, J. (2013). Functional role of the left ventral occipito-temporal cortex in reading. In J. Wu (Ed.), *Biomedical engineering and cognitive neuroscience for healthcare: Interdisciplinary applications* (pp. 192–200). Hershey, PA: Medical Information Science Reference; doi:10.4018/978-1-4666-2113-8.ch020

Queirós, A., Alvarelhão, J., Silva, A. G., Teixeira, A., & Pacheco da Rocha, N. (2013). A conceptual framework for the design and development of AAL services. In M. Cruz-Cunha, I. Miranda, & P. Gonçalves (Eds.), *Handbook of research on ICTs for human-centered healthcare and social care services* (pp. 568–586). Hershey, PA: Medical Information Science Reference; doi:10.4018/978-1-4666-3986-7.ch030

Ravka, N. (2014). Informatics and health services: The potential benefits and challenges of electronic health records and personal electronic health records in patient care, cost control, and health research – An overview. In C. El Morr (Ed.), *Research perspectives on the role of informatics in health policy and management* (pp. 89–114). Hershey, PA: Medical Information Science Reference; doi:10.4018/978-1-4666-4321-5.ch007

Reibling, N., & Wendt, C. (2013). Regulating patients' access to healthcare services. In M. Merviö (Ed.), *Healthcare management and economics: Perspectives on public and private administration* (pp. 53–68). Hershey, PA: Medical Information Science Reference; doi:10.4018/978-1-4666-3982-9.ch005

Reis, C. I., Freire, C. S., Fernández, J., & Monguet, J. M. (2013). Patient centered design: Challenges and lessons learned from working with health professionals and schizophrenic patients in e-therapy contexts. In R. Martinho, R. Rijo, M. Cruz-Cunha, & J. Varajão (Eds.), *Information systems and technologies for enhancing health and social care* (pp. 120–135). Hershey, PA: Medical Information Science Reference; doi:10.4018/978-1-4666-3667-5.ch008

Remmers, H., & Hülsken-Giesler, M. (2013). e-Health technologies in home care nursing: Recent survey results and subsequent ethical issues. In I. Management Association (Ed.), User-driven healthcare: Concepts, methodologies, tools, and applications (pp. 396-420). Hershey, PA: Medical Information Science Reference. doi:10.4018/978-1-4666-2770-3.ch020

Ricketts, M. (2014). Making health information personal: How anecdotes bring concepts to life. In S. Hai-Jew (Ed.), *Packaging digital information for enhanced learning and analysis: Data visualization, spatialization, and multidimensionality* (pp. 1–36). Hershey, PA: Information Science Reference; doi:10.4018/978-1-4666-4462-5.ch001

Rocha, Á. (2013). Evolution of information systems and technologies maturity in healthcare. In J. Tan (Ed.), *Healthcare information technology innovation and sustainability: Frontiers and adoption* (pp. 238–246). Hershey, PA: Medical Information Science Reference; doi:10.4018/978-1-4666-2797-0.ch015

Rodríguez-Gómez, D., & Gairín, J. (2014). Communities of practice in the Catalan public administration: Promoting their improvement. In Y. Al-Bastaki & A. Shajera (Eds.), *Building a competitive public sector with knowledge management strategy* (pp. 383–402). Hershey, PA: Business Science Reference; doi:10.4018/978-1-4666-4434-2.ch018

Rodríguez-Solano, C., Lezcano, L., & Sicilia, M. (2013). Automated generation of SNOMED CT subsets from clinical guidelines. In R. Martinho, R. Rijo, M. Cruz-Cunha, & J. Varajão (Eds.), *Information systems and technologies for enhancing health and social care* (pp. 190–204). Hershey, PA: Medical Information Science Reference; doi:10.4018/978-1-4666-3667-5.ch013

Rosu, S. M., & Dragoi, G. (2014). E-health sites development using open source software and OMT methodology as support for family doctors' activities: A Romanian case study. In M. Cruz-Cunha, F. Moreira, & J. Varajão (Eds.), *Handbook of research on enterprise 2.0: Technological, social, and organizational dimensions* (pp. 72–88). Hershey, PA: Business Science Reference; doi:10.4018/978-1-4666-4373-4.ch004

Roy, N., Das, S. K., & Julien, C. (2013). Resolving and mediating ambiguous contexts in pervasive environments. In I. Management Association (Ed.), User-driven healthcare: Concepts, methodologies, tools, and applications (pp. 630-654). Hershey, PA: Medical Information Science Reference. doi:10.4018/978-1-4666-2770-3.ch031

Russo, M. R. (2014). Emergency management professional development: Linking information communication technology and social communication skills to enhance a sense of community and social justice in the 21st century. In I. Management Association (Ed.), Crisis management: Concepts, methodologies, tools and applications (pp. 651-665). Hershey, PA: Information Science Reference. doi:10.4018/978-1-4666-4707-7.ch031

Sabone, M. B., Mogobe, K. D., & Sabone, T. G. (2013). ICTS and their role in health promotion: A preliminary situation analysis in selected botswana rural communities. In I. Management Association (Ed.), User-driven healthcare: Concepts, methodologies, tools, and applications (pp. 211-224). Hershey, PA: Medical Information Science Reference. doi:10.4018/978-1-4666-2770-3.ch012

Saha, P. (2014). Systemic enterprise architecture as future: Tackling complexity in governments in the cusp of change. In P. Saha (Ed.), *A systemic perspective to managing complexity with enterprise architecture* (pp. 1–70). Hershey, PA: Business Science Reference; doi:10.4018/978-1-4666-4518-9.ch001

Saijo, Y. (2013). Biomedical application of multimodal ultrasound microscope. In J. Wu (Ed.), *Technological advancements in biomedicine for healthcare applications* (pp. 27–35). Hershey, PA: Medical Information Science Reference; doi:10.4018/978-1-4666-2196-1.ch004

Santos, M., Bastião, L., Costa, C., Silva, A., & Rocha, N. (2013). Clinical data mining in small hospital PACS: Contributions for radiology department improvement. In R. Martinho, R. Rijo, M. Cruz-Cunha, & J. Varajão (Eds.), *Information systems and technologies for enhancing health and social care* (pp. 236–251). Hershey, PA: Medical Information Science Reference; doi:10.4018/978-1-4666-3667-5.ch016

Santos, M. F., Portela, F., Miranda, M., Machado, J., Abelha, A., Silva, Á., & Rua, F. (2013). Grid data mining strategies for outcome prediction in distributed intensive care units. In R. Martinho, R. Rijo, M. Cruz-Cunha, & J. Varajão (Eds.), *Information systems and technologies for enhancing health and social care* (pp. 87–101). Hershey, PA: Medical Information Science Reference; doi:10.4018/978-1-4666-3667-5.ch006

Santos, R. J., Bernardino, J., & Vieira, M. (2013). A hypotension surveillance and prediction system for critical care. In M. Cruz-Cunha, I. Miranda, & P. Gonçalves (Eds.), *Handbook of research on ICTs and management systems for improving efficiency in healthcare and social care* (pp. 341–355). Hershey, PA: Medical Information Science Reference; doi:10.4018/978-1-4666-3990-4.ch017

Sanz, P. R., Mezcua, B. R., & Pena, J. M. (2013). ICTs for orientation and mobility for blind people: A state of the art. In M. Cruz-Cunha, I. Miranda, & P. Gonçalves (Eds.), *Handbook of research on ICTs for human-centered healthcare and social care services* (pp. 646–669). Hershey, PA: Medical Information Science Reference; doi:10.4018/978-1-4666-3986-7.ch034

Sarabdeen, J. (2013). Legal issues in e-healthcare systems. In I. Management Association (Ed.), User-driven healthcare: Concepts, methodologies, tools, and applications (pp. 1194-1219). Hershey, PA: Medical Information Science Reference. doi:10.4018/978-1-4666-2770-3.ch060

Sarbadhikari, S. N. (2013). Unlearning and relearning in online health education. In I. Management Association (Ed.), User-driven healthcare: Concepts, methodologies, tools, and applications (pp. 1348-1363). Hershey, PA: Medical Information Science Reference. doi:10.4018/978-1-4666-2770-3.ch067

Sasayama, T., Hamada, S., & Kobayashi, T. (2013). Application of prewhitening beamformer with linear constraints for correlated EEG signal source estimation. In J. Wu (Ed.), *Biomedical engineering and cognitive neuroscience for healthcare: Interdisciplinary applications* (pp. 243–254). Hershey, PA: Medical Information Science Reference; doi:10.4018/978-1-4666-2113-8.ch025

Sato, T., & Minato, K. (2013). Differences in analysis methods of the human uncinate fasciculus using diffusion tensor MRI. In J. Wu (Ed.), *Biomedical engineering and cognitive neuroscience for healthcare: Interdisciplinary applications* (pp. 162–170). Hershey, PA: Medical Information Science Reference; doi:10.4018/978-1-4666-2113-8.ch017

Satoh, J. (2013). Molecular network analysis of target RNAs and interacting proteins of TDP-43, a causative gene for the neurodegenerative diseases ALS/FTLD. In I. Management Association (Ed.), Bioinformatics: Concepts, methodologies, tools, and applications (pp. 964-985). Hershey, PA: Medical Information Science Reference. doi:10.4018/978-1-4666-3604-0.ch052

Satoh, J. (2013). Molecular network analysis of target RNAs and interacting proteins of TDP-43, a causative gene for the neurodegenerative diseases ALS/FTLD. In J. Wu (Ed.), *Technological advancements in biomedicine for healthcare applications* (pp. 314–335). Hershey, PA: Medical Information Science Reference; doi:10.4018/978-1-4666-2196-1.ch032

Schafer, S. B. (2013). Fostering psychological coherence: With ICTs. In M. Cruz-Cunha, I. Miranda, & P. Gonçalves (Eds.), *Handbook of research on ICTs for human-centered healthcare and social care services* (pp. 29–47). Hershey, PA: Medical Information Science Reference; doi:10.4018/978-1-4666-3986-7.ch002

Schmeida, M., & McNeal, R. (2013). Bridging the inequality gap to accessing medicare and medicaid information online: An empirical analysis of e-government success 2002 through 2010. In J. Gil-Garcia (Ed.), *E-government success around the world: Cases, empirical studies, and practical recommendations* (pp. 60–78). Hershey, PA: Information Science Reference; doi:10.4018/978-1-4666-4173-0.ch004

Scholl, J. C., & Olaniran, B. A. (2013). ICT use and multidisciplinary healthcare teams. In M. Cruz-Cunha, I. Miranda, & P. Gonçalves (Eds.), *Handbook of research on ICTs for human-centered healthcare and social care services* (pp. 627–645). Hershey, PA: Medical Information Science Reference; doi:10.4018/978-1-4666-3986-7.ch033

Serenko, N. (2013). The impact of genetic testing and genetic information on ethical, legal and social issues in North America: The framework. In I. Management Association (Ed.), Bioinformatics: Concepts, methodologies, tools, and applications (pp. 1317-1333). Hershey, PA: Medical Information Science Reference. doi:10.4018/978-1-4666-3604-0.ch067

Serrano, M., Elmisery, A., Foghlú, M. Ó., Donnelly, W., Storni, C., & Fernström, M. (2013). Pervasive computing support in the transition towards personalised health systems. In J. Rodrigues (Ed.), *Digital advances in medicine, e-health, and communication technologies* (pp. 49–64). Hershey, PA: Medical Information Science Reference; doi:10.4018/978-1-4666-2794-9.ch003

Shaw, V., & Braa, J. (2013). "Developed in the south": An evolutionary and prototyping approach to developing scalable and sustainable health information systems. In I. Management Association (Ed.), User-driven healthcare: Concepts, methodologies, tools, and applications (pp. 583-607). Hershey, PA: Medical Information Science Reference. doi:10.4018/978-1-4666-2770-3.ch029

Shi, J., Upadhyaya, S., & Erdem, E. (2013). Health information exchange for improving the efficiency and quality of healthcare delivery. In I. Management Association (Ed.), User-driven healthcare: Concepts, methodologies, tools, and applications (pp. 714-736). Hershey, PA: Medical Information Science Reference. doi:10.4018/978-1-4666-2770-3.ch035

Shibata, T. (2013). A human-like cognitive computer based on a psychologically inspired VLSI brain model. In J. Wu (Ed.), *Technological advancements in biomedicine for healthcare applications* (pp. 247–266). Hershey, PA: Medical Information Science Reference; doi:10.4018/978-1-4666-2196-1.ch027

Shimada, S. (2013). Self-body recognition and its impairment. In J. Wu (Ed.), *Biomedical engineering and cognitive neuroscience for healthcare: Interdisciplinary applications* (pp. 156–161). Hershey, PA: Medical Information Science Reference; doi:10.4018/978-1-4666-2113-8.ch016

Shimogonya, Y., Ishikawa, T., Yamaguchi, T., Kumamaru, H., & Itoh, K. (2013). Computational study of the hemodynamics of cerebral aneurysm initiation. In J. Wu (Ed.), *Technological advancements in biomedicine for healthcare applications* (pp. 267–277). Hershey, PA: Medical Information Science Reference; doi:10.4018/978-1-4666-2196-1.ch028

Shimonomura, K. (2013). A neuromorphic robot vision system to predict the response of visual neurons. In J. Wu (Ed.), *Technological advancements in biomedicine for healthcare applications* (pp. 193–199). Hershey, PA: Medical Information Science Reference; doi:10.4018/978-1-4666-2196-1.ch020

Shrestha, S. (2013). Clinical decision support system for diabetes prevention: An illustrative case. In S. Sarnikar, D. Bennett, & M. Gaynor (Eds.), *Cases on healthcare information technology for patient care management* (pp. 308–329). Hershey, PA: Medical Information Science Reference; doi:10.4018/978-1-4666-2671-3.ch017

Siddiqui, Z. S., & Jonas-Dwyer, D. R. (2013). Mobile learning in health professions education: A systematic review. In J. Keengwe (Ed.), *Pedagogical applications and social effects of mobile technology integration* (pp. 193–205). Hershey, PA: Information Science Reference; doi:10.4018/978-1-4666-2985-1.ch011

Silvana de Rosa, A., Fino, E., & Bocci, E. (2014). Addressing healthcare on-line demand and supply relating to mental illness: Knowledge sharing about psychiatry and psychoanalysis through social networks in Italy and France. In A. Kapoor & C. Kulshrestha (Eds.), *Dynamics of competitive advantage and consumer perception in social marketing* (pp. 16–55). Hershey, PA: Business Science Reference; doi:10.4018/978-1-4666-4430-4.ch002

Siqueira, S. R., Rocha, E. C., & Nery, M. S. (2013). Brazilian occupational therapy perspective about digital games as an inclusive resource to disabled people in schools. In I. Management Association (Ed.), User-driven healthcare: Concepts, methodologies, tools, and applications (pp. 918-937). Hershey, PA: Medical Information Science Reference. doi:10.4018/978-1-4666-2770-3.ch046

Sliedrecht, S., & Kotzé, E. (2013). Patients with a spinal cord injury inform and co-construct services at a spinal cord rehabilitation unit. In I. Management Association (Ed.), User-driven healthcare: Concepts, methodologies, tools, and applications (pp. 1054-1072). Hershey, PA: Medical Information Science Reference. doi:10.4018/978-1-4666-2770-3.ch053

Smedberg, Å. (2013). E-health communities for learning healthy habits: How to consider quality and usability. In I. Management Association (Ed.), User-driven healthcare: Concepts, methodologies, tools, and applications (pp. 310-325). Hershey, PA: Medical Information Science Reference. doi:10.4018/978-1-4666-2770-3.ch016

Smedberg, Å., & Sandmark, H. (2013). Dynamic stress management: Self-help through holistic system design. In I. Management Association (Ed.), User-driven healthcare: Concepts, methodologies, tools, and applications (pp. 1469-1486). Hershey, PA: Medical Information Science Reference. doi:10.4018/978-1-4666-2770-3.ch073

Sobol, M., & Prater, E. (2013). Adoption, usage and efficiency: Benchmarking healthcare IT in private practices. In J. Tan (Ed.), *Healthcare information technology innovation and sustainability: Frontiers and adoption* (pp. 145–159). Hershey, PA: Medical Information Science Reference; doi:10.4018/978-1-4666-2797-0.ch009

Sobrinho, Á. A., Dias da Silva, L., Melo de Medeiros, L., & Pinheiro, M. E. (2013). A mobile assistant to aid early detection of chronic kidney disease. In R. Martinho, R. Rijo, M. Cruz-Cunha, & J. Varajão (Eds.), *Information systems and technologies for enhancing health and social care* (pp. 309–323). Hershey, PA: Medical Information Science Reference; doi:10.4018/978-1-4666-3667-5.ch020

Soomlek, C., & Benedicenti, L. (2013). Agent-based wellness indicator. In V. Gulla, A. Mori, F. Gabbrielli, & P. Lanzafame (Eds.), *Telehealth networks for hospital services: New methodologies* (pp. 300–330). Hershey, PA: Medical Information Science Reference; doi:10.4018/978-1-4666-2979-0.ch020

Spadaro, L., Timpano, F., Marino, S., & Bramanti, P. (2013). Telemedicine and Alzheimer disease: ICT-based services for people with Alzheimer disease and their caregivers. In V. Gulla, A. Mori, F. Gabbrielli, & P. Lanzafame (Eds.), *Telehealth networks for hospital services: New methodologies* (pp. 191–206). Hershey, PA: Medical Information Science Reference; doi:10.4018/978-1-4666-2979-0.ch013

Springer, J. A., Beever, J., Morar, N., Sprague, J. E., & Kane, M. D. (2013). Ethics, privacy, and the future of genetic information in healthcare information assurance and security. In I. Management Association (Ed.), Bioinformatics: Concepts, methodologies, tools, and applications (pp. 1405-1423). Hershey, PA: Medical Information Science Reference. doi:10.4018/978-1-4666-3604-0.ch072

Stachura, M. E., Wood, J., Angjellari-Dajci, F., Grayson, J., Astapova, E. V., Tung, H., . . . Lawless, W. (2013). Representing organizational conservation of information: A review of telemedicine and e-health in Georgia. In I. Management Association (Ed.), User-driven healthcare: Concepts, methodologies, tools, and applications (pp. 1220-1235). Hershey, PA: Medical Information Science Reference. doi:10.4018/978-1-4666-2770-3.ch061

Stefaniak, J. E. (2013). Resuscitating team roles within wayburn health system. In A. Ritzhaupt & S. Kumar (Eds.), *Cases on educational technology implementation for facilitating learning* (pp. 130–145). Hershey, PA: Information Science Reference; doi:10.4018/978-1-4666-3676-7.ch008

Sugiura, M. (2013). A cognitive neuroscience approach to self and mental health. In J. Wu (Ed.), *Biomedical engineering and cognitive neuroscience for healthcare: Interdisciplinary applications* (pp. 1–10). Hershey, PA: Medical Information Science Reference; doi:10.4018/978-1-4666-2113-8.ch001

Supnithi, T., Buranarach, M., Thatphithakkul, N., Junsirimongkol, B., Wongrochananan, S., Kulnawan, N., & Jiamjarasrangsi, W. (2014). A self-management service framework to support chronic disease patients' self-management. In M. Kosaka & K. Shirahada (Eds.), *Progressive trends in knowledge and system-based science for service innovation* (pp. 425–450). Hershey, PA: Business Science Reference; doi:10.4018/978-1-4666-4663-6.ch023

Takahashi, K., & Watanabe, K. (2013). Crossmodal interactions in visual competition. In J. Wu (Ed.), *Biomedical engineering and cognitive neuroscience for healthcare: Interdisciplinary applications* (pp. 64–72). Hershey, PA: Medical Information Science Reference; doi:10.4018/978-1-4666-2113-8.ch007

Tamiya, T., Kawanishi, M., Miyake, K., Kawai, N., & Guo, S. (2013). Neurosurgical operations using navigation microscope integration system. In J. Wu (Ed.), *Technological advancements in biomedicine for healthcare applications* (pp. 128–138). Hershey, PA: Medical Information Science Reference; doi:10.4018/978-1-4666-2196-1.ch014

Tan, R., Wang, S., Jiang, Y., Ishida, K., & Fujie, M. G. (2013). Motion control of an omni-directional walker for walking support. In J. Wu (Ed.), *Biomedical engineering and cognitive neuroscience for healthcare: Interdisciplinary applications* (pp. 20–28). Hershey, PA: Medical Information Science Reference; doi:10.4018/978-1-4666-2113-8.ch003

Tanaka, H., & Furutani, M. (2013). Sleep management promotes healthy lifestyle, mental health, QOL, and a healthy brain. In J. Wu (Ed.), *Biomedical engineering and cognitive neuroscience for healthcare: Interdisciplinary applications* (pp. 211–224). Hershey, PA: Medical Information Science Reference; doi:10.4018/978-1-4666-2113-8.ch022

Tang, C., & Carpendale, S. (2013). Human-centered design for health information technology: A qualitative approach. In I. Management Association (Ed.), User-driven healthcare: Concepts, methodologies, tools, and applications (pp. 158-179). Hershey, PA: Medical Information Science Reference. doi:10.4018/978-1-4666-2770-3.ch009

Tang, X., Gao, Y., Yang, W., Zhang, M., & Wu, J. (2013). Audiovisual integration of natural auditory and visual stimuli in the real-world situation. In J. Wu (Ed.), *Biomedical engineering and cognitive neuroscience for healthcare: Interdisciplinary applications* (pp. 337–344). Hershey, PA: Medical Information Science Reference; doi:10.4018/978-1-4666-2113-8.ch035

Targowski, A. (2013). Well-being, wisdom, health, and IT: From the big-picture to the small-picture. In M. Cruz-Cunha, I. Miranda, & P. Gonçalves (Eds.), *Handbook of research on ICTs for human-centered healthcare and social care services* (pp. 1–28). Hershey, PA: Medical Information Science Reference; doi:10.4018/978-1-4666-3986-7.ch001

Tayabali, S., & Martin, C. M. (2013). The need to transform the core values of medical care and health organizations. In R. Biswas (Ed.), *Clinical solutions and medical progress through user-driven healthcare* (pp. 85–92). Hershey, PA: Medical Information Science Reference; doi:10.4018/978-1-4666-1876-3.ch009

Taylor, B. W. (2014). Decision-making and decision support in acute care. In C. El Morr (Ed.), *Research perspectives on the role of informatics in health policy and management* (pp. 1–18). Hershey, PA: Medical Information Science Reference; doi:10.4018/978-1-4666-4321-5.ch001

Teixeira, C., Pinto, J. S., Ferreira, F., Oliveira, A., Teixeira, A., & Pereira, C. (2013). Cloud computing enhanced service development architecture for the living usability lab. In R. Martinho, R. Rijo, M. Cruz-Cunha, & J. Varajão (Eds.), *Information systems and technologies for enhancing health and social care* (pp. 33–53). Hershey, PA: Medical Information Science Reference; doi:10.4018/978-1-4666-3667-5.ch003

Teixeira, L., Saavedra, V., Ferreira, C., & Santos, B. S. (2013). The role of ICTs in the management of rare chronic diseases: The case of hemophilia. In M. Cruz-Cunha, I. Miranda, & P. Gonçalves (Eds.), *Handbook of research on ICTs and management systems for improving efficiency in healthcare and social care* (pp. 635–649). Hershey, PA: Medical Information Science Reference; doi:10.4018/978-1-4666-3990-4.ch033

Teoh, S. Y., Singh, M., & Chong, J. (2013). An overview of e-health development in Australia. In I. Management Association (Ed.), User-driven healthcare: Concepts, methodologies, tools, and applications (pp. 901-917). Hershey, PA: Medical Information Science Reference. doi:10.4018/978-1-4666-2770-3.ch045

Tiwari, S., & Srivastava, R. (2014). Research and developments in medical image reconstruction methods and its applications. In R. Srivastava, S. Singh, & K. Shukla (Eds.), *Research developments in computer vision and image processing: Methodologies and applications* (pp. 274–312). Hershey, PA: Information Science Reference; doi:10.4018/978-1-4666-4558-5.ch014

Tobolcea, I. (2013). The psychosocial impact of ICT efficiency on speech disorders-treatment. In M. Cruz-Cunha, I. Miranda, & P. Gonçalves (Eds.), *Handbook of research on ICTs for human-centered healthcare and social care services* (pp. 70–92). Hershey, PA: Medical Information Science Reference; doi:10.4018/978-1-4666-3986-7.ch004

Toda, T., Chen, P., Ozaki, S., Fujita, K., & Ideguchi, N. (2013). A simple web-based image database system for facilitating medical care in dermatological clinics. In I. Management Association (Ed.), User-driven healthcare: Concepts, methodologies, tools, and applications (pp. 502-513). Hershey, PA: Medical Information Science Reference. doi:10.4018/978-1-4666-2770-3.ch025

Toda, T., Chen, P., Ozaki, S., Fujita, K., & Ideguchi, N. (2013). A simple web-based image database system for facilitating medical care in dermatological clinics. In J. Wu (Ed.), *Technological advancements in biomedicine for healthcare applications* (pp. 385–397). Hershey, PA: Medical Information Science Reference; doi:10.4018/978-1-4666-2196-1.ch038

Trudeau, S. (2013). CoRDS registry: An HIT case study concerning setup and maintenance of a disease registry. In S. Sarnikar, D. Bennett, & M. Gaynor (Eds.), *Cases on healthcare information technology for patient care management* (pp. 197–207). Hershey, PA: Medical Information Science Reference; doi:10.4018/978-1-4666-2671-3.ch012

Tseng, R., & Yi-Luen Do, E. (2013). The role of information and computer technology for children with autism spectrum disorder and the facial expression wonderland (FEW). In A. Gangopadhyay (Ed.), *Methods, models, and computation for medical informatics* (pp. 98–116). Hershey, PA: Medical Information Science Reference; doi:10.4018/978-1-4666-2653-9.ch006

Uchitomi, H., Suzuki, K., Nishi, T., Hove, M. J., Miyake, Y., Orimo, S., & Wada, Y. (2013). Gait rhythm of Parkinson's disease patients and an interpersonal synchrony emulation system based on cooperative gait. In J. Wu (Ed.), *Biomedical engineering and cognitive neuroscience for healthcare: Interdisciplinary applications* (pp. 38–53). Hershey, PA: Medical Information Science Reference; doi:10.4018/978-1-4666-2113-8.ch005

Usher, W., & San Too, L. (2013). E-health knowledge management by australian university students. In I. Management Association (Ed.), User-driven healthcare: Concepts, methodologies, tools, and applications (pp. 938-953). Hershey, PA: Medical Information Science Reference. doi:10.4018/978-1-4666-2770-3.ch047

Van Genderen, E. (2014). Intensive care. In V. Jham & S. Puri (Eds.), *Cases on consumer-centric marketing management* (pp. 53–67). Hershey, PA: Business Science Reference; doi:10.4018/978-1-4666-4357-4.ch006

Van Leuven, N., Newton, D., Leuenberger, D. Z., & Esteves, T. (2014). Reaching citizen 2.0: How government uses social media to send public messages during times of calm and times of crisis. In I. Management Association (Ed.), Crisis management: Concepts, methodologies, tools and applications (pp. 839-857). Hershey, PA: Information Science Reference. doi:10.4018/978-1-4666-4707-7.ch041

Vasudevan, V., & Rao, H. (2013). E-discovery and health care IT: An investigation. In I. Management Association (Ed.), User-driven healthcare: Concepts, methodologies, tools, and applications (pp. 1622-1635). Hershey, PA: Medical Information Science Reference. doi:10.4018/978-1-4666-2770-3.ch080

Vat, K. H. (2013). Conceiving community knowledge records as e-governance concerns in wired healthcare provision. In I. Management Association (Ed.), User-driven healthcare: Concepts, methodologies, tools, and applications (pp. 1093-1111). Hershey, PA: Medical Information Science Reference. doi:10.4018/978-1-4666-2770-3.ch055

Venkatasubramanian, K. K., Nabar, S., Gupta, S. K., & Poovendran, R. (2013). Cyber physical security solutions for pervasive health monitoring systems. In I. Management Association (Ed.), User-driven healthcare: Concepts, methodologies, tools, and applications (pp. 447-465). Hershey, PA: Medical Information Science Reference. doi:10.4018/978-1-4666-2770-3.ch022

Victan, H. (2013). Emerging trends in user-driven healthcare: Negotiating disclosure in online health community organizations. In I. Management Association (Ed.), User-driven healthcare: Concepts, methodologies, tools, and applications (pp. 1589-1606). Hershey, PA: Medical Information Science Reference. doi:10.4018/978-1-4666-2770-3.ch078

Vidal, B., Pereira, J. M., & Santos, G. (2013). SimBody: An interactive simulator for health education. In S. Arnab, I. Dunwell, & K. Debattista (Eds.), *Serious games for healthcare: Applications and implications* (pp. 265–282). Hershey, PA: Medical Information Science Reference; doi:10.4018/978-1-4666-1903-6.ch013

Vivekananda-Schmidt, P. (2013). Ethics in the design of serious games for healthcare and medicine. In S. Arnab, I. Dunwell, & K. Debattista (Eds.), *Serious games for healthcare: Applications and implications* (pp. 91–106). Hershey, PA: Medical Information Science Reference; doi:10.4018/978-1-4666-1903-6.ch005

Wahbeh, A. (2013). Application of handheld computing and mobile phones in diabetes self-care. In S. Sarnikar, D. Bennett, & M. Gaynor (Eds.), *Cases on healthcare information technology for patient care management* (pp. 254–275). Hershey, PA: Medical Information Science Reference; doi:10.4018/978-1-4666-2671-3.ch015

Waidyanatha, N., & Dekker, S. (2013). The RTBP – Collective intelligence driving health for the user. In R. Biswas (Ed.), *Clinical solutions and medical progress through user-driven healthcare* (pp. 70–78). Hershey, PA: Medical Information Science Reference; doi:10.4018/978-1-4666-1876-3.ch007

Waidyanatha, N., Dubrawski, A. M. G., & Gow, G. (2013). Affordable system for rapid detection and mitigation of emerging diseases. In J. Rodrigues (Ed.), *Digital advances in medicine, e-health, and communication technologies* (pp. 271–288). Hershey, PA: Medical Information Science Reference; doi:10.4018/978-1-4666-2794-9.ch015

Walker, D. (2013). How would a ban on prescriber-identifying information impact pharmaceutical marketing? In M. Merviö (Ed.), *Healthcare management and economics: Perspectives on public and private administration* (pp. 141–153). Hershey, PA: Medical Information Science Reference; doi:10.4018/978-1-4666-3982-9.ch011

Wang, B., Yan, T., & Wu, J. (2013). Neuronal function in the cortical face perception network. In J. Wu (Ed.), *Biomedical engineering and cognitive neuroscience for healthcare: Interdisciplinary applications* (pp. 171–182). Hershey, PA: Medical Information Science Reference; doi:10.4018/978-1-4666-2113-8.ch018

Wang, Y., & Wang, Q. (2013). Evaluating the IEEE 802.15.6 2.4GHz WBAN proposal on medical multi-parameter monitoring under WiFi/bluetooth interference. In J. Rodrigues (Ed.), *Digital advances in medicine, e-health, and communication technologies* (pp. 312–325). Hershey, PA: Medical Information Science Reference; doi:10.4018/978-1-4666-2794-9.ch018

Watanabe, T., & Miura, N. (2013). Functional electrical stimulation (FES) control for restoration and rehabilitation of motor function. In J. Wu (Ed.), *Technological advancements in biomedicine for healthcare applications* (pp. 80–93). Hershey, PA: Medical Information Science Reference; doi:10.4018/978-1-4666-2196-1.ch009

Watanabe, Y., Tanaka, H., & Hirata, K. (2013). Evaluation of cognitive function in migraine patients: A study using event-related potentials. In J. Wu (Ed.), *Biomedical engineering and cognitive neuroscience for healthcare: Interdisciplinary applications* (pp. 303–310). Hershey, PA: Medical Information Science Reference; doi:10.4018/978-1-4666-2113-8.ch031

Watfa, M. K., Kaur, M., & Daruwala, R. F. (2013). RFID applications in e-healthcare. In I. Management Association (Ed.), User-driven healthcare: Concepts, methodologies, tools, and applications (pp. 259-287). Hershey, PA: Medical Information Science Reference. doi:10.4018/978-1-4666-2770-3.ch014

Wesorick, B. (2013). Essential steps for successful implementation of the EHR to achieve sustainable, safe, quality care. In A. Moumtzoglou & A. Kastania (Eds.), *E-health technologies and improving patient safety: Exploring organizational factors* (pp. 27–55). Hershey, PA: Medical Information Science Reference; doi:10.4018/978-1-4666-2657-7.ch003

Whitaker, R. (2013). Securing health-effective medicine in practice: A critical perspective on user-driven healthcare. In R. Biswas (Ed.), *Clinical solutions and medical progress through user-driven healthcare* (pp. 35–50). Hershey, PA: Medical Information Science Reference; doi:10.4018/978-1-4666-1876-3.ch005

Wickramasinghe, N., Troshani, I., Hill, S. R., Hague, W., & Goldberg, S. (2013). A transaction cost assessment of a pervasive technology solution for gestational diabetes. In J. Tan (Ed.), *Healthcare information technology innovation and sustainability: Frontiers and adoption* (pp. 109–126). Hershey, PA: Medical Information Science Reference; doi:10.4018/978-1-4666-2797-0.ch007

Wilkowska, W., & Ziefle, M. (2013). User diversity as a challenge for the integration of medical technology into future smart home environments. In I. Management Association (Ed.), User-driven healthcare: Concepts, methodologies, tools, and applications (pp. 553-582). Hershey, PA: Medical Information Science Reference. doi:10.4018/978-1-4666-2770-3.ch028

Willis, E., Wang, Y., & Rodgers, S. (2013). Online health communities and health literacy: Applying a framework for understanding domains of health literacy. In I. Management Association (Ed.), User-driven healthcare: Concepts, methodologies, tools, and applications (pp. 180-196). Hershey, PA: Medical Information Science Reference. doi:10.4018/978-1-4666-2770-3.ch010

Wu, Q., Li, C., Takahashi, S., & Wu, J. (2013). Visual-tactile bottom-up and top-down attention. In J. Wu (Ed.), *Biomedical engineering and cognitive neuroscience for healthcare: Interdisciplinary applications* (pp. 183–191). Hershey, PA: Medical Information Science Reference; doi:10.4018/978-1-4666-2113-8.ch019

Yamaguchi, S., Onoda, K., & Abe, S. (2013). Feedback-related negativity and its clinical implications. In J. Wu (Ed.), *Biomedical engineering and cognitive neuroscience for healthcare: Interdisciplinary applications* (pp. 283–292). Hershey, PA: Medical Information Science Reference; doi:10.4018/978-1-4666-2113-8.ch029

Yan, B., Lei, Y., Tong, L., & Chen, K. (2013). Functional neuroimaging of acupuncture: A systematic review. In J. Wu (Ed.), *Biomedical engineering and cognitive neuroscience for healthcare: Interdisciplinary applications* (pp. 142–155). Hershey, PA: Medical Information Science Reference; doi:10.4018/978-1-4666-2113-8.ch015

Yang, J. (2014). Towards healthy public policy: GIS and food systems analysis. In C. El Morr (Ed.), *Research perspectives on the role of informatics in health policy and management* (pp. 135–152). Hershey, PA: Medical Information Science Reference; doi:10.4018/978-1-4666-4321-5.ch009

Yang, J., Li, Q., Gao, Y., & Wu, J. (2013). Temporal dependency of multisensory audiovisual integration. In J. Wu (Ed.), *Biomedical engineering and cognitive neuroscience for healthcare: Interdisciplinary applications* (pp. 320–326). Hershey, PA: Medical Information Science Reference; doi:10.4018/978-1-4666-2113-8.ch033

Yang, J. J., Liu, J. F., Kurokawa, T., Kitamura, N., Yasuda, K., & Gong, J. P. (2013). Tough double-network hydrogels as scaffolds for tissue engineering: Cell behavior in vitro and in vivo test. In J. Wu (Ed.), *Technological advancements in biomedicine for healthcare applications* (pp. 213–222). Hershey, PA: Medical Information Science Reference; doi:10.4018/978-1-4666-2196-1.ch023

Yang, W., Gao, Y., & Wu, J. (2013). Effects of selective and divided attention on audiovisual interaction. In J. Wu (Ed.), *Biomedical engineering and cognitive neuroscience for healthcare: Interdisciplinary applications* (pp. 311–319). Hershey, PA: Medical Information Science Reference; doi:10.4018/978-1-4666-2113-8.ch032

Yano, Y. (2013). An EMG control system for an ultrasonic motor using a PSoC microcomputer. In J. Wu (Ed.), *Technological advancements in biomedicine for healthcare applications* (pp. 11–17). Hershey, PA: Medical Information Science Reference; doi:10.4018/978-1-4666-2196-1.ch002

Yap, K. Y. (2013). The evolving role of pharmacoinformatics in targeting drug-related problems in clinical oncology practice. In I. Management Association (Ed.), User-driven healthcare: Concepts, methodologies, tools, and applications (pp. 1541-1588). Hershey, PA: Medical Information Science Reference. doi:10.4018/978-1-4666-2770-3.ch077

Yokoi, H., Kato, R., Mori, T., Yamamura, O., & Kubota, M. (2013). Functional electrical stimulation based on interference-driven PWM signals for neuro-rehabilitation. In J. Wu (Ed.), *Technological advancements in biomedicine for healthcare applications* (pp. 180–192). Hershey, PA: Medical Information Science Reference; doi:10.4018/978-1-4666-2196-1.ch019

York, A. M., & Nordengren, F. R. (2013). E-learning and web 2.0 case study: The role of gender in contemporary models of health care leadership. In H. Yang & S. Wang (Eds.), *Cases on formal and informal e-learning environments: Opportunities and practices* (pp. 292–313). Hershey, PA: Information Science Reference; doi:10.4018/978-1-4666-1930-2.ch016

Yu, W. D., & Bhagwat, R. (2013). Modeling emergency and telemedicine health support system: A service oriented architecture approach using cloud computing. In J. Rodrigues (Ed.), *Digital advances in medicine, e-health, and communication technologies* (pp. 187–213). Hershey, PA: Medical Information Science Reference; doi:10.4018/978-1-4666-2794-9.ch011

Yu, Y., Yang, J., & Wu, J. (2013). Cognitive functions and neuronal mechanisms of tactile working memory. In J. Wu (Ed.), *Biomedical engineering and cognitive neuroscience for healthcare: Interdisciplinary applications* (pp. 89–98). Hershey, PA: Medical Information Science Reference; doi:10.4018/978-1-4666-2113-8.ch010

Zaheer, S. (2014). Implementation of evidence-based practice and the PARIHS framework. In C. El Morr (Ed.), *Research perspectives on the role of informatics in health policy and management* (pp. 19–36). Hershey, PA: Medical Information Science Reference; doi:10.4018/978-1-4666-4321-5.ch002

Zalzala, A., Chia, S., Zalzala, L., Sahu, S., Vaghasiya, S., & Karimi, A. (2013). Rural e-health infrastructure development. In I. Management Association (Ed.), User-driven healthcare: Concepts, methodologies, tools, and applications (pp. 870-900). Hershey, PA: Medical Information Science Reference. doi:10.4018/978-1-4666-2770-3.ch044

Zapirain, B. G., & Zorrilla, A. M. (2013). Independent living support for disabled and elderly people using cell phones. In M. Cruz-Cunha, I. Miranda, & P. Gonçalves (Eds.), *Handbook of research on ICTs for human-centered healthcare and social care services* (pp. 379–397). Hershey, PA: Medical Information Science Reference; doi:10.4018/978-1-4666-3986-7.ch020

Zgodavová, K., & Bourek, A. (2013). Potential of web based learning in managing for the sustained success of a healthcare organization based on IMPROHEALTH® project. In I. Management Association (Ed.), User-driven healthcare: Concepts, methodologies, tools, and applications (pp. 1523-1540). Hershey, PA: Medical Information Science Reference. doi:10.4018/978-1-4666-2770-3.ch076

Zhao, X., Zuo, M. J., & Moghaddass, R. (2013). Generating indicators for diagnosis of fault levels by integrating information from two or more sensors. In I. Management Association (Ed.), User-driven healthcare: Concepts, methodologies, tools, and applications (pp. 288-309). Hershey, PA: Medical Information Science Reference. doi:10.4018/978-1-4666-2770-3.ch015

Zhou, X., & Ren, X. (2013). Speed-accuracy tradeoff models of target-based and trajectory-based movements. In J. Wu (Ed.), *Biomedical engineering and cognitive neuroscience for healthcare: Interdisciplinary applications* (pp. 355–368). Hershey, PA: Medical Information Science Reference; doi:10.4018/978-1-4666-2113-8.ch037

Zouag, N., & Driouchi, A. (2014). Trends and prospects of the Moroccan health system: 2010-2030. In A. Driouchi (Ed.), *Labor and health economics in the mediterranean region: Migration and mobility of medical doctors* (pp. 314–336). Hershey, PA: Medical Information Science Reference; doi:10.4018/978-1-4666-4723-7.ch013

Zvikhachevskaya, A., & Mihaylova, L. (2013). Advanced video distribution for wireless e-healthcare systems. In I. Management Association (Ed.), User-driven healthcare: Concepts, methodologies, tools, and applications (pp. 421-446). Hershey, PA: Medical Information Science Reference. doi:10.4018/978-1-4666-2770-3.ch021

Compilation of References

Agresti, A. (2013). *Categorical data analysis* (3rd ed.). Hoboken, NJ: Wiley.

AIR Report. (2014). *Among New STEM Ph.D.'s, Women Likelier to Join Academia, but Men Get More Faculty Jobs*. Retrieved from http://www.air.org/news/press-release/among-new-stem-ph-d-s-women-likelier-join-academia-men-get-more-faculty-jobs

Alumanah, J. N. E. (2005). Access and Use of Information and Communication Technology for the African Girl-child Under Cultural Impediments. In *Proceedings of the International Symposium on Women and ICT: Creating Global Transformation*. New York, NY: ACM. http://doi.org/ doi:10.1145/1117417.1117430

Amenkhienan, C. A., & Kogan, L. R. (2004). Engineering students' perceptions of academic activities and support services: Factors that influence their academic performance. *College Student Journal*, *38*(4), 523–540.

American Association of Medical Colleges. (2014). *2014 Physician Specialty Data Book*. Author.

American Association of Medical Colleges. (2015). *Altering the course: Black Males in Medicine*. Washington, DC: Association of American Medical Colleges.

American Association of University Professors. (2006). *AAUP faculty gender equity indicators 2006*. Retrieved from http://www.aaup.org/NR/rdonlyres/63396944-44BE-4ABA-9815-5792D93856F1/0/AAUPGenderEquityIndicators2006.pdf

American Association of University Women. (2008). *Where the girls are: The facts about gender equity in education*. Washington, DC: Author.

American Association of University Women. (2013). *Improve Girls' and Women's Opportunities in Science, Technology, Engineering, and Math*. Retrieved August 13, 2015, from http://www.aauw.org/files/2013/02/position-on-STEM-education-111.pdf

American Association of University Women. (2013). *Improve girls' and women's opportunities in science, technology, engineering, and math*. Retrieved from http://www.aauw.org/files/2013/02/position-on-STEM-education-111.pdf

American Medical Association. (2006). *Physician Characteristics and Distribution in the U.S., 2006 edition and prior editions*. Retrieved March 8, 2015, from www.ama-assn.org/ama/pub/category/12192.html

Anaya, G., & Cole, D. G. (2001). Latina/o student achievement: Exploring the influence of student-faculty interactions on college grades. *Journal of College Student Development*, *42*(1), 611–622.

APS & IPEDS. (2015). *Percentage of degrees awarded to minorities*. Retrieved from http://www.aps.org/programs/education/statistics/urm.cfm

Arksey, H., & O'Malley, L. (2005). Scoping studies: Towards a methodological framework. *International Journal of Social Research Methodology*, *8*(1), 19–32. doi:10.1080/1364557032000119616

Aronson, J., Quinn, D. M., & Spencer, S. J. (1998). Stereotype threat and academic underperformance of minorities and women. In J. K. Swim & C. Stangor (Eds.), Prejudice: Target's perspective (pp. 83-103). San Diego, CA: Academic Press. doi:10.1016/B978-012679130-3/50039-9

Association of American Medical Colleges. (2011). *Table 5. Distribution of U.S. Medical School Faculty by Degree and Race/Hispanic Origin.* Retrieved March, 7, 2015, from https://www.aamc.org/download/271904/data/11table5.pdf

Association of Women in Science. (2014). *Institutions without family-friendly practices risk losing their best and brightest.* Retrieved from http://awisblog.wordpress.com/2014/06/18/institutions-without-family-friendly-practices-risk-losing-their-best-and-brightest/

Aud, S., Hussar, W., Planty, M., Snyder, T., Bianco, K., Fox, M., & Drake, L. (2010). *The Condition of Education 2010 (NCES 2010-028).* Washington, DC: National Center for Education Statistics, Institute of Education Sciences, U.S. Department of Education.

Ausubel, D. P. (1970). *The use of ideational organizers in science teaching.* Occasional paper 3. The Science Series of the Science and Mathematics Education Information Analysis Center.

Bailyn, L. (2003). Academic careers and gender equity: Lessons learned from MIT. *Gender, Work and Organization, 10*(2), 137–153. doi:10.1111/1468-0432.00008

Baker, S., Tancred, P., & Whitesides, S. (2002). Gender and graduate school: Engineering students confront life after the B. Eng. *The Journal of Engineering Education, 91*(1), 41–47. doi:10.1002/j.2168-9830.2002.tb00671.x

Banatvala, N., & Doyal, L. (1998). Knowing when to say "no" on the student elective: Students going on electives abroad need clinical guidelines. *BMJ: British Medical Journal, 316*(7142), 1404–1405. doi:10.1136/bmj.316.7142.1404 PMID:9572746

Banerjee, A. (2010). Medical electives: A chance for international health. *Journal of the Royal Society of Medicine, 103*(1), 6–8. doi:10.1258/jrsm.2009.090266 PMID:20056663

Barile, J. J., Donohue, D. D., Anthony, E. E., Baker, A. A., Weaver, S. S., & Henrich, C. C. (2012). Teacher-Student Relationship Climate and School Outcomes: Implications for Educational Policy Initiatives. *Journal of Youth and Adolescence, 41*(3), 256–267. doi:10.1007/s10964-011-9652-8 PMID:21404108

Barkin, S. L., Fuentes-Afflick, E., Brosco, J. P., & Tuchman, A. M. (2010). Unintended consequences of the Flexner report: Women in pediatrics. *Pediatrics, 126*(6), 1055–1057. doi:10.1542/peds.2010-2050 PMID:21059716

Barnes, C. W. (2014). *Too Heavy a Yoke.* Eugene, OR: Oregon Cascade Books.

Basford, T. E., Offermann, L. R., & Behrend, T. S. (2014). Do you see what I see? Perptions of gender microaggressions in the workplace. *Psychology of Women Quarterly, 38*(3), 340–349. doi:10.1177/0361684313511420

Bauman, M. D., Howell, L. P., & Villablanca, A. C. (2014). The women in medicine and health science program: An innovative initiative to support female faculty at the University of California Davis School of Medicine. *Academic Medicine, 89*(11), 1462–1466. doi:10.1097/ACM.0000000000000403 PMID:25006704

Beagan, B. (2001). Micro Inequities and Everyday Inequalities: "Race," Gender, Sexuality and Class in Medical School. *The Canadian Journal of Sociology / Cahiers Canadiens De Sociologie, 26*(4), 583-610.

Beck, A. H. (2004). The Flexner report and the standardization of American medical education. *Journal of the American Medical Association, 291*(17), 2139–2140. doi:10.1001/jama.291.17.2139 PMID:15126445

Bellanca, J., & Brant, J. (2010). *21st century skills: Rethinking how students learn.* Bloomington, IN: Solution Tree Press.

Bell, D. (1992). *Faces at the bottom of the well.* New York: Basic Books.

Bell, N. (2008). *Graduate enrollment and degrees: 1997 to 2007.* Washington, DC: Council of Graduate Schools.

Beltrão, K., & Teixeira, M. (2005). Cor e gênero na seletividade das carreiras universitárias. In *Os mecanismos de discriminação racial nas escolas brasileiras.* Rio de Janeiro: IPEA.

Benatar, D. (2003). The second sexism. *Social Theory and Practice, 29*(2), 177–210. doi:10.5840/soctheorpract200329213

Beninger, A. (2014). *High potentials in tech-intensive industries: The gender divide in business roles.* Retrieved from Catalyst website: http://www.catalyst.org/system/files/high_potentials_in_techintensive_industries_the_gender_divide_in_business_roles_1.pdf

Bennett-Conroy, W. (2012). Engaging parents of eighth grade students in parent-teacher bidirectional communication. *School Community Journal, 22*(2), 87–110.

Benokraitis, N. V. (1997). *Subtle sexism: Current practice and prospects for change.* Thousand Oaks, CA: SAGE Publications, Inc.

Benokraitis, N. V., & Feagin, J. R. (1995). *Modern sexism: Blantant, subtle, and covert discrimination* (2nd ed.). Englewood Cliffs, NJ: Prentice Hall.

Berenson, S. B., Slaten, K. M., Williams, L., & Ho, C.-W. (2004). Voices of Women in a Software Engineering Course: Reflections on Collaboration. *Journal of Educational Resources in Computing, 4*(1), 3, es. doi:10.1145/1060071.1060074

Beziat, T. L. R., Issacson, R., & Was, C. (2010, October). *Differences within Female Education Majors on Risk Taking.* Paper presented at the meeting of Mid-Western Educational Research Association, Columbus, OH.

Bhattacharjee, Y. (2007). Academic tenure. MIT colleague quits to protest Sherley dismissal. *Science, 316*(5830), 1405. doi:10.1126/science.316.5830.1405 PMID:17556555

Black, S. E., Devereux, P. J., & Salvanes, K. G. (2010). Small family, smart family? Family size and the IQ scores of young men. *The Journal of Human Resources, 45*(1), 33–58. doi:10.3368/jhr.45.1.33

Blanch, D. C., Hall, J. A., Roter, D. L., & Frankel, R. M. (2008, September01). Medical student gender and issues of confidence. *Patient Education and Counseling, 72*(3), 374–381. doi:10.1016/j.pec.2008.05.021 PMID:18656322

Blickenstaff, J. C. (2005). Women and science careers: Leaky pipeline or gender filter? *Gender and Education, 17*(4), 369–386. doi:10.1080/09540250500145072

Bobbitt-Zeher, D. (2011). Gender discrimination at work: Connecting gender stereotypes, institutional policies, and gender composition of workplace. *Gender & Society, 25*(6), 764–786. doi:10.1177/0891243211424741

Bonilla-Silva, E. (2006). *Racism without racists: color-blind racism and the persistence of racial inequality in the United States.* Lanham, MD: Rowman & Littlefield.

Bonous-Hammarth, M. (2000). Pathways to success: Affirming opportunities for science, mathematics, and engineering majors. *The Journal of Negro Education*, 92–111.

Bourdieu, P. (1999). *A dominação masculina.* Rio de Janeiro: Bertrand Brasil.

Bowen, G. A. (2009). Document analysis as a qualitative research method. *Qualitative Research Journal, 9*(2), 27-40.

Bozeman, S. T., & Hughes, R. J. (2004). Improving the graduate school experience for women in mathematics: The EDGE program. *Journal of Women and Minorities in Science and Engineering, 10*(3), 243–253. doi:10.1615/JWomenMinorScienEng.v10.i3.40

Bradley, K., & Charles, M. (2004). Uneven roads: understandingw omen's status in higher education. In D. Baker, B. Fuller, E. Hannum, & R. Werum (Eds.), *Inequalitieasc rosss ocietiesf:a milies, schools and persisting stratification* (pp. 247–274). Amsterdam: Elsevier.

Brainard, S. G., & Carlin, L. (2001). *A six-year longitudinal study of undergraduate women in engineering and science*. Academic Press.

Brasil, I. P. E. A. (2008). *Retrato das desigualdades de gênero e raça*. Brasília: IPEA.

Breene, L. A. (1992). Women and Computer Science. *Initiatives, 55*(2), 39–44.

Bronfenbrenner, U. (1994). Ecological models of human development. In T. Husen & T. N. Postlethwaite (Eds.), *International Encyclopedia of Education* (2nd ed., Vol. 3, pp. 1643–1647). Oxford, UK: Pergamon Press.

Brown, R. D., Clarke, B., Gortmaker, V., & Robinson-Keilig, R. (2004). Assessing the campus climate for gay, lesbian, bisexual, and transgender (GLBT) students using a multiple perspectives approach. *Journal of College Student Development, 45*(1), 8–26. doi:10.1353/csd.2004.0003

Brown, S. W. (2002). Hispanic students majoring in science or engineering: What happened in their educational journeys? *Journal of Women and Minorities in Science and Engineering, 8*(2), 123–148. doi:10.1615/JWomenMinorScienEng.v8.i2.20

Budden, A. E., Tregenza, T., Aarssen, L. W., Koricheva, J., Leimu, R., & Lortie, C. J. (2007). Double-blind reviews favours increased representation of female authors. *Trends in Ecology & Evolution, 23*(1), 4–6. doi:10.1016/j.tree.2007.07.008 PMID:17963996

Bullock, S. C., & Houston, E. (n.d.). *Perceptions of Racism by Black Medical Students Attending White Medical Schools.*

Buntting, C., Jones, A., McKinley, L., & Gan, M. (2013). *STEM initiatives and issues in New Zealand.* Retrieved from http://www.acola.org.au/PDF/SAF02Consultants/Consultant%20Report%20-%20New%20Zealand.pdf

Bureau of Labor and Statistics. (2010). *Occupational outlook handbook: Overview of the 2008-18 projections*. Washington, DC: Author.

Burge, J. D., & Suarez, T. L. (2005). *Preliminary analysis of factors affecting women and African Americans in the computing sciences*. ACM Press. doi:10.1109/RTCDC.2005.201646

Burge, S. W. (2011). Academic programs: Undergraduate, graduate, and professional. In B. J. Bank (Ed.), *Gender and higher education* (pp. 171–181). Baltimore, MD: The Johns Hopkins University Press.

Burgess, D. J., Joseph, A., van Ryn, M., & Carnes, M. (2012). Does stereotype threat affect women in academic medicine? *Academic Medicine, 87*(4), 506–512. doi:10.1097/ACM.0b013e318248f718 PMID:22361794

Burgess, D., & Borgida, E. (1999). Who women are, who women should be: Descriptive and prescriptive gender stereotyping in sex discrimination. *Psychology, Public Policy, and Law, 5*(3), 665–692. doi:10.1037/1076-8971.5.3.665

Bussey-Jones, J., Bernstein, L., Higgins, S., Malebranche, D., Paranjape, A., Genao, I., & Branch, W. et al. (2006). Repaving the road to academic success: The IMeRGE approach to peer mentoring. *Academic Medicine, 81*(7), 674–679. doi:10.1097/01.ACM.0000232425.27041.88 PMID:16799297

Butler, P. D., Longaker, M. T., & Britt, L. D. (2008). Major deficit in the number of underrepresented minority academic surgeons persists. *Annals of Surgery, 248*(5), 704–711. doi:10.1097/SLA.0b013e31817f2c30 PMID:18948795

Byrd, K. (2015). The relationship between media influence and ethnic identity development among low-income African American and white adolescent girls. *Dissertation Abstracts International. A, The Humanities and Social Sciences, 75*(7-A(E)).

Cabrera, A. F., Nora, A., Terenzini, P. T., Pascarella, E., & Hagedorn, L. S. (1999). Campus racial climate and the adjustment of students to college: A comparison between White students and African-American students. *The Journal of Higher Education*, *70*(2), 134–160. doi:10.2307/2649125

Camp, T. (2012). Computing, We Have a Problem.... *ACM Inroads*, *3*(4), 34–40. doi:10.1145/2381083.2381097

Canes, B. J., & Rosen, H. S. (1995). Following in her footsteps? Faculty gender composition and women's choices of college majors. *Industrial & Labor Relations Review*, *48*(3), 486–504. doi:10.1177/001979399504800308

Caribbean Community (CARICOM) Secretariat. (2011). *CARICOM Member States*. Retrieved May 15, 2015, from http://www.caricom.org/jsp/community/member_states.jsp?menu=community

Carlone, H. B., & Johnson, A. (2007). Understanding the science experiences of successful women of color: Science identity as an analytic lens. *Journal of Research in Science Teaching*, *44*(8), 1187–1218. doi:10.1002/tea.20237

Carolan, B. V. (2012). An examination of the relationship among high school size, social capital, and adolescents' mathematics achievement. *Journal of Research on Adolescence*, *22*(3), 583-595.

Carr, P.L., Gunn, C.M., Kaplan, S.A., Raj, A., & Freund, K.M. (2015). *Inadequate progress for women in academic medicine: findings from the National Faculty Study*. Academic Press.

Carter, S. C., & Brickhouse, N. W. (1989). What makes chemistry difficult? Alternate perceptions. *Journal of Chemical Education*, *66*(3), 223–225. doi:10.1021/ed066p223

Cassell, J. (1998). *The woman in the surgeon's body*. Cambridge, MD: Harvard University Press.

Caygill, R., & Kirkham, S. (2008). *Year 5 Students' mathematics achievement in 2010/11: New Zealand results from the International Mathematics and Science Study (TIMSS)*. Retrieved from http://www.educationcounts.govt.nz/publications/numeracy/TIMSS-200607/34160/9

Ceci, S. J., Williams, W. M., & Barnett, S. M. (2009). Women's underrepresentation in science: Sociocultural and biological considerations. *Psychological Bulletin*, *135*(2), 218–261. doi:10.1037/a0014412 PMID:19254079

Ceci, S., & Williams, W. (2010). Sex differences in math-intensive fields. *Current Directions in Psychological Science*, *19*(5), 275–279. doi:10.1177/0963721410383241 PMID:21152367

Chandler, F. T., & Parsons, S. (1995, April). *A case study of a mother/daughter science club*. Paper presented at the Annual Meeting National Association for Research in Science, San Francisco, CA.

Chan-Malik, S. (2015). Panethnic Religious Institutions. Asian American Religious Cultures, 98.

Chapa, J., & De La Rosa, B. (2006). The problematic pipeline demographic trends and Latino participation in graduate Science, Technology, Engineering, and Mathematics programs. *Journal of Hispanic Higher Education*, *5*(3), 203–221. doi:10.1177/1538192706288808

Charness, G., & Gneezy, U. (2012). Strong evidence for gender differences in risk taking. *Journal of Economic Behavior & Organization*, *83*(1), 50–58. doi:10.1016/j.jebo.2011.06.007

Chen, H. L., Lattuca, L. R., & Hamilton, E. R. (2008). Conceptualizing engagement: Contributions of faculty to student engagement in engineering. *The Journal of Engineering Education*, *97*(3), 339–353. doi:10.1002/j.2168-9830.2008.tb00983.x

Chen, X. (2009). *Students Who Study Science, Technology, Engineering, and Mathematics (STEM) in Postsecondary Education (NCES 2009161)*. Washington, DC: National Center for Education Statistics, Institute of Education Sciences, U.S. Department of Education.

Chen, X. (2013). *STEM Attrition: College Students' Paths Into and Out of STEM Fields (NCES 2014-001)*. Washington, DC: National Center for Education Statistics, Institute of Education Sciences, U.S. Department of Education.

Ching-Hua, M., Ying-Chu, H., & Tzu-Wei, F. (2012). The role of the mother-daughter relationship in Taiwanese college students' career self-efficacy. *Social Behavior & Personality: An International Journal*, *40*(9), 1511–1521. doi:10.2224/sbp.2012.40.9.1511

Chipman, S. F., & Thomas, V. G. (1987). The Participation of Women and Minorities in Mathematical, Scientific, and Technical Fields. *Review of Research in Education*, *14*, 387–430. doi:10.2307/1167316

Chronicle of Higher Education Almanac. (2014). *Diversity*. Retrieved from http://chronicle.com/article/Almanac-2014-Diversity/148173/?sn

Civil Rights Act of 1964 § 7, 42 U.S.C. § 2000e et seq (1964).

Clewell, B. C., & Ginorio, A. B. (1996). Examining women's progress in the sciences from the perspective of diversity. In C.-S. Davis, A. B. Ginorio, C. S. Hollenshead, B. B. Lazarus, & P. M. Rayman (Eds.), *The equity equation: Fostering the advancement of women in the sciences, mathematics, and engineering* (pp. 163–231). San Francisco: Jossey-Bass.

Clifford, M. M. (1988). Failure tolerance and academic risk-taking in ten- to twelve-year-old students. *The British Journal of Educational Psychology*, *58*(1), 15–27. doi:10.1111/j.2044-8279.1988.tb00875.x

Clifford, M. M., & Chou, F. (1991). Effects of payoff and task context on academic risk taking. *Journal of Educational Psychology*, *83*(4), 499–507. doi:10.1037/0022-0663.83.4.499

Colbeck, C. L., Cabrera, A. F., & Terenzini, P. T. (2001). Learning professional confidence: Linking teaching practices, students' self-perceptions, and gender. *The Review of Higher Education*, *24*(2), 173–191. doi:10.1353/rhe.2000.0028

Colby, S. L., & Ortman. (2014). *Projections of the size and composition of the U.S. population: 2014 to 2060*. Current Population Reports.

Cole, D. (2008). Constructive criticism: The role of student-faculty interactions on African American and Hispanic students' educational gains. *Journal of College Student Development*, *49*(6), 587–605. doi:10.1353/csd.0.0040

Cole, D., & Espinoza, A. (2008). Examining the academic success of latino students in science technology engineering and mathematics (STEM) majors. *Journal of College Student Development*, *49*(4), 285–300. doi:10.1353/csd.0.0018

Cole, D., & Griffin, K. A. (2013). Advancing the study of student-faculty interaction: A focus on diverse students and faculty. In *Higher education: Handbook of theory and research* (pp. 561–611). Springer Netherlands. doi:10.1007/978-94-007-5836-0_12

Cole, E. R. (2009). Intersectionality and research in psychology. *The American Psychologist*, *64*(3), 170–180. doi:10.1037/a0014564 PMID:19348518

Colleges, Association of American Medical. (2015). *Altering the Course: Black Males in Medicine*. Washington, DC: Association of American Medical Colleges.

Colom, R., Contreras, M. J., Arend, I., Leal, O. G., & Santacreu, J. (2004). Sex differences in verbal reasoning are mediated by sex differences in spatial ability. *The Psychological Record*, *54*(3), 365–372.

Cook, L. (Producer). (2015a, August 30). U.S. Education: Still Separate and Unequal. *Data Mine Numbers You Can Use*. Retrieved from http://www.usnews.com/news/blogs/data-mine/2015/01/28/us-education-still-separate-and-unequal

Cook, L. (Producer). (2015b, August 30). Why Black Americans Die Younger. *Data Mine Numbers You Can Use*. Retrieved from http://www.usnews.com/news/blogs/data-mine/2015/01/05/black-americans-have-fewer-years-to-live-heres-why?int=a29308&int=a29308

Cooke, C. L., & Dawson, P. L. (2002). Book Reviews - Forged by the Knife: The Experience of Surgical Residency from the Perspective of a Woman of Color. *Signs, 27*(2), 584.

Cooper, J. (2015). *Civil Rights and Equality*. Retrieved from http://cooper.house.gov

Cooper-Patrick, L., Gallo, J. J., Gonzales, J. J., Vu, H. T., Powe, N. R., Nelson, C., & Ford, D. E. (1999). Race, gender, and partnership in the patient-physician relationship. *Journal of the American Medical Association*, *282*(6), 583–589. doi:10.1001/jama.282.6.583 PMID:10450723

Corbett, C., Hill, C., & St. Rose, A. (2008). *Where the girls are: The facts about gender equity in education*. Retrieved from American Association of University Women website: http://www.aauw.org/files/2013/02/Where-the-Girls-Are-The-Facts-About-Gender-Equity-in-Education.pdf

Corp, B. Survey. (2010). *Bayer Facts on Science and Education XIV*. Retrieved from http://www.igert.org/system/content_item_assets/files/579/Bayer_Facts_of_Science_Education_Executive_Summary.pdf?1269877505

Correlation does not imply causation. (2015). Retrieved from http://en.wikipedia.org/wiki/Correlation_does_not_imply_causation

Correll, S. J., Barnard, S., & Paik, I. (2007). Getting a Job: Is there a Motherhood Penalty? *American Journal of Sociology*, *112*(5), 1297–1339. doi:10.1086/511799

Cotten, S. R., & Wilson, B. (2006). Student–faculty interactions: Dynamics and determinants. *Higher Education*, *51*(4), 487–519. doi:10.1007/s10734-004-1705-4

Cox, B. E., McIntosh, K. L., Terenzini, P. T., Reason, R. D., & Quaye, B. R. L. (2010). Pedagogical signals of faculty approachability: Factors shaping faculty–student interaction outside the classroom. *Research in Higher Education*, *51*(8), 767–788. doi:10.1007/s11162-010-9178-z

Crabtree, S. (2013, October 8). *Worldwide, 13% of Employees Are Engaged at Work - Low workplace engagement offers opportunities to improve business outcomes*. Retrieved May 15, 2015, from http://www.gallup.com/poll/165269/worldwide-employees-engaged-work.aspx

Crenshaw, K. W. (1989). Demarginalizing the intersection of race and sex: A Black feminist critique of antidiscrimination doctrine, feminist theory, and antiracist politics. *University of Chicago Legal Forum*, *140*, 139–167.

Crenshaw, K. W. (1994). Mapping the margins: Intersectionality, identity politics, and violence against women of color. In M. A. Fineman & R. Mykitiuk (Eds.), *The public nature of private violence* (pp. 93–118). New York: Routledge.

Crenshaw, K. W. (2014). The structural and political dimensions of intersectional oppression. In P. R. Grzanka (Ed.), *Intersectionality: A Foundations and Frontiers Reader* (pp. 17–22). Boulder, CO: Westview Press.

Cressey, D. (2014). Journals weigh up double-blind peer review. *NATNews*.

Crisp, C., & Cruz, I. (2009). Mentoring college students: A critical review of the literature between 1990 and 2007. *Research in Higher Education*, *50*(6), 525–545. doi:10.1007/s11162-009-9130-2

Cuddy, A. J., Fisk, S. T., & Glick, P. (2004). When Professionals become mothers, warmth doesn't cut the ice. *The Journal of Social Issues*, *60*(4), 701–718. doi:10.1111/j.0022-4537.2004.00381.x

Curtis, J. W. (2013). What roles are we modeling? Gendered academic employment and its consequences. *Peer Review, 15*(3). Retrieved from http://www.aacu.org/peerreview/pr-su13/Curtis.cfm

Cuyjet, M. (2009). Invisible men-almost: The diminution of African American males in higher education. Emerald Group Publishing.

Davis-Kean, P. E. (2005). The influence of parent education and family income on child achievement: The indirect role of parental expectations and the home environment. *Journal of Family Psychology, 19*(2), 294–304. doi:10.1037/0893-3200.19.2.294 PMID:15982107

Day, J. C. (1992). *Population projections of the United States, by age, sex, race, and Hispanic origin: 1992 to 2050*. US Department of Commerce, Economics and Statistics Administration, Bureau of the Census.

DeCastro, R., Griffith, K. A., Ubel, P. A., Stewart, A., & Jagsi, R. (2014, January01). Mentoring and the career satisfaction of male and female academic medical faculty. *Academic Medicine: Journal of the Association of American Medical Colleges, 89*(2), 301–311. doi:10.1097/ACM.0000000000000109 PMID:24362376

Delgado, R., & Stefancic, J. (2001). *Critical race theory: an introduction*. New York: NYU Press.

Dempster, A. P., Laird, N. M., & Rubin, D. B. (1977). Maximum likelihood from incomplete data via the EM algorithm. *Journal of the Royal Statistical Society. Series A (General), 38*(1), 1–38.

Dey, J. G., & Hill, C. (2007). *Behind the pay gap*. Washington, DC: American Association of University Women Educational Foundation.

Diekman, A. B., Brown, E. R., Johnston, A. M., & Clark, E. K. (2010). Seeking congruity between goals and roles: A new look at why women opt out of science, technology, engineering, and mathematics careers. *Psychological Science, 21*(8), 1051–1057. doi:10.1177/0956797610377342 PMID:20631322

DiPrete, T. A., & Eirich, G. M. (2006). Cumulative Advantage as a Mechanism for Inequality: A Review of Theoretical and Empirical Developments. *Annual Review of Sociology, 32*(1), 271–297. doi:10.1146/annurev.soc.32.061604.123127

Duckworth, A. L., Peterson, C., Matthews, M. D., & Kelly, D. R. (2007). Grit: Perseverance and passion for long-term goals. *Journal of Personality and Social Psychology, 9*(6), 1087–1101. doi:10.1037/0022-3514.92.6.1087 PMID:17547490

Duffy, T. P. (2011). The Flexner Report--100 years later. *The Yale Journal of Biology and Medicine, 84*(3), 269–276. PMID:21966046

Durie, M. (1998). *Whaiora: Maori health development* (2nd ed.). Auckland: Oxford University Press.

Durie, M. (2003). *Nga kahui pou launching Maori futures*. Wellington: Huia publishers.

Dyrbye, L. N., Thomas, M. R., & Shanafelt, T. D. (2005, January01). Medical student distress: Causes, consequences, and proposed solutions. *Mayo Clinic Proceedings, 80*(12), 1613–1622. doi:10.4065/80.12.1613 PMID:16342655

Easwaran, E. (1997). *Gandhi the Man: The Story of His Transformation* (3rd ed.). Tomales, CA: Nilgiri Press.

Eckel, C. C., & Grossman, P. J. (2008). Differences in the economic decisions of men and women: Experimental evidence. In C. Plott & V. Smith (Eds.), *Handbook of experimental economics results* (Vol. 1, pp. 509–519). New York: Elsevier. doi:10.1016/S1574-0722(07)00057-1

Education Review Office. (2012). *Science in the New Zealand Curriculum: Years 5 to 8*. Wellington, NZ: Ministry of Education.

Eichler, L. (2012). *You can't get ahead on merit alone.* Retrieved from http://www.theglobeandmail.com/report-on-business/careers/career-advice/you-cant-get-ahead-on-merit-alone/article4226012/

Eliot, L. (2010). The truth about boys and girls. *Scientific American Mind*, *21*(2), 22–29. doi:10.1038/scientificamericanmind0510-22

Elit, L., Hunt, M., Redwood-Campbell, L., Ranford, J., Adelson, N., & Schwartz, L. (2011). Ethical issues encountered by medical students during international health electives. *Medical Education*, *45*(7), 704–711. doi:10.1111/j.1365-2923.2011.03936.x PMID:21649703

Ellington, R. (2006). *Having their say: Eight high-achieving African-American undergraduate mathematics majors discuss their success and persistence in mathematics.* Doctoral dissertation. Retrieved from ProQuest Dissertations and Theses database, Publication No. 305304604.

Ellison, R. (1947). *The invisible man.* New York, NY: Random House, Inc.

Else, H. (2015). 'Sexist' peer review causes storm online. *Times Higher Education.* Retrieved from https://www.timeshighereducation.co.uk/news/sexist-peer-review-causes-storm-online/2020001.article

Elster, J. (1994). *Peças e engrenagens das ciências sociais.* Rio de Janeiro: Relume-Dumará.

Erlick & LeBold. (1977). *Website.* Retrieved from http://www.engr.washington.edu/cwd/Publications/2005-ASEE_Paper.pdf

Espinosa, L. L. (2011). Pipelines and pathways: Women of color in undergraduate STEM majors and the college experiences that contribute to persistence. *Harvard Educational Review*, *81*(2), 209–241. doi:10.17763/haer.81.2.92315ww157656k3u

Fang, D., Moy, E., Colburn, L., & Hurley, J. (2000). Racial and ethnic disparities in faculty promotion in academic medicine. *Journal of the American Medical Association*, *284*(9), 1085–1092. doi:10.1001/jama.284.9.1085 PMID:10974686

Ferreira, M. M. (2003). Gender issues related to graduate student attrition in two science departments. *International Journal of Science Education*, *25*(8), 969–989. doi:10.1080/09500690305026

Ferrini-Mundy, J., & Güçler, B. (2009). Discipline-based efforts to enhance undergraduate STEM education. *New Directions for Teaching and Learning*, *2009*(117), 55–67. doi:10.1002/tl.344

Ferry, T. R., Fouad, N. A., & Smith, P. L. (2000). The role of family context in a social cognitive model for career-related choice behavior: A math and science perspective. *Journal of Vocational Behavior*, *57*(3), 348–364. doi:10.1006/jvbe.1999.1743

Figner, B., & Weber, E. (2011). Who takes risk when and why? Determinants of risk-taking. *Current Directions in Psychological Science*, *20*(4), 211–216. doi:10.1177/0963721411415790

Fisk Vanderbilt Bridge Program. (2010). *Get the preparation you need to earn a Ph.D.* Retrieved from http://www.vanderbilt.edu/gradschool/bridge/descript.htm

Flexner, A. (1910). The Flexner report on medical education in the United States and Canada 1910. New York: Carnegie Foundation.

Foley, D. (1990). *Learning capitalist culture: deep in the heart of Tejas.* Philadelphia, PA: University of Pennsylvania Press.

Foley, D. (2005). Elusive prey: John Ogbu and the search for a grand theory of academic disengagement. *International Journal of Qualitative Studies in Education*, *18*(5), 643–657. doi:10.1080/09518390500224986

Freeman, C. E. (2004). *Trends in educational equity of girls and women: 2004* (NCES Report No. 2005-016). Retrieved from U.S. Department of Education, Institute of Education Sciences, National Center for Education Statistics website: http://nces.ed.gov/pubs2005/2005016.pdf

Frenkel, K. A. (1990). Women and Computing. *Communications of the ACM, 33*(11), 34–46. doi:10.1145/92755.92756

Frohman, H. A. (2015). The nonwhite women surgeon: A rare species. *The Journal of Surgical Research, 72*(6), 1266–1271. PMID:26160130

Furby, L., & Beyth-Marom, R. (1992). Risk taking in adolescence: A decision-making perspective. *Developmental Review, 12*(1), 1–44. doi:10.1016/0273-2297(92)90002-J

Gappa, J. M., Austin, A. E., & Trice, A. G. (2007). *Rethinking faculty work: Higher education's strategic imperative.* San Francisco, CA: John Wiley & Sons, Inc.

Garmon, L. B. (1983). Women in physics: Today and looking toward the future. In B. Lotze (Ed.), *Making contributions: An historical overview of women's role in physics* (pp. 91–119). College Park, MD: American Association of Physics Teachers.

Garnett, K. (1999). Girl Power Clicks Here. *NEA Today, 18*, i2.

George, Y. S., Neale, D. S., Van Horne, V., & Malcolm, S. M. (2001). *In pursuit of a diverse science, technology, engineering, and mathematics workforce.* Paper presented at the American Association for the Advancement of Science.

Germeijs, V., & Verschueren, K. (2009). Adolescents' career decision-making process: Related to quality of attachment to parents? *Journal of Research on Adolescence, 19*(3), 459-483. doi:10.1111/j.1532-7795.2009.00603.x

Giedd, J. N., Raznahan, A., Mills, K. L., & Lenroot, R. K. (2012). Review: Magnetic resonance imaging of male/female differences in human adolescent brain anatomy. *Biology of Sex Differences, 3*(1), 19–27. doi:10.1186/2042-6410-3-19 PMID:22908911

Ginther, D. K., & Kahn, S. (2013). Education and academic career outcomes for women of color in science and engineering. In K. Rapporteur (Ed.), *Seeking solutions: Maximizing American talent by advancing women of color in academia: Summary of a conference* (pp. 71–92). National Academy Press, National Academy of Sciences.

Ginther, D. K., Schaffer, W. T., Schnell, J., Masimore, B., Liu, F., Haak, L. L., & Kington, R. (2011). Race, ethnicity, and NIH research awards. *Science, 333*(6045), 1015–1019. doi:10.1126/science.1196783 PMID:21852498

Glass, J. L., Sassler, S., Levitte, Y., & Michelmore, K. M. (2013). What's so special about STEM? A comparison of women's retention in STEM and professional occupations. *Social Forces, 92*(2), 723–756. doi:10.1093/sf/sot092 PMID:25554713

Goodman, I. F., Cunningham, C. M., Lachapelle, C., Thompson, M., Bittinger, K., Brennan, R. T., & Delci, M. (2002). *The Women's Experiences in College Engineering (WECE) Project final report* (pp. 1–264). Cambridge, MA: Goodman Research Group.

Gouveia, A. J. (1968). Democratização do ensino superior. *Revista Brasileira de Estudos Pedagógicos, 50*(112), 232-244.

Graff, E. J., Babcock, L., & Laschever, S. (2004). Women don't ask: Negotiation and the gender divide. *The Women's Review of Books, 22*(2), 4. doi:10.2307/4024424

Graham, J. W. (2009). Missing data analysis: Making it work in the real world. *Annual Review of Psychology, 60*(1), 549–576. doi:10.1146/annurev.psych.58.110405.085530 PMID:18652544

Grbic, D., & Brewer, R. L. (2012, April01). Which Factors Predict the Likelihood of Reapplying to Medical School? An Analysis by Gender. *Academic Medicine, 87*(4), 449–457. doi:10.1097/ACM.0b013e3182494e54 PMID:22361796

Grossman, J. M., & Porche, M. V. (2014). Perceived gender and racial/ethnic barriers to STEM success. *Urban Education, 49*(6), 698–727. doi:10.1177/0042085913481364

Grzanka. (Ed.). (n.d.). *Intersectionality: A Foundations and Frontiers Reader*. Boulder, CO: Westview Press.

Guimarães, A., & Sérgio, A. (1995). "Raça", racismo e grupos de cor no Brasil. *Estudos Afroasiáticos, 27*, 45–63.

Gurer, D., & Camp, T. (2001). *Investigating the incredible shrinking pipeline for women in computer science*. Final Report–NSF Project, 9812016. Retrieved from http://womendev.acm.org/archives/documents/finalreport.pdf

Gutierrez y Muhs, G., Flores Neiman, Y., Gonzales, C.G., & Harris, A.P. (2012). Presumed Incompetent; the intersections of race and class in academia. Boulder, CO: Utah State University Press.

Gysler, M., Kruse, J. B., & Schubert, R. (2002). *Ambiguity and gender differences in financial decision making: An experimental examination of competence and confidence effects*. Working Paper. Center for Economic Research, Swiss Federal Institute of Technology. Retrieved from http://www.cer.ethz.ch/research/wp_02_23_paper.pdf

Haines, V. A., Wallace, J. E., & Cannon, M. E. (2001). Exploring the gender gap in engineering: A re-specification and test of the hypothesis of cumulative advantages and disadvantages. *The Journal of Engineering Education, 90*(4), 677–684. doi:10.1002/j.2168-9830.2001.tb00659.x

Halpern, D. F., Benbow, C. P., Geary, D. C., Gur, R. C., Hyde, J. S., & Gernsbacher, M. A. (2007). The science of sex differences in science and mathematics. *Psychological Science in the Public Interest, 8*(1), 1–51. PMID:25530726

Hamann, E. T. (2004). Lessons from the interpretation/misinterpretation of John Ogbu's scholarship. *Intercultural Education, 15*(4), 399–412. doi:10.1080/1467598042000313421

Hanson, S. L., Schaub, M., & Baker, D. P. (1996). Gender stratification in the science pipeline: A Comparative Analysis of Seven Countries. *Gender & Society, 10*(3), 271–290. doi:10.1177/089124396010003005

Harlow, R. (2003). 'Race doesn't matter, but . . .:' The effect of race on professors' experiences and emotion management in the undergraduate college classroom. *Social Psychology Quarterly, 66*(4), 348–363. doi:10.2307/1519834

Hasenbalg, C., & Valle Silva, N. (1991). Raça e oportunidades educacionais no Brasil. In *Desigualdades raciais no Brasil contemporâneo*. Belo Horizonte: UFMG/Cedeplar.

Hasenbalg, C., & Valle Silva, N. (1998). *Estrutura social, mobilidade e raça*. Rio de Janeiro: Iuperj/Vértice.

Hasenbalg, C., Valle Silva, N., & Lima, M. (1999). *Cor e estratificação social*. Rio de Janeiro: Contra Capa.

Hassan Al Marzouqi, A., & Forster, N. (2011). An exploratory study of the under-representation of emirate women in the United Arab Emirates' information technology sector. *Equality, Diversity and Inclusion. International Journal (Toronto, Ont.), 30*(7), 544–562.

Haswell, M. M. (2014). *Exploring the influence of departmental culture on perceptions of scholarship and role prioritization of biology professors*. (Doctoral dissertation), Department of Educational Leadership, Central Michigan University, Mt. Pleasant, MI.

Hathaway, R. S., Nagda, B. A., & Gregerman, S. R. (2002). The relationship of undergraduate research participation to graduate and professional education pursuit: An empirical study. *Journal of College Student Development, 43*(5), 614–631.

Hedgepeth, A. (2013). *Balancing work and life: Family-friendly workplace policies*. Retrieved from http://www.aauw.org/files/2013/02/position-on-work-life-balance-112.pdf

Heilman, M. E., Wallen, A. S., Fuchs, D., & Tamkins, M. M. (2004). Penalties for success: Reactions to women who succeed at male gender-typed tasks. *The Journal of Applied Psychology*, *89*(3), 416–427. doi:10.1037/0021-9010.89.3.416 PMID:15161402

Hewlett, S. A. (2008). *The Athena factor: Reversing the brain drain in science, engineering, and technology*. Harvard Business School Watertown.

Hiatt, M. D., & Stockton, C. G. (2003). The impact of the Flexner Report on the fate of medical schools in North America after 1909. *Journal of American Physicians and Surgeons*, *8*(2), 37–40.

Hill, C., Corbett, C., & St Rose, A. (2010). *Why So Few? Women in Science, Technology, Engineering, and Mathematics*. Retrieved from http://www.aauw.org/resource/why-so-few-women-in-science-technology-engineering-mathematics/

Hill, C., Corbett, C., & St. Rose, A. (2010). *Why so few? Women in science, technology, engineering, and mathematics*. Retrieved from American Association of University Women website: http://www.aauw.org/files/2013/02/Why-So-Few-Women-in-Science-Technology-Engineering-and-Mathematics.pdf

Hinze, S. W. (2004). 'Am I being over-sensitive?' Women's experience of sexual harassment during medical training. *Health (London, England: 1997)*, *8*(1), 101-27.

Howard-Vital, M. (1989). African American Women in Higher Education. *Journal of Black Studies*. Retrieved from http://www.jstor.org/discover/2784699?

Huang, G., Taddese, N., & Walter, E. (2000). *Entry and persistence of Women and Minorities in ollege science and engineering education (NCES Rep. No. 2000-601)*. Washington, DC: U.S. Government Printing Office.

Hughes, W. (2000). Perceived gender interaction and course confidence among undergraduate science, mathematics, and technology majors. *Journal of Women and Minorities in Science and Engineering*, *6*(2), 166–167. doi:10.1615/JWomenMinorScienEng.v6.i2.40

Huhman, H. (2012, June). STEM fields and the gender gap: Where are the women? *Forbes*. Retrieved from http://www.forbes.com/sites/work-in-progress/2012/06/20/stem-fields-and-the-gender-gap-where-are-the-women/

Human Development Report. (2009). *Overcoming barriers: Human mobility and development*. Retrieved from http://hdr.undp.org/en/search/node/GEM%202009

Hunter, B. (2006). *Home by Choice: Raising Emotionally Secure Children in an Insecure World*, New York: Multnomah Books.

Hunter, A. B., Laursen, S. L., & Seymour, E. (2007). Becoming a scientist: The role of undergraduate research in students' cognitive, personal, and professional development. *Science Education*, *91*(1), 36–74. doi:10.1002/sce.20173

Hurtado, S., Milem, J. F., Clayton-Pedersen, A. R., & Allen, W. R. (1999). Enacting diverse learning environments: Improving the campus climate for racial/ethnic diversity. ERIC Clearinghouse on Higher Education.

Hurtado, S., & Carter, D. F. (1997). Effects of college transition and perceptions of the campus racial climate on Latino college students' sense of belonging. *Sociology of Education*, *70*(4), 324–345. doi:10.2307/2673270

Hurtado, S., Milem, J. F., Clayton-Pedersen, A., & Allen, W. R. (1998). Enhancing campus climates for racial/ethnic diversity: Educational policy and practice. *The Review of Higher Education*, *21*(3), 279–302. doi:10.1353/rhe.1998.0003

IPENZ. (2010). *Minority groups in engineering education*. Wellington: National Engineering Education Plan Project.

Isaacs, B. (2001). Mystery of the missing women engineers: A solution. *Journal of Professional Issues in Engineering Education and Practice*, *127*(2), 85–91. doi:10.1061/(ASCE)1052-3928(2001)127:2(85)

Jackson, K. (2006). *The Status of the African-American Physicist in the Department of Energy National Laboratories*. Retrieved from http://www.aps.org/publications/apsnews/200205/backpage.cfm

Jagsi, R., Guancial, E. A., Worobey, C. C., Henault, L. E., Chang, Y., Starr, R., & Hylek, E. M. et al. (2006). The "gender gap" in authorship of academic medical literature—a 35-year perspective. *The New England Journal of Medicine*, *355*(3), 281–287. doi:10.1056/NEJMsa053910 PMID:16855268

Jannuzzi, P. M. (2003). Estratificação sócio-ocupacional para estudos de mercado e pesquisa social no Brasil. *Perspectiva, 17*(3-4). Accessed on May 4, 2014, at http://www.scielo.br/Scielo.Php?Script=Sci_Arttext&Pid=S0102-88392003000300024&Lng=En&Nrm=Iso

Jayakumar, U. M. (2008). Can higher education meet the needs of an increasingly diverse and global society? Campus Diversity and Cross-Cultural Workforce Competencies. *Harvard Educational Review*, 615-651.

Jeeyon, L., & Mijin, K. (2015). Parental attachment, career decision self-efficacy, and commitment to the career choice among Korean college students. *Journal of Asia Pacific Counseling*, *5*(1), 39–52. doi:10.18401/2015.5.1.3.

Jeffe, D. B., Yan, Y., & Andriole, D. A. (2012). Do research activities during college, medical school, and residency mediate racial/ethnic disparities in full-time faculty appointments at U.S. Medical schools? *Academic Medicine*, *87*(11), 1582–1593. doi:10.1097/ACM.0b013e31826e3297 PMID:23018339

Jefferson, M. (2015). *Negroland. A Memoir*. Pantheon Books New York.

Jeynes, W. (2013). The data are in: Religious private schools deserve a second look. *Public Discourse*. Retrieved from: http://www.thepublicdiscourse.com/2013/05/10218/

Jeynes, W. H. (2009). The relationship between Biblical literacy, academic achievement, and school behavior among Christian- and public-school students. *Journal of Research on Christian Education*, *18*(1), 36–55. doi:10.1080/10656210902751826

Johnson, A. (2007). Unintended consequences: How science professors discourage women of color. *Science Education*, *91*(5), 805–821. doi:10.1002/sce.20208

Johnson, A. (2010). Privilege, power, difference, and us. In M. S. Kimmel & A. L. Ferber (Eds.), *Privilege, A Reader* (2nd ed.; pp. 77–86). Boulder, CO: Westview Press.

Johnson, D. (2012). Campus Racial Climate Perceptions and Overall Sense of Belonging Among Racially Diverse Women in STEM Majors. *Journal of College Student Development*, *53*(2), 336–346. doi:10.1353/csd.2012.0028

Justia.com. (2015). *Patents by Inventor John W. Jackson*. Retrieved from http://patents.justia.com/inventor/john-w-jackson

Justin-Johnson, C. (2004). *Good fit or chilly climate: An exploration of the persistence experiences of African-American women graduates of predominantly White college science programs*. Doctoral dissertation. Retrieved from ProQuest Dissertations and Theses database, Publication No. 305172242.

Kaatz, A., Gutierrez, B., & Carnes, M. (2014). Threats to objectivity in peer review: The case of gender. *Trends in Pharmacological Sciences*, *35*(8), 371–373. doi:10.1016/j.tips.2014.06.005 PMID:25086743

Kass, R. B., Souba, W. W., & Thorndyke, L. E. (2006). Challenges confronting female surgical leaders: Overcoming the barriers. *The Journal of Surgical Research*, *132*(2), 179–187. doi:10.1016/j.jss.2006.02.009 PMID:16564542

Keenan, K., & Gabovitch, R. (1995). *Evaluating the impact of a freshman seminar program on student development and retention (Rep. No. ED395614)*. ERIC Document Reproduction Service.

Kim, Y. K., & Sax, L. J. (2007). *Different Patterns of Student-Faculty Interaction in Research Universities: An Analysis by Student Gender, Race, SES, and First-Generation Status*. A Student Experience in the Research University (SERU) Project Research Paper. Research & Occasional Paper Series: CSHE. 10.07. Center for Studies in Higher Education.

Kirk, N. (2014). Fostering more women surgeons. UConn Today.

Klein, R. & Ribeiro, S. C. A. (1982). Divisão interna da universidade: posição social das carreiras. *Educação e Seleção*, 5.

Kleinhubbert, G. (2013, September). Lab equality: Sciences struggle to attract young women. *Spiegel Online International*. Retrieved from: http://www.spiegel.de/international/germany/germany-boosts-efforts-to-get-women-in-stem-fields-a-924023.html

Koppelman, K. L. (2014). *Understanding human differences: Multicultural education for a diverse America* (4th ed.). Boston, MA: Pearson Education, Inc.

Kost, L., Pollack, S., & Finkelstein, N. (2009). Unpacking gender differences in students' perceived experiences in introductory physics. *AIP Conference Proceedings*, *1179*(1), 177–180. doi:10.1063/1.3266708

Kristof, A. L. (1996). Person-organization fit: An integrative review of its conceptualizations, measurement, and implications. *Personnel Psychology*, *49*(1), 1–49. doi:10.1111/j.1744-6570.1996.tb01790.x

Ladson-Billings, G. (1998). Just what is critical race theory and what's it doing in a nice field like education? *International Journal of Qualitative Studies in Education*, *11*(1), 7–24. doi:10.1080/095183998236863

Landsman, J., & Lewis, C. (2011). *White teachers/diverse classrooms: Creating inclusive schools, building on students' diversity, and providing true educational equity*. Sterling, VA: Stylus Publishing.

Langen, A. V., & Dekkers, H. (2005). Cross-national differences in participating in tertiary science, technology, engineering and mathematics education. *Comparative Education*, *41*(3), 329–350. doi:10.1080/03050060500211708

Lattuca, L. R., Terenzini, P. T., Harper, B. J., & Yin, A. C. (2010). Academic environments in detail: Holland's theory at the subdiscipline level. *Research in Higher Education*, *51*(1), 21–39. doi:10.1007/s11162-009-9144-9

Laura, P., Lundy-Wagner, V., Drezner, N., Gasman, M., Yoon, S., Bose, E., & Gary, S. (2009). The contribution of HBCUS to the preparation of African American women for STEM careers: A case study. *Research in Higher Education*, *50*(1), 1–23. doi:10.1007/s11162-008-9110-y

Lautenberger, D., Raezer, C., & Bunton, S. (2015). The Underrepresentation of Women in Leadership Positions at U.S. Medical Schools. *Analysis in Brief*, *15*(2), 2.

Lee, J. J. (2013). 6 women scientists who were snubbed due to sexism. *National Geographic*. Retrieved from http://news.nationalgeographic.com/news/2013/13/130519-women-scientists-overlooked-dna-history-science/

Leever, S., Dunigan, M., & Turner, M. (2002). The Power to Change is in Our Hands. *Journal of Computing Sciences in Colleges*, *18*(2), 169–179.

Leggon, C. B. (2006). Women in science: Racial and ethnic differences and the differences they make. *The Journal of Technology Transfer*, *31*(3), 325–333. doi:10.1007/s10961-006-7204-2

Liebschutz, J. M., Darko, G. O., Finley, E. P., Cawse, J. M., Bharel, M., & Orlander, J. D. (2006, January01). In the minority: Black physicians in residency and their experiences. *Journal of the National Medical Association*, *98*(9), 1441–1448. PMID:17019911

Limongi, F., Carneiro, L., Silva, L., & Mancuso, W. (n.d.). *Acesso à Universidade de São Paulo: atributos socioeconômicos dos excluídos e dos ingressantes no exame vestibular.* Work Document 03/02. Accessed on Dec 1, 2013, at http://nupps.usp.br/Downloads/Docs/Dt0203.Pdf

List of African American Scientists and Inventors. (2015). Retrieved from http://en.wikipedia.org/wiki/List_of_African-American_inventors_and_scientists

Loewen, J. W. (2006). *Sundown Towns: A Hidden Dimension of American Racism.* New York: Touchstone publishers.

Lopez, A. M., Jr., & Schulte, L. J. (2002). African American Women in the Computing Sciences: A Group to Be Studied. In *Proceedings of the 33rd SIGCSE Technical Symposium on Computer Science Education* (pp. 87–90). New York, NY: ACM. http://doi.org/ doi:10.1145/563340.563371

Lopez, M. H., & Gonzalez-Barrera, A. (2014, March 6). Women's college enrollment gains leave men behind. *Pew Research Center: Fact Tank.* Retrieved from http://www.pewresearch.org/fact-tank/2014/03/06/womens-college-enrollment-gains-leave-men-behind/

Lord, S. M., Camacho, M. M., Layton, R. A., Long, R. A., Ohland, M. W., & Washburn, M. H. (2009). Who's persisting in engineering? A comparative analysis of female and male Asian, Black, Hispanic, Native American, and White students. *Journal of Women and Minorities in Science and Engineering, 15*(2), 167–190. doi:10.1615/JWomenMinorScienEng.v15.i2.40

Lord, S. M., Layton, R., & Ohland, M. W. (2011). Trajectories of electrical engineering and computer engineering students by race and gender. Education. *IEEE Transactions on, 54*(4), 610–618.

Luckenbill-Edds, L. (2000). The 'Leaky Pipline:' Has It Been Fixed? *The American Society for Cell Biology.* Career Strategy Columns.

Lundberg, C. A., & Schreiner, L. A. (2004). Quality and frequency of faculty-student interaction as predictors of learning: An analysis of student race/ethnicity. *Journal of College Student Development, 45*(5), 549–565. doi:10.1353/csd.2004.0061

MacCorquodale, P., & National Institute of Education (U.S.). (1984). *Self-image, science, and math: Does the image of the "scientist" keep girls and minorities from pursuing science and math?.* Washington, DC: National Institute of Education.

Mahoney, M. R., Wilson, E., Odom, K. L., Flowers, L., & Adler, S. R. (2008). Minority faculty voices on diversity in academic medicine: Perspectives from one school. *Academic Medicine, 83*(8), 781–786. doi:10.1097/ACM.0b013e31817ec002 PMID:18667896

Malcom, S. M., Hall, P. Q., & Brown, J. W. (1976). *The double bind: The price of being a minority woman in science (No. 76-R-3).* Washington, DC: American Association for the Advancement of Science.

Malicky, D. (2003). A literature review on the underrepresentation of women in undergraduate engineering: Ability, self-efficacy, and the "chilly climate". In *Proceedings of the 2003 American Society for Engineering Education Annual Conference & Exposition.* Nashville, TN: American Society for Engineering Education.

Mandela, N. (1975). On perseverance. *Excerpt from letter to Mini Mandela.* Retrieved from http://qz.com/93070/the-wisdom-of-mandela-quotes-from-the-most-inspiring-leader-of-the-20th-century/

Marchant, A., Bhattacharya, A., & Carnes, M. (2007). Can the language of tenure criteria influence women's academic advancement? *Journal of Women's Health, 16*(7), 998–1003. doi:10.1089/jwh.2007.0348 PMID:17903076

Marginson, S., Tytler, R., Freeman, B., & Roberts, K. (2013). *STEM: country comparisons: international comparisons of science, technology, engineering and mathematics (STEM) education. Final report.* Australian Council of Learned Academies.

Margolis, J., & Fisher, A. (2003). *Unlocking the clubhouse: Women in computing*. MIT Press.

Martinez, E. D., Botos, J., Dohoney, K. M., Geiman, T. M., Kolla, S. S., Olivera, A., & Cohen-Fix, O. et al. (2007). Falling off the academic bandwagon: Women are more likely to quit at the postdoc to principal investigator transition. *European Molecular Biology Organization Reports*, *8*(11), 977–981. PMID:17972894

Maryland State Department of Education. (n.d.). *STEM education glossary*. Retrieved from http://mdk12.org/instruction/curriculum/stem/pdf/STEMEducationGlossary.pdf

Ma, X., & Klinger, D. A. (2000). Hierarchical linear modelling of student and school effects on academic achievement. *Canadian Journal of Education*, *25*(1), 41–55. doi:10.2307/1585867

Mayer, A., Blair, J., Ko, M. G., Hayes, S. N., Chang, Y.-H. H., Caubet, S. L., & Files, J. A. (2014). Gender distribution of U.S. medical school faculty by academic track type. *Academic Medicine*, *89*(2), 312–317. doi:10.1097/ACM.0000000000000089 PMID:24362384

McCarren, M., & Goldman, S. (2012). Research leadership and investigators: Gender distribution in the federal government. *The American Journal of Medicine*, *125*(8), 811–816. doi:10.1016/j.amjmed.2012.03.006 PMID:22579138

McCullough, L. (2011). Women's leadership in science, technology, engineering, and mathematics: Barriers to participation. *Forum on Public Policy: A Journal of the Oxford Round Table*, *2011*(2). Retrieved from http://forumonpublicpolicy.com/vol2011.no2/papers2011vol2.html

McGee, E. O., & Martin, D. B. (2011). "You would not believe what I have to go through to prove my intellectual value!" Stereotype management among academically successful Black mathematics and engineering students. *American Educational Research Journal*, *48*(6), 1347–1389. doi:10.3102/0002831211423972

McGowan, J. M. (2000). Multicultural teaching: African American faculty classroom teaching experiences in predominantly White colleges and universities. *Multicultural Education*, *8*, 19–22.

McPherson, E. (2014). Informal learning in science, math, and engineering majors for African American female undergraduates. *Global Education Review*, *1*(4), 96–113.

McQuillan, P. (1998). *Educational Opportunity in an Urban American High School: A Cultural Analysis*. Albany, NY: State University of New York Press.

Meharry Medical College. (2004). *Journal of health care for the poor and underserved*. Nashville, TN: Institute on Health Care for the Poor and Underserved, Meharry Medical College.

Merchant, J. L., & Omary, M. B. (2010). Underrepresentation of underrepresented minorities in academic medicine: the need to enhance the pipeline and the pipe. *Gastroenterology, 138*(1), 19-26 e11-13. doi: 10.1053/j.gastro.2009.11.017

Metz, S. S. (2011). Engineering. In B. J. Bank (Ed.), *Gender and higher education* (pp. 196–202). Baltimore, MD: The Johns Hopkins University Press.

Milem, J. (2003). The educational benefits of diversity: Evidence from multiple sectors. In M. Chang, D. Witt, J. Jones, & K. Hakuta (Eds.), *Compelling interest: Examining the evidence on racial dynamics in higher education* (pp. 126–169). Stanford, CA: Stanford University Press.

Milem, J. F., Chang, M. J., & Antonio, A. L. (2005). *Making diversity work on campus: A research-based perspective*. Washington, DC: Association American Colleges and Universities.

Milgram, D. (2011). How to recruit women and girls to the science, technology, engineering, and math (STEM) classroom. *Technology & Engineering Teacher*, *71*(3), 4–11.

Ministry for Women. (2014). *A snapshot of New Zealand women.* Retrieved from http://women.govt.nz/our-work#sthash.8mv6wmBO.dpuf

Ministry for Women. (2015). *Occupational segregation.* Retrieved from http://women.govt.nz/our-work/economic-independence/paid-and-unpaid-work/occupational-segregation

Ministry of Business Innovation and Employment, & Ministry of Education. (2014). *A nation of curious minds: A national strategic plan for science in society.* Retrieved from http://www.msi.govt.nz/assets/MSI/Update-me/Science-in-society-project/science-in-society-plan.pdf

Ministry of Education. (2014). *Tertiary Education Strategy 2014 - 2019.* Retrieved from http://www.minedu.govt.nz/NZEducation/EducationPolicies/TertiaryEducation/PolicyAndStrategy/TertiaryEducationStrategy2014-2019.aspx

Ministry of Pacific Island Affairs. (2010). *Education and Pacific Peoples in New Zealand.* Retrieved from http://www.stats.govt.nz/browse_for_stats/people_and_communities/pacific_peoples/pacific-progress-health.aspx

Ministry of Research Science and Technology. (1998). *Mathematics in New Zealand: Past, Present and Future.* Wellington: Ministry of Research, Science and Technology.

Mlodinow, L. (2014). *Overcoming Unconscious Bias.* Retrieved from http://www.thebigidea.co.uk/overcoming-unconscious_bias

Moakler, M. Jr, & Kim, M. (2014). College major choice in STEM: Revisiting confidence and demographic factors. *The Career Development Quarterly*, *62*(2), 128–142. doi:10.1002/j.2161-0045.2014.00075.x

Mody, P. N., & Brainard, S. G. (2005). Successful International Initiatives Promoting Gender Equity in Engineering. In *Proceedings of the International Symposium on Women and ICT: Creating Global Transformation.* New York, NY: ACM. http://doi.org/ doi:10.1145/1117417.1117422

Moehlecke, S., & Gabriel, C. T. (2006). Conexões de saberes: uma outra visão sobre o ensino superior. *Revista Contemporânea de Educação da UFRJ.* Retrieved from http://www.educacao.ufrj.br/artigos/n2/numero2-smoehlecke.pdf

Moehlecke, S. (2002). Ação afirmativa: História e debates no Brasil. *Cadernos de Pesquisa, São Paulo*, *117*(117), 197–217. doi:10.1590/S0100-15742002000300011

Moore, J. L. III. (2006). A qualitative investigation of African American males' career trajectory in engineering: Implications for teachers, school counselors, and parents. *Teachers College Record*, *108*(2), 246–266. doi:10.1111/j.1467-9620.2006.00653.x

Moore, R. (1988). *Home School Burnout: What It Is. What Causes It. and How to Overcome It.* Tucson, AZ: Arizona Wolgemuth & Hyatt Pub.

Morrell, C., & Parker, C. (2013) Adjusting micromessages to improve equity in STEM. *Diversity and Democracy*, *16*(2). Retrieved from http://www.aacu.org/diversitydemocracy/vol16no2/morrell_parker.cfm

Mosatche, H. S., Matloff-Nieves, S., Kekelis, L., & Lawner, E. K. (2013). Effective STEM programs for adolescent girls: Three approaches and many lessons learned. *Afterschool Matters*, *17*, 17–25.

Mpofu, C. (2014). *The global dynamics of the medical workforce: A case study of flows from New Zealand to Australia.* Paper presented at the 10th Annual Association of American Medical Colleges Health Workforce Research Conference, Washington, DC.

Muller, C., Katz, S., & Dance, L. J. (1999). Investing in teaching and learning: Dynamics of teacher–student relationship from each actor's perspective. *Urban Education*, *34*(3), 292–337. doi:10.1177/0042085999343003

National Academies. (2010). *Expanding underrepresented minority participation: America's science and technology talent at the crossroads*. Washington, DC: National Academies Press.

National Academy of Engineering. (2004). *The engineer of 2020: Visions of engineering in the new century*. Washington, D.C.: National Academies Press.

National Center for Education in Statistics. (2015). *Degrees conferred by sex and race*. Retrieved from https://nces.ed.gov/fastfacts/display.asp?id=72

National Center for Education Statistics. (2005). *Digest of education statistics*. Washington, DC: National Center for Education Statistics.

National Center for Education Statistics. (2011). *Digest of Education Statistics: 2011. Table 301. Bachelor's degrees conferred by degree-granting institutions, by sex, race/ethnicity, and field of study: 2009-10*. Author.

National Center for Educational Statistics. (2011). *Table 301 Bachelor's degrees conferred by degree-granting institutions, by sex, race/ethnicity, and field of study: 2009-2010* [Data file]. Retrieved from http://nces.ed.gov/programs/digest/d11/tables/dt11_301.asp

National Center for Science and Engineering Statistics. (2013). *Women, Minorities, and Persons with Disabilities in Science and Engineering: 2013*. National Science Foundation. Retrieved from http://www.nsf.gov/statistics/wmpd/2013/pdf/nsf13304_full.pdf

National Science Board. (2006). *America's pressing challenge - building a stronger foundation: A companion to science and engineering indicators - 2006* (NSB-06-02). Retrieved from http://www.nsf.gov/statistics/nsb0602/nsb0602.pdf

National Science Board. (2006). Science and Engineering Indicators 2006. National Science Foundation.

National Science Board. (2014). Science and engineering indicators 2014. Arlington, VA: National Science Foundation (NSB 14-01).

National Science Foundation, Division of Science Resource Statistics. (2004). *Women, minorities, and persons with disabilities in science and engineering: 2004 (NSF 04-317)*. Arlington, VA: Author.

National Science Foundation, Division of Science Resources Statistics. (2011). *Women, minorities, and persons with disabilities in science and engineering: 2011 (NSF 11-309)*. Arlington, VA: Author.

National Science Foundation. (2006). *Investing in America's future: Strategic plan FY 2006–2011* (NSF 06-48). Retrieved from http://www.nsf.gov/pubs/2006/nsf0648/NSF-06-48.pdf

National Science Foundation. (2015). *Women, Minorities, and Persons with Disabilities in Science and Engineering: 2015*. Arlington, VA: National Science Foundation.

National Women's Law Center. (2012). *The next generation of Title IX: STEM–science, technology, engineering, and math*. Retrieved from http://www.nwlc.org/sites/default/files/pdfs/nwlcstem_titleixfactsheet.pdf

Negros na UFBA. (n.d.). Accessed on Dec 1, 2013, at http://www1.folha.uol.com.br/fsp/cotidian/ff23089813.htm#_=_

Nelson, D. (2007). *A National analysis of minorities in science and engineering faculties at research universities*. Retrieved from http://faculty-staff.ou.edu/N/Donna.J.Nelson-1/diversity/Faculty_Tables_FY07/07Report.pdf

Newman, L. A., Pollock, R. E., & Johnson-Thompson, M. C. (2003). Increasing the pool of academically oriented African-American medical and surgical oncologists. *Cancer*, *97*(1Suppl), 329–334. doi:10.1002/cncr.11027 PMID:12491497

Nolen, W. (1968). *The making of a surgeon*. Minneapolis, MN: Mid-List Press.

Nosek, B. A., Smyth, F. L., Sriram, N., Lindner, N. M., Devos, T., Ayala, A., & Greenwald, A. G. et al. (2009). National differences in gender-science sterotypes predict national sex differences in science and math achievement. *Proceedings of the National Academy of Sicneces*, *106*(26), 10593–10597. doi:10.1073/pnas.0809921106 PMID:19549876

Nunez-Smith, M., Ciarleglio, M. M., Sandoval-Schaefer, T., Elumn, J., Castillo-Page, L., Peduzzi, P., & Bradley, E. H. (2012). Institutional variation in the promotion of racial/ethnic minority faculty at US medical schools. *American Journal of Public Health*, *102*(5), 852–858. doi:10.2105/AJPH.2011.300552 PMID:22420820

Nurse, K. (2007). *Science, Technology and innovation in the Caribbean*. Paper presented at the Meeting on Technology Policy and Development in Latin America of the Economic Commission for Latin America and the Caribbean, Santiago, Chile. Retrieved from http://vi.unctad.org/files/papers/nursepaper.pdf

Nye, B., Konstantopoulos, S., & Hedges, L. (2004). How large are teacher effects? *Educational Evaluation and Policy Analysis*, *26*(3), 237–257. doi:10.3102/01623737026003237

O'Shea, M., Heilbronner, N., & Reis, S. (2010). Characteristics of academically talented women who achieve at high levels on a scholastic achievement test-mathematics. *Journal of Advanced Academics*, *21*(2), 234–271. doi:10.1177/1932202X1002100204

Obama, B. (2015). *Selma: Remarks by the President at the 50th Anniversary of the Selma to Montgomery Marches*. Retrieved from https://www.whitehouse.gov/the-press-office/2015/03/07/remarks-president-50th-anniversary-selma-montgomery-marches

O'Brien, L. T., Blodorn, A., Adams, G., Garcia, D. M., & Hammer, E. (2015). Ethnic variation in gender-STEM stereotypes and STEM participation: An intersectional approach. *Cultural Diversity & Ethnic Minority Psychology*, *21*(2), 169–180. doi:10.1037/a0037944 PMID:25244590

Office of Minority Health U.S. Department of Health and Human Services. (2002). *Teaching Cultural Competence in Health Care. A Review of Current Concepts, Policies and Procedures*. Contract Number: 282–98–0029 Task Order #41 Task 2: Synthesis Report. Retrieved from: http://minorityhealth.hhs.gov/assets/pdf/checked/1/em01garcia1.pdf

Ogbu, J. (1974). *The Next Generation: An Ethnography of Education in an Urban Neighborhood*. New York: Academic Press.

Ogbu, J. (1981). School ethnography: A multilevel approach. *Anthropology & Education Quarterly*, *12*(10), 3–29. doi:10.1525/aeq.1981.12.1.05x1281g

Ogbu, J. (2003). *Black American Students in an Affluent Suburb: A Study of Academic Disengagement*. Mahwah, NJ: Lawrence Erlbaum Associates.

Ohland, M. W., Brawner, C. E., Camacho, M. M., Layton, R. A., Long, R. A., Lord, S. M., & Washburn, M. H. (2011). Race, gender, and measures of success in engineering education. *The Journal of Engineering Education*, *100*(2), 225–252. doi:10.1002/j.2168-9830.2011.tb00012.x

Ojokoh, B. A., Adeola, O. S., Isinkaye, F. O., & Abraham, C. (2014). Career Choices in Information and Communication Technology among South Western Nigerian Women. *Journal of Global Information Management*, *22*(2), 48–77. doi:10.4018/jgim.2014040104

Ong, M. (2002). *Against the current: Women of color succeeding in physics*. Doctoral dissertation. Retrieved from ProQuest Dissertations and Theses database, Publication No. 304803810.

Ong, M. (2005). Body projects of young women of color in physics: Intersections of gender, race, and science. *Social Problems*, *52*(4), 593–617. doi:10.1525/sp.2005.52.4.593

Ong, M., Wright, C., Espinosa, L., & Orfield, G. (2011). Inside the double bind: A synthesis of empirical research on women of color in science, technology, engineering, and mathematics. *Harvard Educational Review*, *81*(2), 172–208. doi:10.17763/haer.81.2.t022245n7x4752v2

Ornek, F., Robinson, W. R., & Haugan, M. P. (2008). What makes physics difficult? *International Journal of Environmental and Science Education*, *3*(1), 30–34.

Parks, J. B., & Roberton, M. A. (2004). Attitudes toward women mediate the gender effect on attitudes toward sexist language. *Psychology of Women Quarterly*, *28*(3), 233–239. doi:10.1111/j.1471-6402.2004.00140.x

Pascarella, E. T., & Terenzini, P. T. (2005). *How college affects students: A third decade of research* (Vol. 2). San Francisco: Jossey-Bass.

Peña Dolhun, E., Muñoz, C., & Grumbach, K. (2003). Cross-cultural education in U.S. medical schools: Development of an assessment tool. *Academic Medicine*, *78*(6), 615–622. doi:10.1097/00001888-200306000-00012 PMID:12805042

Peng & Jaffe. (1979). *Women Who Enter Male-dominated Fields of Study in Higher Education. American Educational Research Journal*, *16*, 285–293.

Perez-Felkner, L., McDonald, S., Schneider, B., & Grogan, E. (2012). Female and male adolescents' subjective orientations to mathematics and the influence of those orientations on postsecondary majors. *Developmental Psychology*, *48*(6), 1658–1673. doi:10.1037/a0027020 PMID:22390658

Persaud, A., & Salter, D. W. (2003). Understanding women's perceptions of classroom 'fit' and participation as interactions between psychological and environmental types. *Journal of Classroom Interaction*, *38*(2), 1–10.

Peters, A.-K. (2013). Identity Development of CS and IT Students: What's the Role of Higher Education? In *Proceedings of the Ninth Annual International ACM Conference on International Computing Education Research* (pp. 187–188). New York, NY: ACM. http://doi.org/ doi:10.1145/2493394.2493427

Piotrowski, C., & Hemasinha, R. (2012). Career aspirations of undergraduate mathematics majors: An exploratory study. *College Student Journal*, *46*(1), 46–50.

Pittman, C. T. (2010). Race and gender oppression in the classroom: The experience of women faculty of color with white male students. *Teaching Sociology*, *38*(3), 183–196. doi:10.1177/0092055X10370120

Pittman, C. T. (2012). Racial microaggressions: The narratives of African American faculty at a predominantly white university. *The Journal of Negro Education*, *81*(1), 82–92. doi:10.7709/jnegroeducation.81.1.0082

Planty, M., Hussar, W., Snyder, T., Kena, G., KewalRamani, A., Kemp, J., . . . Dinkes, R. (2009). The Condition of Education 2009 (NCES 2009-081). National Center for Education Statistics, Institute of Education Sciences, U.S. Department of Education.

Pollack, E. (2013, October 3). Why are there still so few women in science? *The New York Times*. Retrieved from http://www.nytimes.com/2013/10/06/magazine/why-are-there-still-so-few-women-in-science.html?_r=0&pagewanted=print

Pololi, L. H., Civian, J. T., Brennan, R. T., Dottolo, A. L., & Krupat, E. (2013). Experiencing the culture of academic medicine: Gender matters, a national study. *Journal of General Internal Medicine*, *28*(2), 201–207. doi:10.1007/s11606-012-2207-1 PMID:22936291

Pololi, L. H., & Jones, S. J. (2010). Women faculty: An analysis of their experiences in academic medicine and their coping strategies. *Gender Medicine*, *7*(5), 438–450. doi:10.1016/j.genm.2010.09.006 PMID:21056870

Pope, J., & Joseph, J. (1997). Student harassment of female faculty of African American descent in the academy. In L. Benjamin (Ed.), *Black women in the academy: Promises and perils* (pp. 252–260). Gainesville: University of Florida Press.

Porter, J. (2014). Why are women leaving science, engineering, and tech jobs? *Fast Company.* Retrieved from: http://www.fastcompany.com/3037075/strong-female-lead/why-are-women-are-leaving-science-engineering-tech-jobs

Prewitt-Freilino, J. L., Caswell, T. A., & Laakso, E. K. (2012). The gendering of language: A comparison of gender, equality in countries with gendered, natural gender, and genderless languages. *Sex Roles, 66*(3-4), 268–281. doi:10.1007/s11199-011-0083-5

Queiroz, D. M. (2001). *Raça, gênero e educação superior.* (Doctoral Thesis). Faculty of Education, Universidade Federal da Bahia, Brazil.

Queiroz, D. M. (2004). O negro e a universidade brasileira. *Historia Actual Online, 1*(3), 73-82, 2004. Accessed on Aug 28, 2006, at http://www.historia-actual.com/hao/volumes/volume1/issue3/esp/v1i3c8.pdf

Queiroz, D. M. (2014). *Quem são os negros? Classificação racial no Brasil: aproximações e divergências.* Accessed on Sept 13, 2014, at http://www.anped.org.br/24/t0344239895038.doc

Ramanan, R. A., Taylor, W. C., Davis, R. B., & Phillips, R. S. (2006, April01). Mentoring matters: Mentoring and career preparation in internal medicine residency training. *Journal of General Internal Medicine, 21*(4), 340–345. doi:10.1111/j.1525-1497.2006.00346.x PMID:16686809

Rankin, S. R., & Reason, R. D. (2005). Differing perceptions: How students of color and White students perceive campus climate for underrepresented groups. *Journal of College Student Development, 46*(1), 43–61. doi:10.1353/csd.2005.0008

Redmond, K., Evans, S., & Sahami, M. (2013). *A large-scale quantitative study of women in computer science at Stanford University.* ACM Press. doi:10.1145/2445196.2445326

Renault, M. (2015). *Conspicuously invisible: Women of color in university sciences.* Education · Minneapolis. Retrieved from http://www.mprnews.org/story/2015/02/21/women-of-color-in-science

Rheingans, P., Brodsky, A., Scheibler, J., & Spence, A. (2011). The Role of Majority Groups in Diversity Programs. *Trans. Comput. Educ., 11*(2), 11:1–11:15. http://doi.org/10.1145/1993069.1993075

Riegle-Crumb, C., & King, B. (2010). Questioning male advantage in STEM: Examining disparities in college majors by gender and race/ethnicity. *Educational Researcher, 39*(9), 656–664. doi:10.3102/0013189X10391657

Rifkin, T. (2014). *Public community college faculty.* Retrieved from http://www.aacc.nche.edu/Resources/aaccprograms/pastprojects/Pages/publicccfaculty.aspx

Ritchie, J., & Lewis, J. (Eds.). (2003). *Qualitative research practice: A guide for social science students and researchers.* London: Sage.

Roberts, J., Gardiner, B., Gilbert, J., & Vaughan, K. (2008). *Trading choices: young people's career decisions and gender segregation in the trades.* Retrieved from http://women.govt.nz/documents/trading-choices-young-people%E2%80%99s-career-decisions-and-gender-segregation-trades-2008

Rodriguez, J. E., Campbell, K. M., & Pololi, L. H. (2015). Addressing disparities in academic medicine: What of the minority tax? *BMC Medical Education, 15*(1), 6. doi:10.1186/s12909-015-0290-9 PMID:25638211

Roger, A., & Duffield, J. (2000). Factors underlying persistent gendered option choices in school science and technology in Scotland. *Gender and Education, 12*(3), 367–383. doi:10.1080/713668300

Ro, H. K., & Loya, K. I. (2015). The effect of gender and race intersectionality on student learning outcomes in engineering. *The Review of Higher Education*, *38*(3), 359–396. doi:10.1353/rhe.2015.0014

Ro, H. K., Terenzini, P. T., & Yin, A. C. (2013). Between-college effects on students reconsidered. *Research in Higher Education*, *54*(3), 253–282. doi:10.1007/s11162-012-9269-0

Rose, A. S., & Hill, C. (2013). *Women in community colleges: Access to success*. Retrieved from http://www.aauw.org/files/2013/05/women-in-community-colleges.pdf

Rossi, A. S. (1965). Women in science: Why so few? Social and psychological influences restrict women's choice and pursuit of careers in science. *Science*, *148*(3674), 1196–1202. doi:10.1126/science.148.3674.1196 PMID:17748114

Rossiter, M. (n.d.). *Has Feminism Changed Science?* Retrieved from muse.jhu.edu/journals/bulletin_of_the_history_of.../77.1rossiter.html

Roue, L. C. (2007). Young women's perceptions of technology and engineering: Factors influencing their participation in math, science and technology. Menomonie, WI: University of Wisconsin--Stout.

Rozek, C. S., Hyde, J. S., Svoboda, R. C., Hulleman, C. S., & Harackiewicz, J. M. (2015). Gender differences in the effects of a utility-value intervention to help parents motivate adolescents in mathematics and science. *Journal of Educational Psychology*, *107*(1), 195–206. doi:10.1037/a0036981

Rury, J. L. (2009). Education and social change (3rd ed.). New York, NY: Routledge.

Sant'Anna, W. (2001). Novos marcos para as relações étnico/raciais no Brasil: uma responsabilidade coletiva. In Anais de Seminários Regionais Preparatórios Para Conferência Mundial Contra Racismo, Discriminação Racial, Xenofobia e Intolerância Correlata. Brasília: Ministério da Justiça e Secretaria de Estado dos Direitos Humanos.

Santos, J. C. (1998). *Por uma teoria da redemocratização do ensino superior: perspectivas e implicações*. Rio de Janeiro: Fórum Educacional.

Satterfield, D., Heisel, S., & Budischak, S. (2013). *Retaining women in STEM careers: Graduate students as the building blocks of change.* Unpublished manuscript. Retrieved from http://www.nsf.gov/news/special_reports/gradchallenge/images/winners/entries/second-place-parasite-ladies.pdf

Savitt, T. (2006). Abraham Flexner and the black medical schools. 1992. *Journal of the National Medical Association*, *98*(9), 1415. PMID:17019906

Sax, L. J. (2008). *The gender gap in college: Maximizing the developmental potential of women and men.* San Francisco, CA: Jossey-Bass.

Schiebinger, L. (2001). *Has Feminism Changed Science?* Cambridge, MA: Harvard University Press.

Schmader, T., Forbes, C., Zhang, S., & Berry, W. (2009). A metacognitive perspective on the cognitive deficits experienced in intellectually threatening environments. *Personality and Social Psychology Bulletin*, *35*(5), 584–596. doi:10.1177/0146167208330450 PMID:19228598

Schwitzer, A. M., Griffin, O. T., Ancis, J. R., & Thomas, C. R. (1999). Social adjustment experiences of African American college students. *Journal of Counseling and Development*, *77*(2), 189–197. doi:10.1002/j.1556-6676.1999.tb02439.x

Science, Technology and Innovation Council. (2013). *2012 State of the Nation Canada's Science, Technology and Innovation System: Aspiring to Global Leadership*. Ottawa, ON, Canada: Science, Technology and Innovation Council Secretariat.

Scragg, G., & Smith, J. (1998). *A study of barriers to women in undergraduate computer science*. ACM Press; doi:10.1145/273133.273167

Settles, I. H., Cortina, L. M., Steward, A. J., & Malley, J. (2007). Voice matters: Buffering the impact of negative climate for women in science. *Psychology of Women Quarterly*, *31*(3), 270–281. doi:10.1111/j.1471-6402.2007.00370.x

Setton, M. G. J. (2002). A divisão interna do campo universitário: Uma tentativa de classificação. *Revista Brasileira de Estudos Pedagógicos, Brasília*, *80*(196), 451–471.

Seymour, E., & Hewitt, N. M. (1997). *Talking About Leaving: Why Undergraduates Leave the Sciences*. Boulder, CO: Westview Press.

Shain, C. H. (2002). *Revisiting the problem of engineering school persistence in African-American women students*. Doctoral dissertation. Retrieved from ProQuest Dissertations and Theses database, Publication No. 304798888.

Shakeshaft, C. (1986). A gender at risk. *Phi Delta Kappan*, *67*(7), 499–503.

Sheldon, S. B., & Epstein, J. L. (2005). Involvement Counts: Family and Community Partnerships and Mathematics Achievement. *The Journal of Educational Research*, *98*(4), 196–206. doi:10.3200/JOER.98.4.196-207

Sheltzer, J. M., & Smith, J. C. (2014). Elite male faculty in the life sciences employ fewer women. *Proceedings of the National Academy of Sciences of the United States of America*, *111*(28), 10107–10112. doi:10.1073/pnas.1403334111 PMID:24982167

Shettle, C., Roey, S., Mordica, J., Perkins, R., Nord, C., Teodorovic, J., . . . Brown, J. (2007). The Nation's Report Card [TM]: America's High School Graduates. NCES 2007-467. National Center for Education Statistics.

Siss, A. (2003). *Afro-brasileiros, cotas e ações afirmativas: razões históricas*. Niterói: Penesb.

Smedley, A. (2007). *Race in North America: Evolution of a worldview*. Boulder, CO: Westview Press.

Smyth, F. L., & Nosek, B. A. (2015). On the gender–science stereotypes held by scientists: Explicit accord with gender-ratios, implicit accord with scientific identity. *Frontiers in Psychology*, 6. PMID:25964765

Snyder, T. D., & Dillow, S. A. (2015). *Digest of Education Statistics 2013* (NCES 2015-011). Retrieved from U.S. Department of Education, Institute of Education Sciences, National Center for Education Statistics website: http://nces.ed.gov/pubs2015/2015011.pdf

Solorzano, D., Ceja, M., & Yosso, T. (2000). Critical race theory, racial microaggressions, and campus racial climate: The experiences of African American college students. *The Journal of Negro Education*, 60–73.

Soper, S. C. (2015, June 18). What It's Like as a "Girl" in the Lab, Op-Ed. *The New York Times*. Retrieved from http://www.nytimes.com/2015/06/18/opinion/what-its-like-as-a-girl-in-the-lab.html?_r=0

Sosnowski, N. H. (2002). *Women of color staking a claim for cyber domain: Unpacking the racial/ gender gap in science, mathematics, engineering and technology (SMET)*. Doctoral dissertation. Retrieved from ProQuest Dissertations and Theses database, Publication No. 275796259.

Southern Education Foundation. (2005). *Igniting potential: Historically black colleges and universities and science, technology, engineering and mathematics*. Atlanta, GA: Author.

Stahlberg, D., Braun, F., Irmen, L., & Sczesny, S. (2007). Representation of the sexes in language. In K. Fiedler (Ed.), *Social communication* (pp. 163–187). New York, NY: Pscyhology.

Statistics Canada. (2012, May 3). *Full-time teaching staff at Canadian universities, by rank and sex*. Retrieved May 15, 2015, from http://www.statcan.gc.ca/tables-tableaux/sum-som/l01/cst01/educ68a-eng.htm

Statistics New Zealand. (2004). Degrees of difference: The employment of university qualified immigrants in New Zealand. Wellington, New Zealand: Author.

Statistics New Zealand. (2005a). *Focusing on women.* Wellington: Statistics New Zealand.

Statistics New Zealand. (2005b). *School achievement in maths and science.* Retrieved from http://www2.stats.govt.nz/domino/external/web/nzstories.nsf/092edeb76ed5aa6bcc256afe0081d84e/6b69fd321fb4557fcc256b1800067224?OpenDocument

Statistics New Zealand. (2006a). *Distribution of NZ population aged 15 and over by highest qualification and field of study 2006.* Wellington: Statistics New Zealand.

Statistics New Zealand. (2006b). *Distribution of the working age population by labour force status, highest qualification and ethnic group 2006 - percentages.* Wellington: Statistics New Zealand.

Statistics New Zealand. (2006c). *Domestic equivalent full-time students enrolled by field of study, level of study, age group and gender, 2013.* Wellington: Statistics New Zealand.

Statistics New Zealand. (2013). Historical population estimates tables. *Estimates and projections.* Retrieved from http://www.stats.govt.nz/browse_for_stats/population/estimates_and_projections/historical-population-tables.aspx

Steele, C. M., & Aronson, J. (1995). Stereotype threat and the intellectual test performance of African Americans. *Journal of Personality and Social Psychology, 69*(5), 797–811. doi:10.1037/0022-3514.69.5.797 PMID:7473032

Steele, J., James, J. B., & Barnett, R. C. (2002). Learning in a man's world: Examining the perceptions of undergraduate women in male-dominated academic areas. *Psychology of Women Quarterly, 26*(1), 46–50. doi:10.1111/1471-6402.00042

Steinecke, A., & Terrell, C. (2010). Progress for whose future? The impact of the Flexner Report on medical education for racial and ethnic minority physicians in the United States. *Academic Medicine, 85*(2), 236–245. doi:10.1097/ACM.0b013e3181c885be PMID:20107348

Stewart, P. (2012). *After decades, revolving door remains for black, latino scholars in the academy.* Retrieved from Diverse: Issues in Higher Education website: http://diverseeducation.com/article/17227/

Stockton, M., & Hiatt, C. (2003). The Impact of the Flexner Report on the Fate of Medical Schools in North America After 1909. *Journal of American Physicians and Surgeons, 8*(2), 4.

Stockton, M. H. C. (2003). The Impact of the Flexner Report on the fate of medical schools in North America After 1909. *Journal of American Physicians and Surgeons, 8*(2), 4. Retrieved from http://www.jpands.org/vol8no2/hiatt.pdf

Stodolsky, S. S., Salk, S., & Glaessner, B. (1991). Student views about learning math and social studies. *American Educational Research Journal, 28*(1), 89–116. doi:10.3102/00028312028001089

Stout, J. G., Dasgupta, N., Hunsinger, M., & McManus, M. A. (2011). STEMing the tide: Using ingroup experts to inoculate women's self-concept in science, technology, engineering, and mathematics (STEM). *Journal of Personality and Social Psychology, 100*(2), 255–270. doi:10.1037/a0021385 PMID:21142376

Strauss, A. L., & Corbin, J. M. (1998). *Basics of qualitative research: Techniques and procedures for developing grounded theory.* Thousand Oaks, CA: Sage Publications.

Strelnick, A. H., Lee-Rey, E., Nivet, M., & Soto-Greene, M. L. (2008). Diversity in academic medicine no. 2 history of battles lost and won. *The Mount Sinai Journal of Medicine, New York, 75*(6), 499–503. doi:10.1002/msj.20080 PMID:19021213

Sue, D. W. (2010). *Microaggressions in everyday life: Race, gender, and sexual orientation.* Hoboken, NJ: John Wiley & Sons.

Sue, D., Capodilupo, C. M., Tonino, G. C., Bucceri, J. M., Holder, A. M. B., Nadal, K. L., & Esquilin, M. (2007). Racial microaggressions in everyday life. *The American Psychologist*, *62*(4), 271–286. doi:10.1037/0003-066X.62.4.271 PMID:17516773

Sullivan, L. W., & Suez Mittman, I. (2010). The state of diversity in the health professions a century after Flexner. *Academic Medicine*, *85*(2), 246–253. doi:10.1097/ACM.0b013e3181c88145 PMID:20107349

Sullivan, O. R. (2002). *African American women scientists and inventors. Black stars.* New York: Wiley.

Swim, J. K., Mallett, R., & Stangor, C. (2004). Understanding subtle sexism: Detection and use of sexist language. *Sex Roles*, *41*(3/4), 117–127. doi:10.1023/B:SERS.0000037757.73192.06

Talamantes, E., Mangione, C. M., Gonzalez, K., Jimenez, A., Gonzalez, F., & Moreno, G. (2014). Community college pathways: Improving the U.S. physician workforce pipeline. *Academic Medicine*, *89*(12), 1649–1656. doi:10.1097/ACM.0000000000000438 PMID:25076199

Tanenbaum, C. (2014). *Early Academic Career Pathways in STEM: Do Gender and Family Status Matter*? Retrieved from http://www.air.org/resource/early-academic-career-pathways-stem-do-gender-and-family-status-matter

Tavares, I., Braga, M. L. S., & Lima, B. S. (2015). *Análise sobre a participação de negras e negros no sistema científico.* Retrieved from http://www.cnpq.br/documents/10157/1f95db49-f382-4e22-9df7-933608de9e8d

Terenzini, P. T., & Reason, R. D. (2005, November). *Parsing the first year of college: A conceptual framework for studying college impacts.* Annual meeting of the Association for the Study of Higher Education, Philadelphia, PA. Education, Philadelphia, PA.

Terenzini, P. T., & Reason, R. D. (2010, June). *Toward a more comprehensive understanding of college effects on student learning.* Paper presented at the Annual Conference of the Consortium of Higher Education Researchers (CHER), Oslo, Norway.

The Journal of Blacks in Higher Education. (2006). *Doctoral Degree Awards to African Americans Reach another All-Time High.* Retrieved from http://www.jbhe.com/news_views/50_black_doctoraldegrees.html

The Royal Society of New Zealand. (2014). *Women in Science – a Royal Society Overview.* Retrieved from http://www.royalsociety.org.nz/expert-advice/papers/yr2005/women-in-science/

Thompson, W. E., Pattillo, R. A., Stiles, J. K., & Schatten, G. (2014). Biomedical research's unpaid debt: NIH's initiative to support and implement fairer competition for minority students is a welcome step to redress the exploitation of African Americans by science. *EMBO Reports*, *15*(4), 333–337. doi:10.1002/embr.201338274 PMID:24652854

Toldson, I.A., & Cooper, G. (2014). *Historically Black Colleges and Universities Data Dashboard.* Academic Press.

Townsend, C. M. (2012). *Sabiston Textbook of Surgery*. Philadelphia, PA: WB Saunders Co.

Trauth, E. M., Quesenberry, J. L., & Yeo, B. (2008). Environmental Influences on Gender in the IT Workforce. *SIGMIS Database*, *39*(1), 8–32. doi:10.1145/1341971.1341975

Treadwell, H. M., Braithwaite, R. L., Braithwaite, K., Oliver, D., & Holliday, R. (2009). Leadership development for health researchers at historically Black colleges and universities. *American Journal of Public Health*, *99*(S1), S53–S57. doi:10.2105/AJPH.2008.136069 PMID:19246669

Trix, F., & Psenka, C. (2003). Exploring the color of glass: Letters of recommendation for female and male medical faculty. *Discourse & Society*, *14*(2), 191–220. doi:10.1177/0957926503014002277

Turk-Bicakci, L. (2014). *Women, Blacks Most Likely to Leave STEM Careers, New Research by AIR Finds.* Retrieved from http://www.air.org/news/press-release/women-blacks-most-likely-leave-stem-careers-new-research-air-finds

Tziner, A., Loberman, G., Dekel, Z., & Sharoni, G. (2012). The influence of the parent offspring relationship on young people's career preferences. *Revista de Psicología del Trabajo y de las Organizaciones*, *28*(2), 99–105. doi:10.5093/tr2012a8

U.S. National Library of Medicine. (2014). *National Library of Medicine Profiles in Science Rosalind Franklin paper.* Retrieved from http://profiles.nlm.nih.gov/ps/retrieve/Narrative/KR/p-nid/183

United States Department of Health and Human Services. (2015). *Medically underserved areas/populations.* Retrieved from http://www.hrsa.gov/shortage/mua/index.html

USP tem 78,7% de calouros brancos e 2,4% de pretos, diz pesquisa. (n.d.). Accessed on Dec 1, 2013, at http://g1.globo.com/educacao/noticia/2013/05/usp-tem-787-de-calouros-brancos-e-24-de-pretos-diz-pesquisa.html

Varma, R., & Hahn, H. (2007). Gender differences in students' experiences in computing education in the United States. *International Journal of Engineering Education*, *23*(2), 361–367.

Varma, R., Prasad, A., & Kapur, D. (2006). Confronting the "socialization" barrier: Crossethnic differences in undergraduate women's preference for IT education. In J. M. Cohoon & W. Aspray (Eds.), *Women and information technology: Research on underrepresentation* (pp. 301–322). Cambridge, MA: MIT Press. doi:10.7551/mitpress/9780262033459.003.0010

Vogt, K. E. (2005). *Asian American women in science, engineering, and mathematics: Background contextual and college environment influences on self-efficacy and academic achievement.* Doctoral dissertation. Retrieved from ProQuest Dissertations and Theses database, Publication No. 304996659.

Von Hippel, P. T. (2009). How to impute interactions, squares, and other transformed variables. *Sociological Methodology*, *39*(1), 1–27. doi:10.1111/j.1467-9531.2009.01215.x

Vygotsky, L. S. (1978). *Mind and society: The development of higher mental processes.* Cambridge, MA: Harvard University Press.

Wagner, J. M., Fleming, A. E., Moynahan, K. F., Keeley, M. G., Bernstein, I. H., & Shochet, R. B. (2014). Benefits to faculty involved in medical school learning communities. *Medical Teacher*, 1–6. doi:10.3109/0142159X.2014.947940 PMID:25159341

Wagner, T. (2008). *The global achievement gap.* New York, NY: Basic Books.

Washington, J. (2011). *STEM Education and Jobs: Declining Numbers of Blacks Seen in Math, Science, Jesse Washington.* Retrieved from http://www.huffingtonpost.com/2011/10/24/stem-education-and-jobs-d_n_1028998.html?view=print&comm_ref=false Ll.

Watson, J. D. (1969). *The double helix.* New York, NY: The Penguin Group.

Watts, R. (2007). *Women in Science: A Social and Cultural History.* London: Academic Press.

Wenneras, C., & Wold, A. (1997). Nepotism and sexism in peer-review. *Nature*, *387*(6631), 341–343. doi:10.1038/387341a0 PMID:9163412

Wentzel, K. (1998). Social relationships and motivation in middle school: The role of parents, teachers, and peers. *Journal of Educational Psychology*, *90*(2), 202–209. doi:10.1037/0022-0663.90.2.202

West, J. D., Jacquet, J., King, M. M., Correll, S. J., & Bergstrom, C. T. (2013). The role of gender in scholarly authorship. *PLoS ONE*, *8*(7), e66212. doi:10.1371/journal.pone.0066212 PMID:23894278

White House Council on Women and Girls. (2014). *White House Council on Women and Girls: recent agency accomplishments*. Washington, DC: Government Printing Office Catalog.

Whitley, R. (2005). Book Review: Cultural Diversity, Mental Health and Psychiatry: The Struggle Against Racism. *Transcultural Psychiatry*, *42*(3), 507–509. doi:10.1177/136346150504200310

Whitt, E. J., Edison, M. I., Pascarella, E. T., Terenzini, P. T., & Nora, A. (1998). *Influences on Students' Openness to Diversity and Challenge in the Second and Third Years of College*. Paper presented at the Annual Meeting of the Association for the Study of Higher Education.

Whorf, B. L. (1941). The relation of habitual thought and behavior to language. In J. B. Carroll (Ed.), *Language, thought, and reality: Selected writings of Benjamin Lee Whorf* (pp. 134–159). Cambridge, MA: MIT Press.

Wilcox, C. W., Wells, J. V., Haddad, G., & Wilcox, J. K. (2014). The Changing Democtratic Functions of Historically Black Colleges and Universities. *New Political Science*, *36*(4), 556–572. doi:10.1080/07393148.2014.954793

Wilkinson, P. (2003). *The damage the bullet has done. Rolling Stone.*

Williams, J. C., Phillips, K. W., & Hall, E. V. (2014). *Double jeopardy? Gender bias against women of color in science*. Retrieved from http://www.uchastings.edu/news/articles/2015/01/double-jeopardy-report.pdf

Williams, J. C., Phillips, K. W., & Hall, E. W. (2014). *Double Jeopardy? Gender Bias against Women of Color in Science*. Retrieved from www. Lifeworkflow.org

Williams, M. A., McGuire, D., & Rizzo, M. (2012). *Practical ethics in clinical neurology: A case-based learning approach*. Philadelphia: Wolters Kluwer Health/Lippincott Williams & Wilkins.

Willis, P. (1977). *Learning to Labor: How Working Class Kids Get Working Class Jobs*. New York: Columbia University Press.

Willis, P. (1981). *Learning to labor: how working class kids get working class jobs*. New York: Teachers College Press.

Wilson, D. E., & Balotin, J. (1999). Has the well run dry? Priming the diversity pump in PhD programs. *Journal of the Association for Academic Minority Physicians*, *10*(2), 27–33. PMID:10826006

Wong, E. Y., Bigby, J., Kleinpeter, M., Mitchell, J., Camacho, D., Dan, A., & Sarto, G. (2001). Promoting the advancement of minority women faculty in academic medicine: The National Centers of Excellence in Women's Health. *Journal of Women's Health & Gender-Based Medicine*, *10*(6), 541–550. doi:10.1089/15246090152543120 PMID:11559451

Wong, R. L., Sullivan, M. C., Yeo, H. L., Roman, S. A., Bell, R. H. Jr, & Sosa, J. A. (2013). Race and surgical residency: Results from a national survey of 4339 US general surgery residents. *Annals of Surgery*, *257*(4), 782–787. doi:10.1097/SLA.0b013e318269d2d0 PMID:23001076

Wood, S. L. (2002). *Becoming a Woman Engineer in the Community of Practice: Validity and Value in Engineering-Education Research*. Academic Press.

Word Press. (2014). Collected Young Minds. *Journey On*. Retrieved from http://collectedyoungminds.org/art-and-culture/journey-on/

World Economic Forum. (2014). *The Global Gender Gap Report 2014*. Retrieved from http://www.weforum.org/reports/global-gender-gap-report-2014

Worthington, R. L., Navarro, R. L., Loewy, M., & Hart, J. (2008). Color-blind racial attitudes, social dominance orientation, racial-ethnic group membership and college students' perceptions of campus climate. *Journal of Diversity in Higher Education*, *1*(1), 8–19. doi:10.1037/1938-8926.1.1.8

Wright, S. M., & Carrese, J. A. (2003). Serving as a physician role model for a diverse population of medical learners. *Academic Medicine*, *78*(6), 623–628. doi:10.1097/00001888-200306000-00013 PMID:12805043

Xu, Y. J. (2008). Gender disparity in STEM disciplines: A study of faculty attrition and turnover intentions. *Research in Higher Education*, *49*(7), 607–624. doi:10.1007/s11162-008-9097-4

Young, S., Fisher, W., & Vrongistinos, K. (2009). Calibrating a measure of gender differences in motivation for learning technology. *Journal of Instructional Psychology*, *36*(3), 259–272.

Zweben, S. (2014, November 30). *Computing Degree and Enrollment Trends from the 2010/11 Taulbee Survey*. Computing Research Association. Retrieved from http://cra.org/uploads/documents/resources/taulbee/CS_Degree_and_Enrollment_Trends_2010-11.pdf

Zweben, S., & Bizot, B. (2014, May). *2013 Taulbee Survey*. Retrieved May 15, 2015, from http://cra.org/resources/crn-archive-view-detail/2013_taulbee_survey/

About the Contributors

Ursula Thomas is the Director of Field Experience and Assessment at Georgia Perimeter College. Her research agenda includes issues of cultural mediation and its effects on instructional choices, the power of teacher educator research on diversity in the classroom, views of social justice in the early childhood classroom, and preservice teachers' disposition on professionalism and diversity in teacher preparation programs. Dr. Thomas's research has resulted in more than 30 publications in journals, book chapters, and conference proceedings. She is the editor of "Culture or chaos in the village: The journey to cultural fluency "(2011).

Jill Drake is a mathematics education professor at the University of West Georgia in Carrollton, Georgia. Dr. Drake has served the mathematics education community as a teacher, professor, and consultant for the past 25 years.

* * *

Tara L. R. Beziat is an assistant professor of educational psychology at Auburn University at Montgomery. Her research interests include metacognition, teaching strategies and assessment practices.

Brooke A. Burks, Ph.D., is an assistant professor of Secondary Education at Auburn University at Montgomery. She serves as Secondary Intern Coordinator and has trained secondary teachers as well as faculty colleagues in Writing Across the Curriculum. Dr. Burks has published and presented research related to education, writing, and technology, which are her key research interests. Dr. Burks is the author of the short story collection Tawanda's Quest.

Jennifer Byrom received her M.Ed. in Biology Education from Auburn University Montgomery (AUM) in 2000. Since then, she has earned her M.Ed. in Instruction Technology from AUM in 2014 and will earn her Ed.S. in Instructional Technology from AUM in May 2016. She has worked in the field of Instructional Support, primarily STEM courses, since 2000. She is currently a math specialist at Columbia Southern University, a fully-online private university with over 20,000 students worldwide.

Patricia Costa Pereira da Silva received her PhD in Education from Fluminense Federal University in 2016. She works as an Educator at Social Work School of Rio de Janeiro Federal University since 2009. Her research interests are sociology of education, higher education, racism, racial relations, blackness and affirmative actions.

Melissa Dean is a respected leader in STEM education, having led for years the development of formal and informal curriculum and programs for K-12 schools and science museums in Alabama and Louisiana. She is highly experienced in curriculum development, writing, training and implementation and served as the Primary Investigator for the National Science Foundation grant-funded program, Engaging Youth through Engineering. She has led a number of teacher development programs and works closely with STEM teachers in the Mobile County Public School System which serves 60,000 students annually. Ms. Dean received her bachelors of science from Louisiana State University in Shreveport and is currently working toward her doctorate in Instructional Design and Development at the University of South Alabama in Mobile, AL.

Bonita Flournoy is currently the Site Director for Year UP Greater Atlanta. She formerly served as the Dean for the Division of Science, Mathematics, and Health Professions and Professor of Chemistry and Science Education at Atlanta Metropolitan State College. She has held tenured positions at Columbus State University and Clark Atlanta University and has prepared scientists and science teachers for over 23 years. She received her doctorate degree from the University of Georgia. She has served in administrative positions for over six years in positions of Chair, Dean, and Interim Vice President for Academic Affairs. In these positions, she has been responsible for maintaining quality and standards-based academic programming, increasing the retention, progression, and graduation of students, while ensuring that the corps of faculty are of high quality and accountable to student-centered academic programming. Her research in science education focuses on learning styles of students in science with their problem solving strategies in chemistry. She has also investigated the synthesis and characterization of modified nucleosides in biochemical systems.

Mechelle Gittens has worked and carried out research in software engineering since 1995. She is currently a Lecturer (Assistant Professor) at the University of the West Indies – Cave Hill Campus. Mechelle has a doctorate from the University of Western Ontario (UWO), where she was the first black woman to graduate with a degree in Computer Science, and where continues active collaboration with colleagues and is now a Research Adjunct Professor. Her work is in lifestyle management technologies, software quality, testing, empirical software engineering, software reliability, and project management, and she has published at several international forums in these areas. Dr. Gittens' industrial work has included work with IBM Canada as a Research Staff Member and then Consultant for IBM Software Group. She has also consulted for the Institute for Government Information Professionals in Ottawa and done software development for an Internet start-up. Gittens' research has led to work with the IBM Centre Advanced Studies (CAS) where she is a Faculty Fellow. On the academic side, she was also a faculty member in Computer Science at the University of Waterloo from 2002 to 2004; and more recently, she taught Software Quality for the Consortium for Graduate Education in Software Engineering (ConGESE) for UWO at IBM. She has acted as the elected president of a professional association whose membership includes representation from management in software quality in the Canada's Technology Triangle in Kitchener/Waterloo. Her current research is in customer-based software quality where she works to understand how customers use software in order to create intelligence for more reflective quality assurance, development and installation. She is also expanding her interest in Quality as she explores a new frontier in Mobile Applications for Quality of Life.

Djana Harp, MD, MSCR, is Co-Director of the Third Year Obstetrics and Gynecology Clerkship. She is an Associate Professor in the Department of Obstetrics and Gynecology at Morehouse School of Medicine.

Jamil Harp, MD, is Co-Director of the Third Year Obstetrics and Gynecology Clerkship. She is board-certified in Urogynecology and Obstetrics and Gynecology and serves as an Assistant Professor in the Department of Obstetrics and Gynecology at Morehouse School of Medicine.

Helen Jackson has formerly worked as a research physicist for three of the National Research Laboratories and is now part time faculty at Wright State University. She also has her own consulting company, ElectroGenesis. She received her Ph.D. in Engineering Physics at the Air Force Institute of Technology, specializing in radiation effects on semiconductors. Her research benefited military Air Defense. Her experimental, modeling and simulation efforts of radiation damaged semiconductors was used to obtain reliability and lifetime information, leading to more optimal device design and performance. She patented a method to achieve radiation hardness for military sensors. She previously did medical physics as a Vanderbilt University research associate and developed a prototype for radiation dosimetry for cancer. She did modelling used in nuclear nonproliferation with the Department of Energy while also at Fisk University. She is very active in mentoring the underrepresented in STEM and is a member of many humanitarian and civic organizations.

Japera Johnson is an Instructor in the Department of Community Health and Preventive Medicine and a Health Policy Leadership Fellow at the Satcher Health Leadership Institute, Morehouse School of Medicine in Atlanta, GA.

Robert Kurtz MD FACS, is Associate Professor of Surgery at the SUNY Downstate Medical Center in Brooklyn, NY. He has been the director of service for trauma and surgical critical care at the Kings County Hospital in Brooklyn and senior clinical advisor to the chief medical officer of the New York City Health and Hospitals Corporation, the largest municipal healthcare system in the country. At his alma mater, the Massachusetts Institute of Technology, he co-chairs the MIT industrial advisory committee on minority education. He also holds a gubernatorial warrant on the New York state commission on corrections, which oversees and strives to improve prison medical care statewide. His parents, militant anti-racists, active in the struggle for civil rights, raised him in that tradition, and his interest in it is deep and abiding.

Susan F. Martin is project mentor/director of Noyce Pathway to Science and Noyce Pathway to Mathematics programs and is Coordinator of Secondary Education at the University of South Alabama College of Education. Dr. Martin is an associate professor of English Language Arts and English for Speakers of Other Languages, as well as being a former secondary Biology teacher. Her research focus is in accommodating English Language Learners and other special populations through the content area classroom.

Kristin McCombs is a PhD candidate in Educational Psychology/Instructional Technology and an adjunct instructor of psychology and education.

Kadian McIntosh is a Postdoctoral Research Fellow for the Arizona Medical Education Research Initiative at the University of Arizona, a collaboration between the College of Education and the College of Medicine which focuses on holistic admissions, access and success in the health sciences, climate assessment, diversity, and assessing student learning and development. Her research interests also include examining effects of organizational structures and policies, faculty-student interaction, understanding experiences and outcomes of underrepresented students in STEM, and using actionable institutional research to inform strategic planning and data-driven policies.

Charles Mpofu is a health sciences lecturer at Auckland University of Technology, New Zealand. In addition to tertiary education his research interests are in the areas of migrant health and social inequalities and hence his interest in contributing a chapter on sexism and racism in STEM fields. He has previously used the social critical theories to examine issues of medical licensure in New Zealand and other Western English-Speaking countries.

Hyun Kyoung Ro is an assistant professor of Higher Education and Student Affairs at Bowling Green State University in Ohio. Her research interests include college effects, diversity, and women and underrepresented minority faculty and students in science, technology, engineering, and mathematics.

Ruth Shim, MD, MPH, is Vice Chair of Education and Faculty Development and Chief of Outpatient Psychiatry Services in the Department of Psychiatry at Lenox Hill Hospital in New York City, part of the North Shore-Long Island Jewish Health System. She is an Associate Professor in the Department of Psychiatry at Hofstra/North Shore-LIJ School of Medicine.

Clyde Wilcox is Professor of Government at Georgetown University. He has written extensively on religion and politics, race politics, gender politics, GLBT politics, interest groups, campaign finance, and the politics of science fiction. He has written more than 30 books and scores of journal articles, and lectured in dozens of countries. He provides diplomatic training for the U.S. foreign service, and has provided similar training in Jordan, Poland, Mexico and the UAE. He is currently teaching at Georgetown's Doha campus in Qatar.

Judith Wilcox (formerly Judith Gwathmey) was Associate Professor of Medicine and Physiology at Harvard Medical School and is now professor at Boston University School of Medicine. She is currently a charter member of the National Institutes of Health, Vascular Cell and Molecular Biology study section. She was founder, CEO and CSO of Gwathmey, Inc., a pre-clinical research biotechnology company. She is known internationally for her research on calcium channeling and heart failure. Wilcox is also a highly sought after motivational speaker and mentor. She is the recipient of the Presidential Mentoring Award in Science, Technology, Engineering and Mathematics, and the Small Business Administration Tibbetts Award for Excellence in Research.

Index

H

I

J

L

M

N

O

P

Q

R

S

www.ingramcontent.com/pod-product-compliance
Ingram Content Group UK Ltd.
Pitfield, Milton Keynes, MK11 3LW, UK
UKHW050951290926
15470UKWH00005B/1757

9 781522 50174